Advances in Radiation Protection

EURO COURSES

A series devoted to the publication of courses and educational seminars organized by the Joint Research Centre Ispra, as part of its education and training program.
Published for the Commission of the European Communities, Directorate-General Telecommunications, Information Industries and Innovation, Scientific and Technical Communications Service.

The EUROCOURSES consist of the following subseries:

- Advanced Scientific Techniques
- Chemical and Environmental Science
- Energy Systems and Technology
- Environmental Impact Assessment
- Health Physics and Radiation Protection
- Computer and Information Science
- Mechanical and Materials Science
- Nuclear Science and Technology
- Reliability and Risk Analysis
- Remote Sensing
- Technological Innovation

HEALTH PHYSICS AND RADIATION PROTECTION

Volume 1

The publisher will accept continuation orders for this series which may be cancelled at any time and which provide for automatic billing and shipping of each title in the series upon publication. Please write for details.

Advances in Radiation Protection

Edited by

Martin Oberhofer
*Commission of the European Communities,
Joint Research Centre (JRC), Ispra Establishment,
Ispra, Italy*

KLUWER ACADEMIC PUBLISHERS
DORDRECHT / BOSTON / LONDON

Based on the lectures given during the Eurocourse
'Advanced Seminar on Selected Topics in Radiation Protection'
held at the Centro de Formação Técnica, Lisbon, Portugal, October 23–27, 1989,
in collaboration with the Laboratorio Nacional de Engenharia e Tecnologia Industrial (LNETI), Lisbon, Portugal

ISBN 0-7923-1232-5

Publication arrangements by
Commission of the European Communities
Directorate-General Telecommunications, Information Industries and Innovation,
Scientific and Technical Communications Service, Luxembourg

EUR 13454

Published by Kluwer Academic Publishers,
P.O. Box 17, 3300 AA Dordrecht, The Netherlands.

Kluwer Academic Publishers incorporates the publishing programmes of
D. Reidel, Martinus Nijhoff, Dr W. Junk and MTP Press.

Sold and distributed in the U.S.A. and Canada
by Kluwer Academic Publishers,
101 Philip Drive, Norwell, MA 02061, U.S.A.

In all other countries, sold and distributed
by Kluwer Academic Publishers Group,
P.O. Box 322, 3300 AH Dordrecht, The Netherlands.

Printed on acid-free paper

Printed in the Netherlands

Advances in Radiation Protection

This book highlights the most outstanding achievements in the field of radiation protection within the last years. International experts in health physics deal with the latest scientific findings justifying the present ICRP (International Commission on Radiation Protection) philosophy and proposals of ICRU (International Commission on Radiation Quantities and Units). The role of the Commission of the European Communities (CEC) in this context is outlined.

After long discussions new operational quantities for radiation protection purposes have been established in order to overcome some drawbacks of the index-quantities used up to now and, considering that these are foreseen to be applied in day-to-day health physics practices in the near future, the issues of their actual introduction are emphasized. New developments in the field of beta- and neutron-dosimetry are presented taking into account the importance given to both within the last decade. This holds also for high dose measuring techniques of which the ALANIN-method is described. Recent progress in solid state dosimetry and in microdosimetry is discussed and future trends are shown up.

After an overview of thermoluminescent dosemeter (TLD) uses in medicine (diagnosis and therapy), the second part of the book is dedicated to radiation protection problems in medical and industrial radiation applications with emphasis on dose-assessment and quality control in radiological practices as well as on radiation protection aspects and intervention in case of radiological incidents and/or accidents.

Finally selected topics in the field of protection against the hazards from non-ionizing radiations (NIR), such as radio-microwave, laser and ultraviolet (UV) radiation and also ultrasound are reviewed with special attention to recent research results considering the safe operation of nuclear resonance magnetic equipment.

Contents

Acknowledgements

The editor of this volume is very much indebted to the Departamento de Protecão e Seguranca Radiologica of the Laboratorio Nacional de Engenharia e Tecnologia Industrial (LNETI), Lisbon, Portugal, represented by its Director, Dr. Julio Pistacchini Galvão. In particular, he is indebted to Dr. João José Vaz Carreiro, Coordinator of the Radiological Protection Service of LNETI, for the co-operation in organizing the "Advanced Seminar on Selected Topics in Radiation Protection" from 23-27 October 1989 in Lisbon, the lectures of which are presented in these proceedings.

Thanks are also due to the Centro de Formacão Tecnica of LNETI having made available its staff and facilities for the performance of the seminar. Mr. Gomes of the Centro assisted seminar organisers, lecturers and participants in the seminar in a very personal way, for which he deserves much credit.

Particular acknowledgement is given to Mr. S.G. Porrini in his function as Head of the Ispra-Courses Organization and his staff for their continuous efforts in making the Seminar a success.

Also the editor appreciates very much the work of the Ispra Publication Service which cared for the re-typing and print-ready finalizing of the manuscripts, as well as proof-reading of his collegue Mr. Hans Penkuhn of the Technology Safety Institute of the Joint Research Centre, Ispra.

Furthermore, the editor would like to thank Kluwer Academic Publishers for their patience with regards to the production of the final versions of the manuscripts and care taken in obtaining the first volume in the Health Physics and Radiation Protection Series of the EURO-Courses Publications.

Last and not least, the editor thanks all the lecturers in the seminar and authors of this publication for their contributions, dedicated to a field which is gaining more and more attention from the authorities and the public.

Ispra, October 1990

Dr. Martin Oberhofer
Seminar Manager
and Editor

Preface

From October 23 to 29, 1989 the Joint Research Centre Ispra organized, in collaboration with the Laboratorio Nacional de Engenharia e Tecnologia Industrial (LNETI), Departamento de Protecão e Seguranca Radiologia in Lisbon, Portugal, an advanced seminar on selected topics in radiation protection. The idea of the Seminar was to highlight the most outstanding achievements in the field of radiation protection within the last few years. Internationally konwn experts in health physics dealt with the latest scientific findings justifying the present ICRP (International Commission on Radiation Protection) philosophy and proposals of the ICRU (International Commission on Radiation Quantities and Units).

In this publication all the texts presented at the Seminar are reproduced with the intention of making its contents available to a wider group of interested persons than those having attended the Seminar.

The initial contribution is dedicated to radiation risk and the question of whether it needs to be re-valuated as a consequence of more recent results of radiation research.

In this context the activity of the Commission of the European Communities within the framework of its radiation protection programme is also very important, as well as its attitude with regard to research in the radiation protection field. In two papers this programme is outlined and the research priorities are presented.

The request for optimization in radiation protection is one consequence of ICRP's modern position. For years its optimization principle has been followed and practised in various health physics areas. Experience gained so far is analysed in another contribution.

After long discussions new operational quantities for radiation protection purposes have been established in order to overcome some drawbacks of the index-quantities used up to now and, considering that these are foreseen to be applied in day-to-day health physics practices in the near future, the issue of their actual introduction is emphasized in the chapter on new operational quantities in radiation protection.

New developments in the field of beta and neutron dosimetry are presented taking into account the importance given to both within the last decade.

This holds also for high dose measuring techniques of which the ALANIN-method is described.

Recent progress in solid state dosimetry and in microdosimetry is discussed and future trends appear in a corresponding article.

After an overview of thermoluminescent dosemeter (TLD) uses in medicine (diagnostic and therapy) the second part of this publication is dedicated to radiation protection problems in medical and industrial radiation applications. Emphasis is placed on dose-assessment and quality control in radiological practices, as well as to radiation protection aspects and intervention in case of radiological incidents and/or accidents, such as the one in Goiania, Brazil, 1987.

Finally the state of the art in the field of protection against the hazards from non-ionizing radiations (NIR), such as radiofrequency-, microwave-, laser- and ultraviolet (UV)- radiation and also ultrasound is reviewed with special attention to recent research results with regard to the safe operation of nuclear resonance magnetic equipment.

The Seminar was addressed to research workers in the radiation protection field, to health physicists engaged in operational health physics, to radiation protection officers in hospitals and in industry, and to radiation protection experts in governmental agencies. The same group of people is addressed by this book on a much wider scale.

Ispra, October 1990

M. Oberhofer
Editor

List of lecturers

R. CIBRIAN
Commission of the European Communities
D.G. XI: Environment, Nuclear Safety and Civil Protection
Radiation Protection Division
Rue de la Loi, 200
B-1049 Bruxelles - Belgium

M. GRANDOLFO
Istituto Superiore di Sanità
Laboratorio delle Radiazioni
Viale Regina Elena, 298
I-00161 Roma - Italy

B. LINDELL
National Institute of Radiation Protection
Box 60204
S-10401 Stockholm - Sweden

C. MACCIA
INSERM U 240
B.P. 48
F-92263 Fontenay aux Roses Cedex - France

T.O. MARSHALL
National Radiological Protection Board
Chilton, Didcot
Oxon, OX11 ORQ - United Kingdom

B.M. MOORES
Integrated Radiological Services Ltd. (IRS)
Radiation Protection Service
42 Rodney Street
Liverpool - United Kingdom

M. OBERHOFER
Commission of the European Communities
DG XII - Joint Research Centre Ispra
I-21020 Ispra (VA) - Italy

D.F. REGULLA
GSF Forschungszentrum für Unwelt und Gesundheit München
D-8042 Neuherberg
Federal Republic of Germany

H. SCHIBILLA
Commission of the European Communities
DG XII: Science, Research and Development
Rue de Loi, 200
B-1049 Bruxelles - Belgium

Th. SCHMITZ
Institut für Medizin
Kernforschungsanlage Jülich
Postfach 1913
D-5170 Jülich - Federal Republic of Germany

J. VAZ CARREIRO
Laboratório Nacional de Engenharia e Tecnologia Industrial (LNETI)
Departamento de Protecção e Segurança Radiológica
Estrada Nacional, 10
P-2685 Sacavém - Portugal

B.F. WALL
National Radiological Protection Board
Chilton, Didcot
Oxon, OXII ORQ - United Kingdom

Welcome Address

Julio Pistacchini Galvão
Director of the Radiological Protection and Safety Department
Laboratório Nacional de Engenharia e Tecnologia Industrial (LNETI)
Lisbon, Portugal

First, I should like to thank the Ispra Courses Organisation for having selected Lisbon for this seminar, the installation of LNETI, and for having invited me to present the welcome address, which I gladly do, in my own name as well as on behalf of Prof. Veiga Simão, president of the LNETI.

It is really our pleasure in wishing all those coming from abroad an excellent stay in our old city of Lisbon, of which we are so proud of.

The programme of the seminar is very demanding and free time is rather short. Nevertheless, my advice to you is to profit from your stay in Lisbon as much as you can. I am sure you will enjoy it.

I also would like that you might have time to visit the LNETI facilities, both here at Lumiar, where you are, and in our research centre at Sacavem, where the laboratories of the Radiological Protection and Safety Department are located.

As the programme of the seminar probably will not allow such a visit, I will keep this address short in order to have time to show you a brief video about the LNETI.

Having in mind the aim of this seminar, I should like to recall here what Prof. Bo Lindell said in one of his papers presented at the Seventh International Congress of the International Radiation Protection Association in Sydney, Australia, last April 1988. In his paper, entitled "How safe can we be?" Prof. Lindell, whom we the radiological protection experts so much admire, and from whom, for so many years, I learned so much in the meetings of the Radiological Protection and Public Health Committee of the OECD, wrote: "To live safely does not mean to live without risk. We wish to take some risks in order to enjoy particular benefits".

Not intending to engage myself in the public controversy of whether nuclear energy is safe or not, the truth is, that in most of the CEC countries we have power reactors running, providing adequate and cost- effective energy for maintaining the level of standard of living of CEC industrialized countries. This would not be possible, as for any peaceful use of nuclear energy, if the general public would not have approved it.

As Prof. Lindell said: "to enjoy benefits, we wish to take some risks. Living safely does not mean living without risk". How safe can we be through radiological protection? could equally be the title of this seminar.

Apart from this, in agreement with article 2a of the Euratom Treaty the Commission of the European Communities has the obligation and competence to "promote research and ensure the dissemination of technical information" also in the field of radiation protection. This, I believe, is only possible through advanced seminars, like this one conceived with the idea to highlight the most outstanding achievements in the field of radiation protection within the last years and addressed to research workers as well as to radiation protection officers in hospitals, industry and governmental agencies.

We wish you a good and profitable working week and that you will enjoy our video.

Lisbon, 23 October 1989

RE-VALUATION OF RADIATION RISKS?

B. LINDELL
National Institute of Radiation Protection
Box 60204
S - 10401 Stockholm
Sweden

1. Introduction

This lecture is intended to be a review of the current picture of radiation risks. In 1977, the United Nations Scientific Committee on the Effects of Atomic Radiation (UNSCEAR) issued a report which included assessments of the risk of cancer and hereditary effects of ionizing radiation. The same year, the International Commission on Radiological Protection (ICRP) published basic radiation protection recommendations in its Publication 26. In 1980, the Committee on the Biological Effects of Ionizing Radiations of the U.S. National Academy of Sciences published its most recent review (the "BEIR III") on the effects of low levels of ionizing radiation. Since then there has been a decade of continued studies and assessments which have just been published or are about to be published. UNSCEAR made new risk assessments in its report of 1988, a "BEIR V" is about to be published, and ICRP is finalizing a revision of its 1977 recommendations, perhaps to be published in 1990.

Rather than merely discussing re-evaluation of earlier risk assessment, these lecture notes give a review of the current view on radiation risks, with special emphasis on new findings and risk models.

2. Which Effects Do we Protect Against?

At the cellular level, radiation can either kill or modify cells. Cell killing is then usually the result of a failure to reproduce into viable new cells. For each individual cell, the outcome is stochastic (of random nature) and the probability of any particular cell being killed increases with the radiation dose. Therefore, the total number of cells killed in an irradiated tissue also increases with the dose. For the result to be clinically observable, a sufficient number of

M. Oberhofer (ed.), Advances in Radiation Protection, 1–15.

cells must be killed. This cannot happen unless the radiation dose is high enough, i.e. exceeds a certain threshold value. However, if the dose is higher, the harmful result is inevitable. Such biological harm (e.g. acute radiation sickness, skin erythema, and observable cataract) is therefore called *deterministic*, which is a better term than the previous "non-stochastic", since the primary effects at the cellular level are, in fact, of stochastic nature.

Some cells may have their DNA affected but still be able to reproduce. This will give rise to clones of cells with modified properties. If these cells deviate significantly from the normal cells in the irradiated tissue, they will be eliminated by the body's defence mechanisms. Only in the very rare case that these mechanisms fail, the modified cells may propagate. If these cells are somatic cells, the end result will be a tumour, if they are germ cells involved in the conception of a new individual, the DNA modification becomes hereditary.

Since, for each individual, radiation-induced tumours and hereditary harm appears to occur at random, these effects are called *stochastic*.

There is a third type of effect, namely developmental harm after radiation exposure of an embryo or foetus, which is sometimes less easy to classify as deterministic or stochastic. They are therefore usually discussed separately.

3. Developmental Effects

3.1. General

Immediately after conception, i.e. before or just after the embryo has been implanted on the uterine wall, the number of cells is small. Cell killing is then equal to death of the embryo, an essentially stochastic effect that will only be observed as an absent pregnancy. Cell damage within this period can therefore not result in harm to any individual at birth or thereafter.

During organogenesis, i.e. during the first months of pregnancy when cell multiplication is high, cell killing may give rise to malformations or growth disturbances, depending on the stage of pregnancy when the exposure occurs. These effects are mostly deterministic and animal experiments indicate at least practical thresholds. In man, a threshold of the order of 100 mGy has been suggested.

3.2 Mental Retardation

The human brain is mainly formed during the 8th to 15th week after conception. During this period of growth it is quite sensitive to radiation and severe mental retardation has been observed among those who were exposed in utero within the period by the atomic bombs in Hiroshima and Nagasaki.

At a low-LET dose of 1 Gy during the sensitive period, about 40 percent of those exposed show severe mental retardation. The question is whether this is a deterministic effect with a threshold dose or a stochastic effect without threshold.

The answer may be found in the fact that less severe mental impairment has also been observed. Changes in intelligence test scores seem to be dose-related, indicating a general shift of the whole IQ distribution, by about 30 IQ points per Gy (cf. Figure 1).

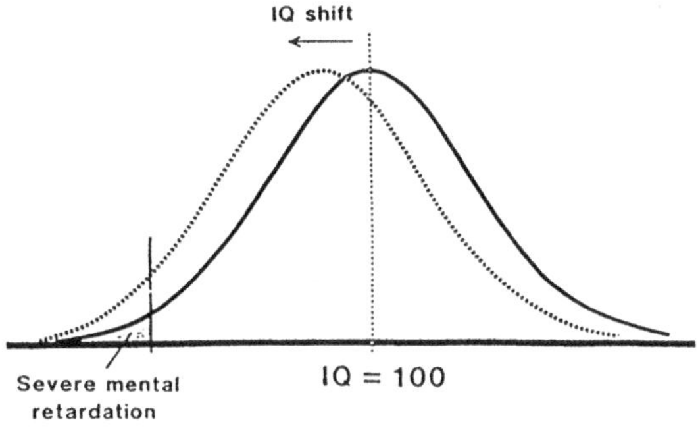

Figure 1. Shift of IG distribution with dose.

For this effect, there is no obvious threshold dose. One possible conclusion would be that, for a population, there would also be no threshold for the increase in the fraction of mentally retarded, but, at low doses, the extra cases would be those who before the exposure were just at the border of being classified as retarded and there would be no risk for those normal or bright to be pushed over that border. There would therefore be a threshold dose which would depend on the expected IQ without exposure, but everyone, including the brightest would be marginally less intelligent after radiation exposure within the sensitive period and for this there would be no threshold. However, whether this interpretation is correct remains to be seen.

4. Deterministic Effects

The number of cells killed by radiation would be expected to increase with the dose in a continuous manner. The shape of this increase (cf.

Figure 2) may be somewhat different for different individuals. The dose which is sufficient to cause manifest tissue damage by killing a sufficient number of cells would therefore by different for different individuals (the individual threshold dose) and the dose-response curve for a population of individuals would be sigmoid from the lowest to the highest individual threshold doses.

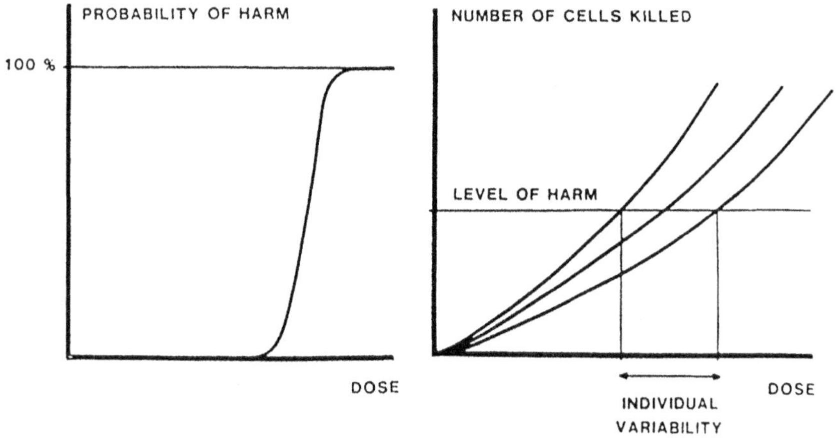

Figure 2. Dose response curve (left) and dose-effect curves (right).

4.1 Repair Mechanisms

If each small dose increment ΔD causes a proportional probability $k \Delta D$ of cell damage that will kill the cell, the probability of survival after a dose D will be exp(-kD) and the probability that the cell will be killed is then 1 - exp(-kD).

However, if one type of change in the cell is not enough to kill it but n changes (each with the same probability k per unit dose) are needed, then the probability of the cell being killed is (1 - exp(-kD)) to the power of n, and the probability of survival is

$$S = 1 - (1 - \exp(-kD))^n$$

This expression gives curves like the one in Figure 3. If the diagram is plotted in a lin-log scale, the curve becomes linear first at high doses. If the linear part is extrapolated to low doses, it will intercept the survival probability axis at S = n. Then n is referred to

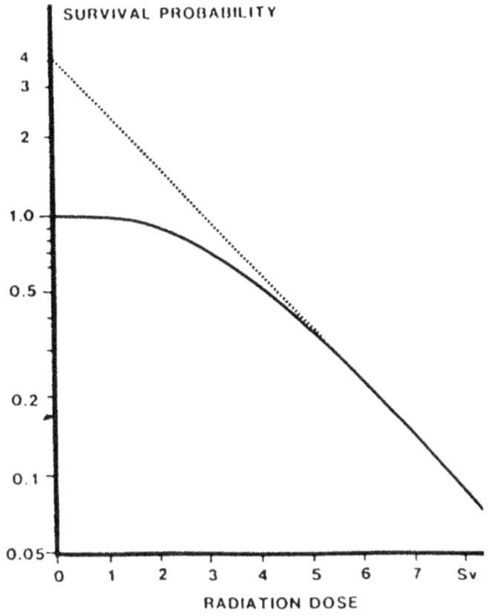

Figure 3. Probability of survival.

as the "extrapolation number". Experiments on cell cultures yield curves like the one in Figure 3, and the extrapolation number for mammalian cells is usually found to be about 4. The value of k is about 0.5 to 1 $gray^{-1}$.

This is often interpreted as an evidence of a requirement of n changes in the cell, and therefore of n primary physical events, within a short period of time. If the dose rate is low, some repair may occur and a higher dose will be needed to give the same probability of death to the cell. With high LET radiation, the dense ionization can cause several primary events simultaneously and the survival curve is exponential from zero dose. For low-LET radiation, repair seems to reduce the effectiveness of the dose when the dose rate is lower than about 100 mGy per minute.

The cell survival probability as described above will determine the number of surviving cells in a cell culture but not in functioning tissues. In those, there is additional repopulation of cells and complex interactions when the normal equilibrium is disturbed.

Much is known about the sensitivity of the various tissues and body organs to large doses of ionizing radiation. The bone marrow, the gonads and the lens of the eye are the most sensitive organs. The threshold for temporary sterility in the male in the case of a single dose of low-LET radiation is about 150 mGy and for permanent sterility about 3.5 Gy. For females, permanent sterility is caused by doses in the range of 2.5 to 6 Gy, older women being more sensitive. Cataract is caused by doses exceeding 5 Gy, but already doses in the range 0.5-2 Gy may cause detectable opacities. Annual doses below 100 mSv do not seem to cause any manifest bodily harm even if continued over a lifetime. This means that deterministic effects (not including developmental effects on the foetus) will not be caused by any radiation exposure controlled at the present dose limits unless the dose distribution in the body is consistently very uneven.

4.2 Acute Radiation Sickness

At high doses deterministic effects may lead to the death of the exposed individual. In the case of whole-body exposures, the primary cause of death is then loss of bone marrow stem cells resulting in a reduction of the number of vital blood cell. The immediate cause of death may be an infection three to four weeks after the exposure. The chance of survival is often described by the "LD_{50}", i.e. the dose at which 50 per cent of the exposed individuals die within two months (60 days). After two months, the risk of dying of bone marrow destruction is negligible. The LD_{50} in the case of loss of bone marrow function is in the range of 3-5 Gy. It depends on the health of the exposed individuals and of the medical care, particularly a successful isolation from infections. The usefulness of bone marrow transplantation is disputed and it is usually recommended only at doses near the LD_{50}. At lower doses it is considered an unjustifiable risk and at higher doses it may not help because other tissues may have been lethally harmed. It is then first damage to the gastrointestinal epithelium which is life-threathening. This damage gives an LD_{50} of between 5 and 15 Gy and will cause death within 10-20 days. At even higher doses, damage to the central nervous system will cause death within 1 to 5 days.

4.3 Deterministic Effects of High-LET Radiation

For stochastic effects, the dose equivalent in various organs and tissues is supposed to indicate the risk. In that case, the absorbed dose (D) is multiplied by a "quality factor" (Q) to give the dose equivalent (H):

$$H = Q \cdot D$$

While the absorbed dose is given in Gray, the dose equivalent is given in Sievert (both being special names of the basic unit, Joule/kilogram). For deterministic effects, the dose equivalent is not an equally good indicator of the risk, because the relative biological effectiveness (RBE) of high-LET radiation in relation to low-LET radiation is usually lower than the corresponding values of Q. The degree of deterministic harm from high-LET radiation is greater per unit absorbed dose than from low-LET radiation, but per unit dose equivalent it is lower.

With regard to the most significant source of high-LET radiation, radon daughter products in mines and indoor air, the situation is further complicated by the dosimetric problem, namely to identify the relevant cells and calculate the dose in these from the very poorly penetrating alpha particles. However, deterministic effects are rarely considered in relation to this source.

5. Stochastic Effects

The stochastic effects are *cancer* and *hereditary effects*. As in the case of cell killing, it is quite likely that more than one primary event is needed to initiate the damage at the cellular level. This becomes particularly important with regard to the stochastic effects since these may arise from damage in only one cell, leading to a clone of affected cells. There may then be the complication that some of the required cellular events have already been caused by other mutagenic agents than radiation, which have what has been called a "rad-equivalent effect".

5.1 Hereditary Effects

Hereditary effects of ionizing radiation are well known for plants and animals and radiation has, in fact, been used as a tool to produce mutations for purposes of research and in plant breeding. However, observations of such effects on man are not yet conclusive. This does not mean that they are unlikely to exist. On the contrary, there is no reason to believe that man is an exception; it only means that the risk is too low to permit observations in the exposed groups which have been studied.

Assessment of the risk of hereditary effects of radiation on man must therefore be inferred from observations on animals, e.g. the mouse. Two main methods have been employed, the "direct method" and the "doubling dose method". The direct method is based on observations on the risk of some particular mutations in animals and a number of assumptions on the total number of genes, differences in gene structures between man and animals and some other relevant factors. It is only natural that this method, which depends heavily on such uncertain assumptions cannot yield a very reliable result.

The doubling dose method assumes proportionality between the radiation-induced harm and the background of "naturally" occurring hereditary harm in the population, i.e. that a given dose will increase the background by an amount which is proportional to the dose. The so-called "doubling dose" is the dose which, if received by everyone in each generation, is assumed to double the normal mutation rate when a new equilibrium is reached.

If the doubling dose is D*, then a dose of D per generation, will lead to a new equilibrium situation where the mutation rate has increased by the fraction D/D*.

However, in radiation protection we are concerned about the total commitment of harm from each single dose rather than the effect in a new equilibrium situation. After a single dose, the number of mutant genes will increase in the first generation of offspring, but will then decrease for each generation until it is back at its value before the exposure. It can be shown that the integrated effect over all generations is the same as the number of extra mutant genes in each generation at equilibrium if the dose had continued to be received in every generation. For this reason, the assessed equilibrium situation can be used as the basis for the estimate of the total effect of a single exposure. For this method to be useful, it must be assumed that there is at present an equilibrium between spontaneous mutations and the elimination of mutant genes by selection in each generation.

For protection purposes, we are less interested in the mutations than in the harm which is actually expressed. It must be assumed that there is a direct proportionality between these.

The assessment of hereditary harm by ICRP in 1977 was that the total number of cases of severe hereditary harm in the first two generations of offspring would average one per 100 man Sv to the reproductive individuals, regardless of the dose distribution between the sexes. However, in radiation protection collective dose assessments are usually made for the total population and risk estimates relate to an average individual and not to a reproductive individual. The risk must then be reduced to 30/70 (the ratio of the mean reproductive age to the life expectancy at birth), i.e. to about 40% of the value for the reproductive population. The probability of severe hereditary harm in the first two generations of offspring was therefore assessed at 0.4% per Sv. The total probability over all generations was estimated at twice that value.

The UNSCEAR 1988 report, referring to a more comprehensive UNSCEAR report on genetic effects 1986, gave a revised estimate for the reproductive population, namely 0.3% per Sv for the first two generations and four times that value for all generations. With the reduction to 40% for a population of all ages, the values would be about 0.12% per

Sv and 0.5% per Sv, respectively.

This means that the current estimate of the severe hereditary harm is lower than the estimate from 1977, particularly for the harm in the first two generations. One explanation is that the ICRP estimate from 1977, as well as the estimate by UNSCEAR the same year, included a type of harm which was not quantitatively estimated by UNSCEAR in 1986 and 1988, namely diseases of complex etiology, multifactorial diseases. This is the most striking change from the assessment in 1977, although the present estimate of the risk of chromosomal anomalies is substantially lower than in 1977 which reduces the proportion of harm expressed in the first two generations.

Multifactorial disorders are those where mutations may play a role but not be exclusively responsible. In 1977 it was assumed that the doubling dose for these was 1 Sv and that there was a 5% mutational component. Since these disorders were then believed to affect only a few per cent of all newborn, the estimated risk was a minor part of the total risk of hereditary harm.

In the latest UNSCEAR assessments, the natural prevalence of these disorders is estimated at near 70% (which does not mean that 70% of all newborn children are affected, because some may be affected by several disorders of this type). With the doubling dose method, the risk estimate would increase in proportion to the assumed prevalence. If one still assumes a doubling dose of 1 Sv and a mutation component of 5%, the multifactorial disorders would be expected to contribute about 0.3% per Sv to the risk for expression in the first two generations and referring to an average individual in a population of all ages. The UNSCEAR risk coefficient of 0.1% per Sv for the other, and severe, effects would then have to be raised to 0.4% per Sv, a value which is incidentally equal to the ICRP 1977 value.

The UNSCEAR experts refused to make any such quantitative estimate, since they did not feel that they had enough information on which to base assumptions on the doubling dose and the mutation component. However, it is obvious that disorders which have a 70% prevalence can hardly all be considered severe and it would be misleading to add this uncertain risk directly to the estimates of the truly severe hereditary harm.

5.2 Cancer

Cancer may arise as a clone of "asocial" cells originating in one primarily affected cell. After a single dose of radiation, there is a latent period before the disease can be clinically detected. There may be several reasons for this. One simple reason is that it takes time for a sufficient number of cell divisions to form a detectable tumour.

A basal cell carcinoma with a volume of 1 ml may contain about 10^9 tumour cells. To create so many cells will take about 30 cell divisions. The time period between cell divisions varies between different tissues and only a small fraction of the time is used for the actual division process.

Another reason why cancer will not appear immediately is that the primary cell damage caused by ionizing radiation may be only one of the *initiating* effects needed to start the build-up of a clone of cancer cells, i.e. a similarity to the requirement of perhaps 4 primary events to kill the cell. A further requirement is that the first cancer cells succeed to survive in spite of immune reactions. Therefore, some *promoting* condition may also be necessary. The mechanisms for cancer induction are not yet fully understood and there is room for much speculation.

If several, say n, initiating effects are needed and the order of these effects is irrelevant (which it may not be), the accumulated probably of a tumour cell can be calculated under simplified assumptions, as recently suggested by M. Holmberg. If we assume that, without radiation exposure, the probability per unit time of the i:th initiating event is q_i and that q_i is very small, then the probability that n such effects will have occurred at age t will be the product of the probabilities $q_i t$ for each effect, i.e.

$$P(t) = (q_1 q_2 \ldots q_n) t^n$$

This would mean that the probability of having cancer of a given type would increase with the n:th power of age, an assumption which has some support in cancer statistics. If one of the events i is caused by a single radiation dose, with a probability of r per unit dose, then the probability R(t) of cancer actually occurring because of dose D would be

$$R(t) = r\, D\, (q_1 q_2 \ldots q_{n-1})\, t^{n-1}$$

This means that the cancer risk from radiation would be proportional to P(t)/t, i.e. the background risk divided by age. The risk per unit time (e.g. the annual risk) would be proportional to the age to the power of (n-2). In the case that n = 2 (only two initiating events needed), the annual risk would be constant and independent of age. If n = 1, the accumulated risk would be independent of time, as it would be if only one initiating event were needed.

In reality, the above assumptions may be wrong or oversimplified. For example, promoting conditions may increase by age, in which case R(t) might well be proportional to t^n and the annual risk of radiation-induced cancer proportional to the background rate. However, even though what has been said here is very speculative, it illustrates the basic problems.

How does this compare with actual observations? Except for leukaemia, there is yet not enough information to tell how the radiation-induced cancer risk will vary with time.

The main sources of epidemiological information on the cancer risk after exposures to ionizing radiation are the survivors of the bomb explosions over Hiroshima and Nagasaki, some groups of patients who have been exposed in radiological examinations or treatments, and personnel in radiological work. An additional source which may provide some information in the future is the persons who lived in the immediate environment of the Chernobyl nuclear power plant and received relatively high radiation doses. The best information may be obtained from the studies in Japan since the exposed groups in Hiroshima and Nagasaki received whole body exposures and included all ages at the time of exposure. However, the majority of these individuals are still alive, and to the extent that the risk of radiation-induced cancer increases with the general background frequency of cancer, most cases of cancer are still, i.e. in 1989 when this is being written, to be expected.

5.3 Leukaemia

Leukaemia is the exceptional malignant condition where (a) the latent period is short, in some cases only a few years, and (b) the annual risk does not seem to increase with time but rather disappear after a "plateau" period with constant annual risk. The short latent period in comparison to other types of cancer would be consistent with the assumption that fewer primary events would be needed in this case.

This observation supports the so-called "absolute" or "additive" model, which describes a minimum latent period (which for myeloid leukaemia is about two years) with no risk, followed by a plateau with constant annual excessive risk, again followed by no extra annual risk. In the case of leukaemia the annual risk seems to decline after a peak after about 6 years and is significantly reduced after 20 years.

All observations are at relatively high doses and dose rates. The annual risk of leukaemia after a single high dose (about 1 Sv) in the active bone marrow seems to be between 0.0001 and 0.0006 per Sv, depending on sex and age at the time of exposure. The risk at low doses will be discussed later.

5.4 "Solid" Cancer

"Solid" cancer, i.e. all other cancers than leukaemia, show longer latent periods, on the average possibly 30-40 years. The minimum latent period for these tumours is 5-10 years. This would indicate that more primary events are needed than in the case of leukaemia and that one might expect the annual risk to increase with time over the plateau and that the plateau may be longer and perhaps exceed the normal life expectancy. This is also what has been observed over recent years, when a sufficiently long time period has passed since the exposure (i.e. since 1945 in Japan).

On this basis, the so-called "relative" or "multiplicative" model has been suggested. With this model, the annual risk of excess cancer is assumed to increase in proportion to the background rate, after the minimum latent period.

Models have been introduced to make it possible to estimate the total risk of a given dose. Since most of the exposed individuals are still alive, the total risk will depend on how many additional cancers can be expected to occur in the future. In 1977, UNSCEAR and ICRP based such estimates on the assumption of the additive model. However, recent data indicate that the annual risk of solid cancers is neither constant or decreasing but, in fact, increasing with time, i.e. as the individual gets older.

In its 1988 report, therefore, UNSCEAR also made use of the multiplicative model. This was rather because it seemed to describe what happened than because of any assumption on the underlying mechanism of cancer induction.

With the multiplicative model, each radiation dose would add an extra annual risk of cancer, which would always be a fraction of the background, "natural" cancer risk. The magnitude of the fraction would depend on the dose.

With the multiplicative model, the total cancer risk, anticipating cases of cancer not yet observed, is estimated at twice the risk assessed by means of the additive model.

There is also another reason for a higher risk estimate now than in 1977. Since that time, the estimate of the doses received by the Japanese survivors has been changed. It is now believed that the neutron component of the dose was earlier much overestimated. This makes the gamma radiation almost alone responsible for the cancer induction. The change in the dosimetry therefore also increases the risk estimate.

In summary, the total risk of lethal cancer, including leukaemia, for an average individual and for high doses is now estimated to be higher than what was believed in 1977. It was then assessed at about 2.5% per Sv. With the additive model it is now estimated to be about 4.5% per Sv on similar assumptions but with the new dose assessment. With the multiplicative model, on the other hand, the estimate is about twice as high. On the other hand, the cancers are then more likely to come late in life, so the mathematical expectation of years lost will be about the same with the two models.

It should be added that neither the additive nor the multiplicative model satisfactorily describes the cancer risk among those who were exposed at young age and among which excess cancers have only recently been observed.

The types of cancer for which there is an indication of excess risk among the Japanese survivors as well as in other groups that have been studied are, in addition to leukaemia, primarily cancers of breast, lung and stomach, but also of bladder, colon, ovary and oesophagus, and multiple myeloma.

5.5 Risk at Low Doses and Dose Rates

There are risk estimates down to doses of 0.1-0.2 Sv but the uncertainties are large. In order to assess the risk at the low doses and dose rates which are usually of concern in radiation protection below the ICRP dose limits, assumptions on the dose-response relationship has to be made.

There is a general agreement that there is no reason to assume any dose threshold for stochastic effects such as cancer and hereditary effects. Microdosimetric considerations as well as results from experiments on animals make it reasonable to believe that the dose-response relation is linear at low doses and with a slope that may depend on the dose rate. At higher doses, it is assumed that the relationship becomes quadratic, and at even higher doses there will be a decrease in the cancer risk because of competing cell death.

The risk coefficient (probability per unit dose of lethal cancer) of relevance in radiation protection is described as the slope of the dose response relation at the point determined by the individuals' inevitable exposure to natural background radiation and, perhaps, medical x-ray examinations. No one is at zero dose over a lifetime, so the question of interest is not what the risk is at very low doses but what extra cancer risk a small change in the already received (and expected) dose will cause, c.f. Figure 4.

The slope of a line from the origin to the point of actual observations gives the risk coefficient at high doses. The slope of interest is the slope of the linear part of the curve. It can be estimated as the risk coefficient at high doses divided by a "reduction factor". In the 1988 UNSCEAR report, the value of the reduction factor was believed to fall in the range 2-10.

In the 1977 risk estimates, a reduction factor of 2 was used, i.e. the risk coefficient for low doses (and dose rates) was estimated to be 2.5/2 = 1.25% per Sv, usually presented as "of the order of 1% per Sv"

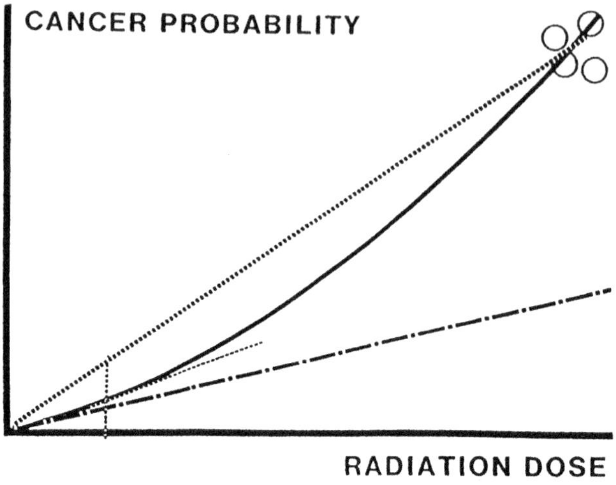

Figure 4. Dose dependence of cancer probability.

considering the great uncertainties. Even though UNSCEAR, on scientific grounds, did not wish to give a more precise estimate of the reduction factor, it is difficult to draw the conclusion that, in spite of the higher risk estimates at high doses, the estimate for low doses would now be lower than earlier. There is, in fact, no significant additional information on which to base a guess. If the same reduction factor as in 1977, i.e. 2, is used, the risk coefficient at low doses would be about 2% per Sv with the additive model and 4-5% per Sv with the multiplicative model. ICRP will have to make an assumption for the purposes of radiation protection when the Commission revises its main recommendations. May one guess that this will lead to a low dose risk coefficient that will fall in the region 2-4 per Sv?

It should be remembered that these risk coefficients relate to lethal cancer and that the total risk of incidence rather than death of cancer is about twice as high.

6. References

Holmberg, M. "Hypothesis: Cancer among A-bomb survivors in relation to a radiation-induced mutational event". National Institute of Radiation Protection, Stockholm, Sweden. To be published.

International Commission on Radiological Protection. "Recommendations of the ICRP". ICRP Publication 26. Annals of the ICRP, Vol. 1, No. 3, 1977.

International Commission on Radiological Protection. "Nonstochastic Effects of Ionizing Radiation". ICRP Publication 41. Annals of the ICRP, Vol. 14, No. 3, 1984.

International Commission on Radiological Protection. "Developmental Effects of Irradiation on the Brain of the Embryo and Fetus. ICRP Publication 49. Annals of the ICRP, Vol. 16, No. 4, 1986.

Shimizu, Y., Kato, H. and Schull, W.J. "Life Span Study Report 11, Part 2: Cancer mortality in the years 1950-85 based on the recently revised doses (DS86)". Radiation Effects Research Foundation, Technical Report RERF TR 5-88.

Schull, W.J., Otake, M. and Yoshimaru, H. "Effect on intelligence test score of prenatal exposure to ionizing radiation in Hiroshima and Nagasaki: A comparison of the T65DR and DS86 dosimetry systems". Radiation Effects Research Foundation, Technical Report RERF TR 3-88.

United Nations Scientific Committee on the Effects of Atomic Radiation (UNSCEAR). Reports to the UN General Assembly, with scientific Annexes. United Nations, N.Y. 1977, 1982, 1986 and 1988.

U.S. National Academy of Science "The Effects on Populations of Exposure to Low Levels of Ionizing Radiation". (BEIR III). NAS, Washington, 1980.

ACTIVITY OF THE COMMISSION OF THE EUROPEAN COMMUNITIES WITHIN THE FRAMEWORK OF ITS RADIATION PROTECTION PROGRAMME

R. CIBRIAN
Commission of the European Communities
D.G. XI: Environment, Nuclear Safety and Civil Protection
Radiation Protection Division
Rue de la Loi, 200 - 1049 Bruxelles - Belgium

1. Introduction

The actions of the European Community in the nuclear field are based on the Euratom Treaty of 1957. The main objective of this Treaty is to provide the conditions necessary for the growth of nuclear industries. The number of nuclear installations and the range of radioactive applications has expanded notably since the entry into force of the Treaty. Today nearly one million persons in the Community are regularly checked for exposure to ionizing radiation in the course of their work. On the other hand, in terms of potential exposure to radioactivity, the whole population of the Community lives in effect in the vicinity of a nuclear installation.

The protection of health and the environment is becoming of paramount importance to the citizens of the Community. There is increasing concern over the risks of exposure to radiation. The future of nuclear activities depends largely on people being assured that they and their environment are adequately protected against the dangers of radiation.

Article 2(b) of the Euratom Treaty provides that one of the main tasks of the Community shall be to establish uniform safety standards to protect the health of workers and of the general public and ensure that they are applied. Chapter III of the Euratom Treaty deals with health and safety, and it is by virtue of the provisions of Chapter III that, since 1959 (1), the Council of Ministers of the Community has adopted a series of directives laying down basic safety standards for radiation protection. The basic standards currently in force are defined in the Council Directive 80/836/Euratom of July 1980 (2) as

(1) OJ L 11/21 of 20.02.59.
(2) OJ L 246/1 of 17.09.80.

M. Oberhofer (ed.), Advances in Radiation Protection, 17–28.

amended by the Directive 84/467/Euratom of September 1984 (3).

After the Chernobyl accident the Commission undertook a work programme focused on the development of Community measures for an extended application of Chapter III of the Euratom Treaty. The objective of this work programme was to reinforce and where appropriate to complete the instruments available to the Community in the area of radiation protection.

This report presents on the one hand the specific actions initiated in the framework of the work programme, and on the other the permanent activities which were enlarged, reoriented, or just updated afterwards.

2. The Community Regulatory Framework

The regulatory framework for achieving the radiation protection goals of the Community is built around a number of duties incumbent of the Commission, the Council of Ministers and Member States and involving the European Parliament and the Economic and Social Committee. This framework comprises:
- Basic Standards
- Harmonisation at Community level of national provisions for radiation protection
- Monitoring of radioactivity in the environment
- Emergency planning and preparedness including public information
- Transport and disposal of radioactive materials including radioactive wastes.

3. Basic Standards

The Euratom Treaty states that Basic Standards for the protection of the health of workers and the general public shall lay down dose limits, levels of exposure and principles for the health surveillance of workers. The Community has since its inception based its approach to radiation protection on scientific principles in the light of the recommendations of the International Commission on Radiological Protection (ICRP). This aims at achieving conformity between the Euratom Directives and the radiation protection provisions of non-Community countries. The Community has, moreover, continuously furthered progress in the field of radiation protection by maintaining an extensive programme of research.

---

(3) OJ L 265/4 of 05.10.84.

The Directives specify three main principles of radiation protection:

- every type of activity resulting in exposure to ionizing radiation should be justified by the advantages which it produces;
- all exposures shall be kept as low as reasonably achievable (ALARA);
- the annual dose limits laid down for the various population groups (exposed workers, apprentices, students and members of the public) shall not be exceeded.

The dose limits fixed in the present Directives should be interpreted in the light of these principles: they provide upper limits as opposed to targets. The annual effective dose limit is 50 mSv (5 rems) for exposed workers and 5 mSv (0.5 rem) for members of the public. However, in 1985 the ICRP observed that, for members of the public, an annual dose limit of 5 mSv a year is only permissible for a limited period of years, and that the average annual effective dose over a lifetime must not exceed 1 mSv per year.

The Directives also indicate methods of dose evaluation and make provisions governing the operational protection of workers and the population. Workers are classified into sub-categories according to the degree of exposure; provisions are made for exposure assessment and medical surveillance, as well as for information about health risks and preventive measures. For the protection of the population, examination and testing of protective arrangements and the assessment of doses received are required, both in normal circumstances and in the event of an accident.

3.1. Modifications and Updating

The European Commission is responsible for keeping the Basic Standard Directives up to date and in line with the most recent scientific data and ICRP recommendations. Therefore, the Commission is closely following scientific developments in this area and especially the work of the ICRP and of the UNSCEAR. A working group has been set up by the group of experts referred to in Article 31 of the European Treaty to help prepare an eventual revision of the Community Basic Safety Standards.

The Commission is cooperating in the work of other relevant international organisations to help maintain the international consensus in this field.

3.2. Protection of External Workers

The Community Basic Safety Standards cover both regular workers and workers from external subcontractors who are also exposed. The Community Directives make no distinction between these two categories.

Nevertheless, the Commission has been aware that in day-to-day practice difficulties may arise for an appropriate dosimetric control of external workers and following a number of resolutions of the European Parliament it adopted on 28 July 1989, a proposal for a Council directive on specific practical arrangements for the protection of external workers. In agreement with Article 31 of the Euratom Treaty the adopted text should be mandatorily informed by the Economic and Social Committee. This opinion is expected for mid-November 1989.

The aim of the proposal is to extend the primary employer's responsibility in respect of the radiation protection of external workers to the operator of the installation where they actually work. This is achieved via a radiation protection exposure certificate which should be used every time a category A external worker intervenes in a control zone of an installation.

3.3. Transfrontier Shipments of Radioactive Waste

The Commission is also preparing a proposal for amending the Basic Safety Standards to include the administrative control of transfrontier shipments of radioactive waste following the conclusions drawn by the European Parliament and the Commission itself from the 1988 Transnuklear-Mol incident. This proposal will extend the Community provisions concerning transfrontier shipments of other hazardous wastes to radioactive wastes. The proposed system requires prior notification and approval for all transfers of radioactive waste originating in the Community, including national transfers and exports.

3.4. Protection of Persons Undergoing Medical Examination

The Basic Safety Standards for radiation protection also relate to problems connected with the use of ionizing radiation for diagnostic and therapeutic purposes.

Apart from natural background radiation, medical exposure to radiation is at present by far the major source of exposure to ionizing radiation. This problem is reviewed repeatedly by the World Health Organisation and the UNSCEAR which have recommended measures to prevent unnecessary excessive medical exposure.

On the other hand, ionizing radiation has enabled great progress to be made in the diagnostic, therapeutic and preventive aspects of medicine, particularly in nuclear medicine, high-energy therapy and computer-controlled tomography.

The aim of Council Directive 84/466/Euratom of September 1984 (4)

---

(4) OJ L 265/1 of 05.10.84.

laying down measures for the radiation protection of persons undergoing medical examination or treatment is to harmonize national practices as regards the training of medical and auxiliary staff and the monitoring of the safety of the technical equipment used so as to improve the protection of patients while in no way jeopardizing the benefits of ionizing radiations in the early recognition, diagnosis or treatment of disease.

The Directive addresses the problem of the growing uses of ionizing radiation and the proliferation of techniques by ensuring that users have the necessary competence and experience to avoid inappropriate uses and unnecessary proliferation of radiological installations.

The Directive demands that any ionizing radiation used in medical procedures be effected under the responsibility of fully entitled and trained doctors or dental practitioners. Assistants must receive instruction and training appropriate to their work.

Competent authorities have to draw up an inventory of medical and dental radiological equipment and nuclear medical installations. All installations in use are subject to strict criteria of acceptability and surveillance with regard to radiological protection and quality control of appliances.

The Directive contains an annex with practical recommendations for Member States to discourage the unnecessary proliferation of equipment for radiotherapy, radiodiagnosis and nuclear medicine.

3.5. Control of the Implementation of Directives Laying Down the Basic Standards

Member States have, in general, made efforts to implement the changes in the Basic Standards. These were first laid down by the Council in 1959 and have been revised and supplemented several times afterwards.

Nevertheless, the Commission had to intervene to expedite the full application of the most recent revisions and, to this end, initiated infringement proceedings against 10 Member States for failure to communicate their implementing legislation. Following these proceedings, a reasoned opinion was sent to one Member State, and 3 others are in preparation.

Since the entry into force of the 1980 version of the Directive laying down the basic standards, the Commission has delivered 22 opinions under Article 33 of the Euratom Treaty on the provisions laid down by law, regulation or administrative action in the Member States to bring their legislation in line with the Directive.

The Commission has found that there have also been delays implementing Council Directive 84/466/Euratom laying down measures for the protection of persons undergoing medical examination in the Member

States and has initiated infringement proceedings against 10 Member States for failure to communicate national implementing regulations. The Commission is also concerned to ensure that this Directive is uniformly applied throughout the Community. The optimisation of image quality and patient exposure in diagnostic radiology requires technical development and for this reason the Commission organised a scientific seminar on this subject in September 1988.

4. Harmonisation Activities: Foodstuffs Intervention Levels

One of the top priority actions undertaken by the Community immediately after the Chernobyl accident was to establish a regulatory system laying down maximum permitted levels of radioactivity in foodstuffs. In those days concern over foodstuffs from contaminated areas led to the imposition by national authorities of restrictions on internal trade and on imports.

A number of important Community provisions in this area have been approved by the Council. The provisions currently in force are set out in Council Regulation 3955/87 of 22 December 1987 (5), which supplemented and adjusted the regulations adopted hitherto. This regulation will remain in force until the end of December 1989.

The system that will be applicable in case of future accidents is laid down by Regulation 3954/87 of 22 December 1987 (6), as modified by Council Regulation 2219/89 of 18 July 1989 (7) which provides that, in the event of an accident, the Commission may impose predetermined maximum permitted contamination levels. This emergency action may then be adjusted according to the exact nature and scope of the incident concerned.

At the time of the adoption of the December 1987 regulations the Council also adopted a resolution (8) inviting the Commission to:

a) propose another Council regulation aiming at applying the Community limits to foodstuffs exported from the Community;
b) endeavour to bring about international consensus on the base of the Community levels.

Council Regulation (CEE) 2218/89 adopted in July 1989 (9) establishes that foodstuffs and feedingstuffs with radioactive contamination greater than the maximum levels which would be admissible within the

---

(5) OJ L 371 of 30.12.1987.
(6) OJ L 371 of 30.12.1987.
(7) OJ L 211 of 22.07.1989.
(8) OJ C 352 of 30.12.1987.
(9) OJ L 211/4 of 22.07.1989.

Community cannot be exported.

At international level, the joint World Health Organization/Food and Agriculture Organization Codex Alimentarius has put forward a proposal for radioactivity limits on food in international trade. In pursuit of the mentioned Council Resolution, the Commission is coordinating the Community position on this proposal and is participating in discussion in the Codex Alimentarius with a view to getting levels adopted by the Codex Commission which are suitable and coherent with those of the Community.

The Community regulatory system setting the maximum permitted levels of radioactivity in foodstuffs is now a coherent instrument in which the shortcomings and inconsistencies which emerged after the Chernobyl accident have been corrected.

5. Environmental Impact of Radioactivity

The Commission has developed various activities related to the monitoring and evaluation of the impact of radioactivity on the environment.

5.1. Publication of Data Sent to the Commission Under Article 36 of the Euratom Treaty

The Commission periodically publishes reports on the levels of radioactivity in the European environment on the basis of data sent by the Member States. The report for 1984, 1985 and 1986 is currently being prepared and will be published in 1989.

After the Chernobyl accident the Commission set up a computerized data bank, named REM (Radioactivity Environmental Monitoring) for the collection of environmental radioactive contamination results in the Community.

The bank has recently been enlarged to include selected data measured before the Chernobyl accident, and submitted to the Commission under the provisions of Art. 36 of the Euratom Treaty.

6. Emergency Planning and Preparedness

6.1. Rapid information System in the Case of a Radiological Emergency: the 1986 IAEA Convention and the Community System

Another major Community development following the Chernobyl accident has been the setting up of rapid exchange information systems after a nuclear accident.

The Commission already announced in a Communication to Council in August 1986 (10) a proposal to establish a system designed for radiological emergency purposes. Following preparatory work in which the Commission was involved, the IAEA General Conference adopted in September 1986 a convention on early notification of a nuclear accident. This Convention was immediately signed by all Community Member States and came into force in October of the same year. On proposal of the Commission the Council decided in December 1987 (11) that the Community would accede to this Convention after all Member States concerned have ratified it. But the Commission has declared to the IAEA that it will already act in conformity with the Convention.

At Community level the Commission considered that the scope of the IAEA Convention could be expanded in order to better meet the Community information needs. The Council, following a Commission proposal of April 1987 (12), decided to establish a specific Community system which, while compatible with that of the IAEA, expands the scope of the emergencies to which it applies, gives a more exact definition of the triggering mechanism, widens the range of information to be supplied and ensures diffusion of the information received to all Community Member States.

6.2. Protection and Assistance in the Event of a Nuclear Accident or a Radiological Emergency

The Commission also announced its intention to submit to the Council a proposal for the implementation of a Community system of mutual assistance in the event of an emergency (13).

The Chernobyl accident showed that it would be useful to have a system of mutual assistance on as wide an international scale as possible. The IAEA convention on mutual assistance in the case of a nuclear accident or radiological emergency has so far been signed by 11 Member States. The Commission favours the accession of the Community to this Convention and a formal proposal is before the Council (14).

The reasons supporting the proposal for accession concern the Commission's Joint Research Centre (JRC). They reflect the view that international cooperation in this field is particularly required and show the Commission's willingness to cooperate as closely as possible with the IAEA.

---

(10) COM (86) 434 final.
(11) COM (86) 760 final.
(12) COM (87) 135 final.
(13) COM (86) 327 final.
(14) COM (86) 760 final.

7. Information Activities

7.1. The Directive of the Provision of Information to the General Public

The Directive laying down the basic safety standards does not contain specific provisions on the question of informing people living in the vicinity of nuclear installations. In the Community, general provisions of this kind are contained in the Council Directive on the major-accident risks of certain industrial activities, which is known as the Seveso Directive.

The Commission has taken the view that the Basic Standards should be supplemented to fill this gap and in June 1988 adopted a proposal for a Council Directive on informing the population in the event of a radiological emergency (15).

The proposed Directive introduces two separate categories of information to be provided to the public at risk:

1) information to be provided under normal conditions;
2) information to be provided in a real emergency.

The proposal specifies the particular nature of the information which should be provided in each case. The proposal has been examined in the instances of the Council in recent months and its adoption is expected for November 1989.

7.2. Standing Conference on Health and Safety in the Nuclear Age

In October 1987, the Commission held the first meeting of the Standing Conference on Health and Safety in the Nuclear Age, its aim being to provide the public, via the media, with as much objective information as possible on topical aspects of nuclear safety and radiation protection. The first meeting was devoted to informing the public about the potential hazards of ionizing radiation from all sources.

Representatives of scientific groups, the media, consumer and environmental protection organisations, the two sides of industry and the national authorities attended the Conference. This was the first Community-level meeting at which all groups concerned were able to exchange views on the health problems associated with ionizating radiation.

The proceedings from the first meeting have been published and a second meeting will be held in Brussels on 5-6 December 1989. The topic of the second meeting will be: "Informing the public about improvements in emergency planning and nuclear accident management".

---

(15) COM (88) 809 final.

7.3. Information of Workers

The activities of the Commission to enhance the information of workers exposed to ionizing radiations are implemented at different levels.

Activities supporting the actions of national authorities are the following:

International seminars on information and training of workers, the last one having been organised on November 1988 in Luxembourg.

The 1992 perspective was contemplated in the conclusions of this seminar, which emphasised the following points:

(i) Technical training in radiation protection should become a reality in secondary school.

(ii) An evaluation at Community level of training requirements for radiation protection experts is advisable.

(iii) The Commission should made a study to determine if it is appropriate to bring closer the different training systems.

(iv) The Commission was invited to collect all the information concerning the main training actions in order to create a directory to be used by radiation protection officers.

(v) The Commission should promote an exchange of views on the problem of training requirements for emergency intervention officers.

In addition, the information activities of the Commission include seminars for trade union representatives on radiation protection matters and production of audiovisual tapes to enhance the awareness of exposed workers to the risks of radiation.

8. Dosimetry

8.1. Occupational Doses at Nuclear Power Plants

Another area where the Commission carries out important activities is the dosimetry area.

The Commission has, for a long time, had regular contacts with some of the most representative European nuclear power plants, not only within the Community but also in Sweden, Finland and Switzerland. For comparison purposes only PWRs and BWRs have been considered, because those reactor types are by far the most representative ones.

A standard questionnaire is used to canvass the collective doses of different occupational posts. To interpret correctly these values it is necessary to consider them together with the characteristics of the reactor involved, such as:

- the reactor age,

- dose rates in the primary circuit piping,
- programmed and unprogrammed shut-downs.

The values collected up to now constitute an important databank which is regularly used for specific intercomparison studies: each year the Commission organises a workshop attended by radiation protection officers from nuclear power plants. In these workshops information obtained from the databank is presented to the participants providing in this way an occasion for an exchange of experiences and evaluations on the current radiation protection practices at power plant level.

The databank is accessible to the participants installations. In the reports published by the Commission the use of plant reference codes is aimed at preserving the confidentiality of the information sources. These dosimetric surveys are regularly published by the Commission.

Since 1988 the Nuclear Energy Agency (NEA) of the OECD cooperates with the Commission with a view to setting up its own dosimetric databank.

To minimize interference with operators, the Commission supports the use of a standardised survey questionnaire. The one in use now by the Commission could be a good reference model.

8.2. Intercomparison Programmes

Since 1964 the services of the Commission have organised and implemented technical programmes which allow different institutions and dosimetry services to calibrate both operational and experimental dosimeters.

The main goal of these technical exercises is to assure that dose evaluations are in agreement with the limits and values laid down in the Euratom Basic Safety Standards.

These activities also allow to compare measures made in Member States laboratories with different methods.

There are in of the order of 15 intercomparison programmes in operation. They are carried out under experimental conditions which try to reproduce as close as possible the conditions in the field.

These programmes embrace all types of radiations but in particular they focus on those where dose evaluations are more difficult: neutrons, X-rays, alpha- and beta-radiations.

After being discussed in seminars, the results of these intercomparison programmes are published by the Commission.

9. Concluding Remarks

Since the early days of the Euratom Community its Commission has collaborated closely and fruitfully with the ICRP and the IAEA, and has based its actions in the radiological protection field on the recommendations of these bodies.

It is in the Community system presented above that the Member States of the Community have worked together to translate the principles stated by the ICRP into regulations and practices acceptable and applicable throughout the Community. If the multiple applications of radioactivity in the medical, agricultural and industrial field are to be accepted by the public the Commission considers necessary to lay down a prudent regulatory framework and to apply it strictly.

The reinforcement of international cooperation in the radiological protection field continues to be of high priority.

The harmonisation of standards at international level will increase the confidence of the citizens and help prevent inconsistencies such as those which appeared after the Chernobyl accident in the contamination limits for foodstuffs.

The serious accident that occurred in Chernobyl in April 1986 raised the problems of radiological protection in Europe in the most concrete fashion. Public opinion, particularly in Europe, reacted intensely.

Confronted with this accident, the European Commission considered necessary to adopt a responsible position, avoid extreme reactions and draw all the lessons from the accident. In order to do so, the Commission:

- analysed the origins and causes of the accident;
- clarified existing responsibilities in the nuclear and radiological protection sphere at national level, at Community level and at the international level;
- exercised the powers which the Treaties, and particularly the Euratom Treaty, confer on the Community including the power to undertake new initiatives.

The Commission considers as very important the continuation of the international cooperation efforts in order to guarantee a high level of nuclear safety and radiological protection everywhere in the world.

The Commission will do its best to fullfil these undertakings with the means at its disposal.

THE RADIATION PROTECTION RESEARCH PROGRAMME OF THE COMMISSION OF THE EUROPEAN COMMUNITIES

H. SCHIBILLA
Commission of the European Communities
DG XII: Science, Research and Development
Rue de la Loi, 200 - 1049 Bruxelles
Belgium

1. Introduction

All research actions of the CEC are integrated in the framework programme for Community activities in the field of research and technological development. One of the eight goals of the current 1987-1991 framework programme is

Quality of Life

This heading covers
- Health
- Radiation Protection
- Environment.

It is the concept of the CEC that in the industrialized world radiation protection is one of the factors which must be considered when the quality of life is to be guaranteed.

Research on radiation protection is legally anchored in the EURATOM Treaty which entered into force on 1. January 1958. By this Treaty the European Atomic Energy Community (EAEC) was given the task to "contribute to the raising of the standard of living in the Member States and to the development of relations with the other countries by creating the conditions necessary for the speedy establishment and growth of nuclear industries".

The Treaty also gave the EAEC the responsibility to study the harmful effects of radiation on living organisms. The EURATOM Treaty endowed the Commission in 1958 with the following instruments:
- A Joint Nuclear Research Centre
- Research and Training Programmes which can in part be carried out by contracts.

The first contract on radiation protection research was signed in

M. Oberhofer (ed.), Advances in Radiation Protection, 29–50.

1959, and the general purpose of the various radiation protection programmes has not been changed during these 30 years:

- Protect the public and workers under "normal situations"
- Deal with possible accidents
- Provide the scientific basis for Safety Standards.

However, the question is legitimate, whether there is still a need for a *Community* research programme on radiation protection, since every individual Member State has its own interests in radiation protection. The answer can be found in the following list of programme objectives:

1. Contribution to the scientific basis for the updating of Basic Safety Standards
2. Independent assessment of risks
3. Rapid referral of Community-wide problems arising in radiation protection
4. Stimulation and organization of cooperation on a Community scale is cost-saving and allow Member States to draw on the collective knowledge accumulated in the Community
5. Creation of a network of European scientists.

Furthermore Table 1 shows the various research area which are emphasized in the individual Member States and illustrates the importance of the Community Research Programme for stimulation and organization of cooperation on a Community scale. Every Member State has the possibility to optimize its own main expertise, having at the same time access to the expertise in another field, optimized by another Member State.

Another question can now be raised: whether there is still a need for radiation protection *research* and what the CEC is contributing to this research?

2. The CEC Radiation Protection Research Programme

2.1. Need for Community Radiation Protection Research

Living organisms have always been exposed to radiation from natural sources and have adapted to this situation. Medical and industrial developments have added other sources of radiation which have caused increasing concern to the public. More and more demands arise for a better knowledge of possible harmful consequences and greater protection of man and his environment.

Public acceptance of radiation technologies in industry and medicine will depend on the understanding of what radiation is and how it acts, what hazards it presents under operational conditions and after accidents, and how the benefits of radiation applications are perceived. When an emergency situation arises, authorities must be able

TABLE 1. Radiation protection research areas and priorities in Member States (without Luxembourg).

Subject \ Country	B	D	DK	F	I	IRL	NL	GB	GR	E	P
Dosimetry		++	(+)	++	+		+	++		+	
Ecology:											
terrestrial	++	++		++	+	+	+	++	+	+	++
marine		+	++	+	++	+	+	++		+	++
Accidents	+	++		++	+		++	+		+	
Non-stochastic/Rad. carcinogenesis:											
molecular	++	++	++	+	+	++	+	++	++	+	
animal	++	++		++	+		+	++			
Epidemiologie		++	(+)	+	+			+		+	
Embryologie	+	++		(+)				+		+	
Nuclide Metabolism	+	+		++	+			++		+	
Radiation genetics:											
molecular	+	+	++	+	+		++	+	++	+	
mammalian	++	++		+	+		++	++		+	+
Risk assessment	+	+		++	++		++	++	+	++	
Medical Exposure		++		++	+	+	+	++	+	+	+

++ high emphasis;
+ special interest;
(+) implicated.

to rely on a sound and comprehensive scientific basis of radiation protection in order to make rational decisions that will safeguard the health of individuals, protect the environment, and are also acceptable to the public. Such decisions which involve important social and economic considerations require a scientific assessment of risks and benefits for which research in radiation protection must provide the basis.

Such research must therefore define the extent of, and the ways by which exposure to radiation can occur. It must study the consequences of irradiation to man including the treatment of over-exposure, and it must assess, in a comparative way, the risks radiation can pose to man and his environment. Research must also contribute to an understanding of how radiation accidents occur, how they can be prevented and how their consequences can be reduced and managed.

In the Member States of the CEC about 30% of the electrical energy is produced with nuclear techniques and medical techniques using ionizing radiation are widely employed and are essential to good health care. Almost one million workers are potentially exposed to radiation as a result of their employment.

In this context it is instructive to have a look on the amount of radiation to which the EC population is actually exposed (Figure 1).

The subdivisions of the "cake" of annual radiation exposure vary as a function of the estimations of natural expcsure and also, to some extent, of the methods of calculations. These data can be complemented by the data for the exposure after the Chernobyl accident (Figure 2(*)).

The professional scientific community has consequently important reasons for vigorously pursuing radiation protection studies. These can be grouped under the following headings.

2.1.1. Cancer Risks. It has recently been recognized that the cancer risk estimates are significantly higher than those estimated about a decade ago and on which actual radiation protection standards are largely based. This necessitates reconsideration of standards for workers and for the public, re-evaluation of risk-benefit equations and a reconsideration of potential epidemiological studies that might be necessary to cast further light on cancer risks.

(*) (Ref. "A preliminary assessment of the radiological impact of the Chernobyl reactor accident on the population of the European Community", M. Morrey, J. Brown, J.A. Williams, M.J. Crick, J.R. Simmonds, M.D. Hill, National Radiological Protection Board, United Kingdom, EUR-11523, January 1987).

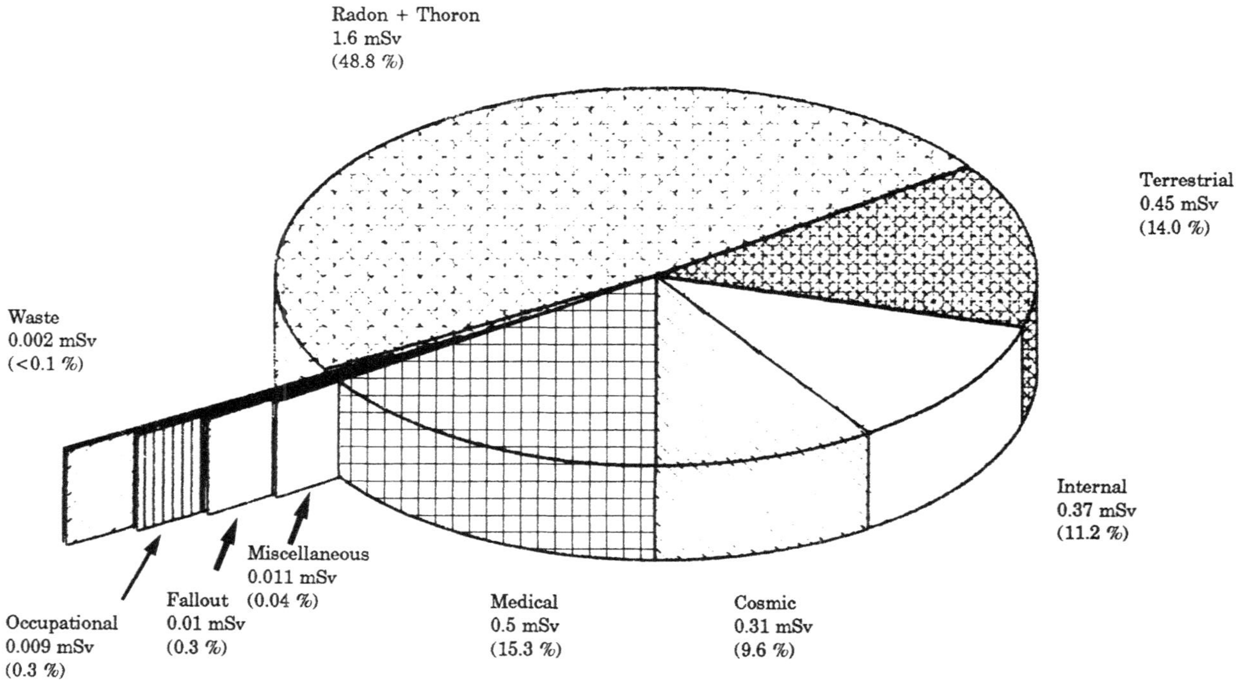

The total dose is 3.3 mSv. In the first year after the Chernobyl accident, this dose increased by 0.0002-0.35 mSv for the average adult in the different EC Member States.

Figure 1. Annual radiation exposure of the EC population under normal conditions.

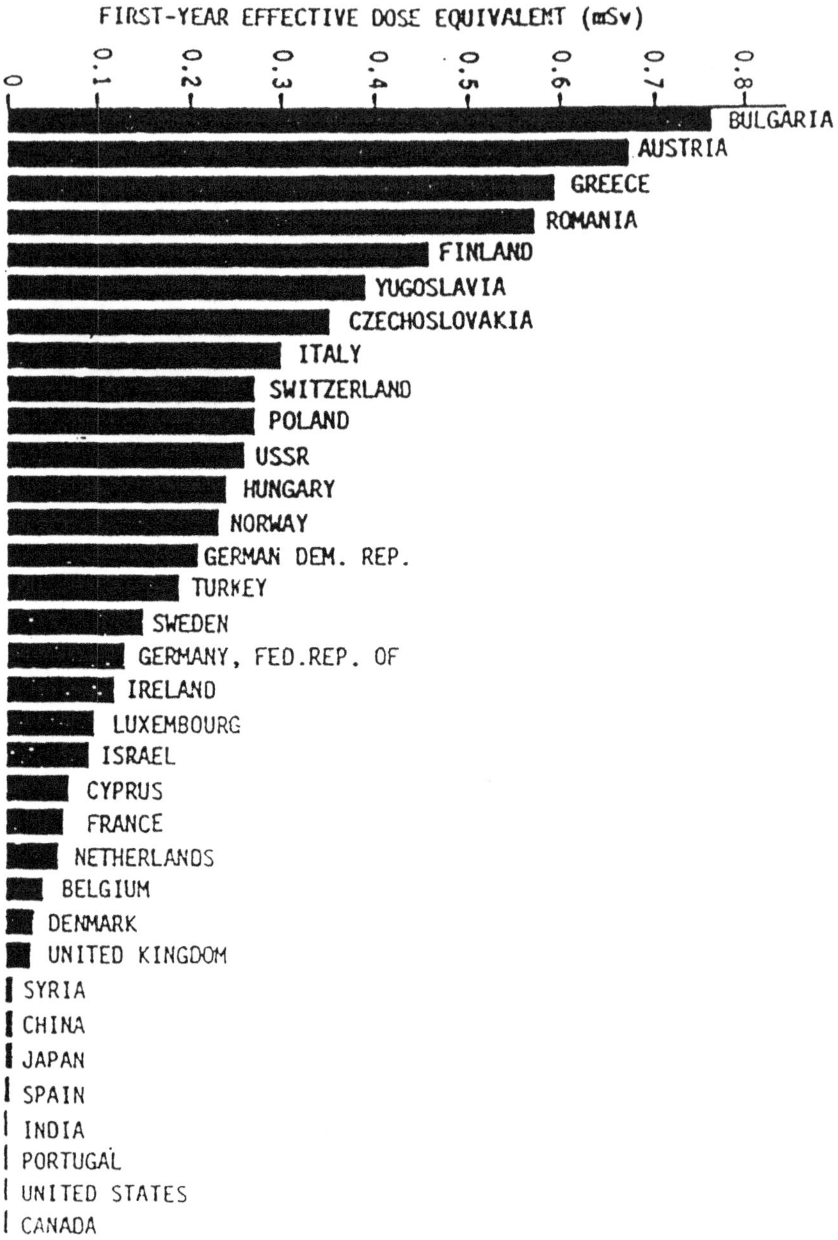

Figure 2. Country average first-year committed effective dose equivalent from the Chernobyl accident.

2.1.2. Genetic Effects. Risks of genetic damage must be considered under a new aspect: multifactorial diseases of genetic origin, which might appear only later in life, seem to be much larger than we recognized before.

2.1.3. Irradiation of embryo. Mental retardation, and some general decrease of IQ, has been found as being a particularly likely consequence of irradiation during a specific "high risk" period of gestation of the human embryo: during the 8-15 weeks (and with less sensitivity 16-25 weeks) during which the radiation risk is considerable. Although the possibility of a threshold for this effect is real and perhaps even likely, experimental work has so far failed to establish such a threshold.

2.1.4. Nuclear Accidents. The Chernobyl accident has demonstrated the need for additional efforts to clarify pathways for radionuclides after a serious accident. The capability to predict effects of dispersion and fall-out in all areas of the Community, and at different seasons of the year, must be strengthened. The capability to predict consequences of radioactive releases must include such phenomena as secondary contamination, for example, of soil, of lakes and forest areas, which are now being observed after the Chernobyl accident. Research must be intensified on transfer factors for radionuclides, such as Caesium through the biological systems to man, so that improved models will be available. The research must also provide background information for emergency preparedness in different European regions and with adherence to the IAEA convention on mutual assistance.

The need to be able to reduce radioactive levels in animals forming part of the foodchain, and to have a scientific basis for acceptable radioactive levels in foodstuffs was also demonstrated by the Chernobyl accident.

2.1.5. Radon in dwellings. In recent years radon has been recognized as a general problem of "industrialized" countries in which the indoor environment is heated and cooled during much of the year. Levels have been measured to be substantial and, in Europe, may account for about 10% of the 150,000 annual deaths due to lung cancer. This projects radon into the arena of *severe* public health hazards, and the countries of the European Community to a greater or lesser degree share this problem. The evaluation of the radon problem, the detection of the more significantly irradiated regions and homes, and mitigation of the more serious exposures is a considerable radiation problem of the latter part of the 20th century. It also emphasizes the critical importance of further radiobiological and epidemiological research of the radon problem.

2.1.6. Protection of Workers. A related but distinct problem concerns the exposure of radiation workers "at work" to all sources, including natural. The International Labour Organization (ILO) has recently declared that *all* such sources at work must be taken into account in evaluating workers' exposure. This will particularly concern the exposure of some workers to radon (and to a lesser degree to other natural and perhaps some man-made sources) while at work. It will necessitate a wide-spread programme of radon (and other radiation) measurement in thousands of work places not previously recognized as potential sources of exposure.

2.1.7. Radiological Examinations. Up to 15% of the total radiation exposure of members of the Community population and 80% of man-made exposure results from medical and dental radiology. Dose reduction must therefore remain in high priority and new methods of reducing dose or alternative procedures of imaging must constantly be sought. This subject area will be treated by a separate paper.

2.1.8. Effects of Low Doses. Research in radiation biology is of major importance, not only in establishing risk estimates at low doses, but also to advance our understanding of basic mechanisms. First of all, it is critical that we understand how to estimate the cancer risk at low dose and dose rate, compared with the high dose and dose rate to which Japanese atomic bomb survivors were exposed. Currently, the "dose rate effectiveness factor" is determined mainly from animal cell culture experiments in the laboratory since too few human data are available. Secondly, recent studies are beginning to point out ways to an elucidation of the molecular processes involved in radiation-induced cancer. Specific cellular or viral genes associated with the process of cancer induction (oncogenes) are now being characterized.

It is necessary to add two items to this list which represent different tasks in radiation protection: Training and public information.

2.1.9. Training. The Community Radiation Protection Research Programme has been evolving over some 30 years and many of its originators are approaching the end of their professional careers. Now it is essential to fund training programmes that will ensure: (1) continuity of existing knowledge, which has been carefully built up over a period of years, by a selected but still relatively small community of scientists, (2) the background to pursue research avenues that will advance knowledge in future, and (3) the availability of up-to-date knowledge and well trained scientific and medical specialists for advice in the case of accidents such as that at Chernobyl.

2.1.10. Public Information. There is a growing public awareness in Europe of the need for protection of the environment from chemical pollutants, carbon dioxide, and from radiation. Any Community policy concerning information on radiation protection must be based on sound scientific grounds. If the results of scientific research are conveyed to the public in an honest, comprehensive and easily understandable way, the public can acquire greater sense of responsibility and partnership in dealing with radiation problems.

2.2. Thirty years of CEC Radiation Protection Research

The first task of the Community Radiation Protection Research Programme 30 years ago was to establish an inventory of ongoing work and to compile a review of the existing state of knowledge. Subjects were identified which were of common interest to the Member States and which are now reflected in Table 1.

The Programme changed its name from Biology - Health Protection Programme to Radiation Protection Programme, and changed the scientific priorities. We are now at the threshold of the 7th Multiannual Programme.

It is the right moment to outline some highlights of achievements and to indicate the main orientations of the 1990-1991 Programme.

2.2.1. Highlights of Achievements of the 1985-1989 Radiation Protection Research Programme. The running Programme is subdivided into 6 sectors:
A Radiation dosimetry and its interpretation
B Behaviour and control of radionuclides in the environment
C Non-stochastic effects of ionizing radiation
D Radiation carcinogenesis
E Genetic effects of ionizing radiation
F Evaluation of radiation risks and optimization of protection.

Every sector has several subsectors and for only some of them achievements will be presented here (More details can be found in the Report EUR 12145, Evaluation of the Radiation Protection Programmes 1980-1984 and 1985-1989).

Sector A: Radiation Dosimetry and its interpretation
This is a research area which is involved in all the other scientific sectors of the Programme. Its achievements are therefore important for the whole Programme:

Microdosimetry (assessment of doses or dose distribution at the cellular and molecular level).
The elucidation of the possible effects of low doses, low dose rates

and radiation of different radiation quality required a strong development of microdosimetric methods and concepts, and the European Community is leading in this field thanks mainly to the CEC Programme:

- microdosimetry counters have been developed which are suitable for practical applications in radiation environments,
- track structures have been calculated for radiation of different quality and for internal emitters and are increasingly taking into account the composition of the involved tissues,
- biophysical models have been developed on the basis of these calculations and first efforts have been made to take account of factors such as repair.

Sector B: Behaviour and control of radionuclides in the environment

The Programme emphasized the dominant role that the chemical form (speciation) of a radionuclide has on its transfer in the environment.

Radioactive contamination and decontamination of animals

As a result of the observations after the Chernobyl accident and to a certain extent already before, transfer studies to animals have been adapted to more realistic situations in the agricultural and rural environment and, in particular, to natural and semi-natural ecosystems.

- The transfer of caesium through the soil-plant has been shown to depend on its chemical form; differences have been found between Chernobyl particulate Cs, Chernobyl ionic Cs and atomic weapons fallout Cs. Whereas formerly cattle was mainly studied for Cs transfer, other domestic animals have now been considered to a greater extent, and differences in uptake and retention between species have been characterized.
- Decontamination studies on animals contaminated with caesium have shown that treatment with Prussian blue can be very effective, removing up to 80% of the radioactivity, whereas treatment with bentonite gave variable results dependent on the type of soil on which the animals were grazing.

Sector C: Non-stochastic effects of ionizing radiation

In this sector those radiation effects are studied, for which the severity varies with dose and for which a threshold may therefore occur.

a) Treatment of victims of radiation accidents

   The group of contractors dealing with the study of the pathophysiology and treatment of radiation accident victims are among the most experienced world-wide. During the past few years, progress has been made by

- improving the isolation methods for hemopoietic stem cells for bone marrow transplantation in insufficiently matched donors and hosts,
- developing new physiological, biochemical and cytogenetic methods to assess the degree of damage, and develop the optimal therapeutic strategy after accidental local exposure of skin and underlying tissues.

b) Effects on the developing organism

It was notably recognized that:

- chromosome aberrations in the embryo prior to implantation can be transmitted to the foetal stages,
- a low dose rate exposure in utero has a much smaller effect on brain development (brain weight) than high dose rate.

Sector D: Radiation carcinogenesis

In this sector stochastic somatic effects of ionizing radiation are studied, for which the probability of an effect occurring rather than its severity depends on dose, without a threshold.

The spectrum of studies on radiation carcinogenesis is broad, ranging from the possible role of genetic predisposition for cancer; attempts to elucidate some of the molecular mechanisms underlying leukemogenesis and carcinogenesis, including the role of viral oncogenes and their cellular counterparts ("proto-oncogenes").

An oncogene is a latent gene that can be switched on or reinforced by radiation and might transform a cell into a cancer cell.

Among the outstanding achievements the following merit particular attention:

a) Development of in-vitro cell transformation systems and studies of oncogenes in radiation-induced cancer

- Several new cell transformation systems relevant for the human situation and/or for epithelial tissues have been developed and studied, in particular for thyroid, skin and bone cells. The latter system has been investigated in parallel with oncogene, animal and human investigations.
- The retro viruses, (responsible for the transmission of certain oncogenes) involved in radiation-induced osteosarcoma and thymic lymphoma have been further characterized: their life cycle during the latency phase of cancer, their gene products and their genetic alterations have been studied.

b) Epidemiology of thorotrast and radium treated patients

These two large studies have been supported by the programme from their beginning and now approach their final phase. The most important recent results are:

- thorotrast (thorium dioxide gels) injected for angiographic purposes some 30 to 40 years ago cannot only cause liver cancer, liver cirrhosis and leukemia, but also reduce life span in a non-specific manner and possibly cause pancreas tumour. The lung carcinoma thought to arise from the action of exhaled thoron seems, however, not to occur;
- the incidence of osteosarcoma from high doses of Ra-224 falls substantially at late times after exposure. There has been a considerable late incidence of non-stochastic damage in these patients, in particular cataracts and tooth breakage. The patients exposed to low doses of radium so far have not shown the expected incidence of osteosarcoma but a few other tumours related to connective tissues have been observed.

Sector E: Genetic effects of ionizing radiation

Carcinogen-induced alterations of chromosomal DNA (deoxyribonuclic acid) in replication-competent target cells as well as their enzymatic repair are of crucial importance in the process of carcinogenesis. The sequence-specific formation of structural DNA modifications and the differential capacity of cells for the repair of specific DNA lesions, are, therefore, highly relevant determinants of carcinogenic risk. For these reasons, many of the research topics contained in Sector E - Radiation Genetics - may equally well be considered as basic contributions to Sector D - Radiation Carcinogenesis. This is particularly the case for all studies concerned with radiation-induced mutagenesis and chromosomal alterations in higher cells (including germ cells) and with the characterization and cloning of eukaryotic DNA repair genes. Likewise, all attempts to establish ultrasensitive, rapid cytological tests based on technologies of modern molecular genetics may prove to be highly relevant in relation to the definition of carcinogenic risk following low dose irradiation.

a) Cloning of repair genes

The first human repair gene to be cloned (ERRC-1) is able to correct a DNA repair defect in Chinese hamster cells sensitive to UV radiation. The correction of DNA repair defects in human cells by transfection has been very difficult to achieve because human cells do not take up the transfected DNA fragments. The transfer of these repair genes is an important step on the way to identifying the genes, their products and their mode of action in the cell.

b) Mutation frequency in mammals

Measurements of mutation frequency in mice using dominant (cataract) mutations and a set of six recessive marker genes have indicated a lower mutation frequency than that derived in previous studies in the USA using a different set of seven recessive marker genes. These recent results from the programme will contribute to the derivation of more precise genetic risk estimates for man.

Sector F: Evaluation of radiation risks and optimization of radiological protection

Like Dosimetry this sector of research concerns the whole Programme and presents in a certain way the global evaluation and application of the data established by the research projects (Figure 3).

Over the last ten years it has become increasingly evident that more and more judgements have to be made on the expenditure and efforts that should be spent on radiation protection relative to the advantages and disadvantages such as protection brings. Radiation protection may have costs in terms of money directly spent upon it and in terms of lost output. It has benefits in terms of increased health or life quality.

The biggest areas of research effort have been in modelling the dispersion and impact of risk, associated with accidental and planned releases of radiation into the ecosystem and assessing the risks from radon and its daughters in people's homes.

a) Dispersion models

Large and complicated models of atmospheric dispersion of radionuclides from sites of origin have been developed, involving international co-operation. They have been supplemented with modelling work on aquatic and terrestrial dispersion, on movement of radionuclides through plants and into the foodchain, and on shielding provided by buildings in urban deposition.

b) Radon

Studies of radon have examined the factors influencing concentration levels in homes (building construction, ventilation, etc.), take up of radon by those exposed to it in aerosol form, risks from given levels of exposure and methods by which exposures might be reduced. The major consensus based upon physical and dosimetric considerations concerned the effective dose equivalent which was found to be more closely related to the time integral of the radon gas concentration. A reference value of 50 $\mu Sv \cdot y^{-1}$ per average $Bq \cdot m^{-3}$ radon gas concentration was derived for domestic environments.

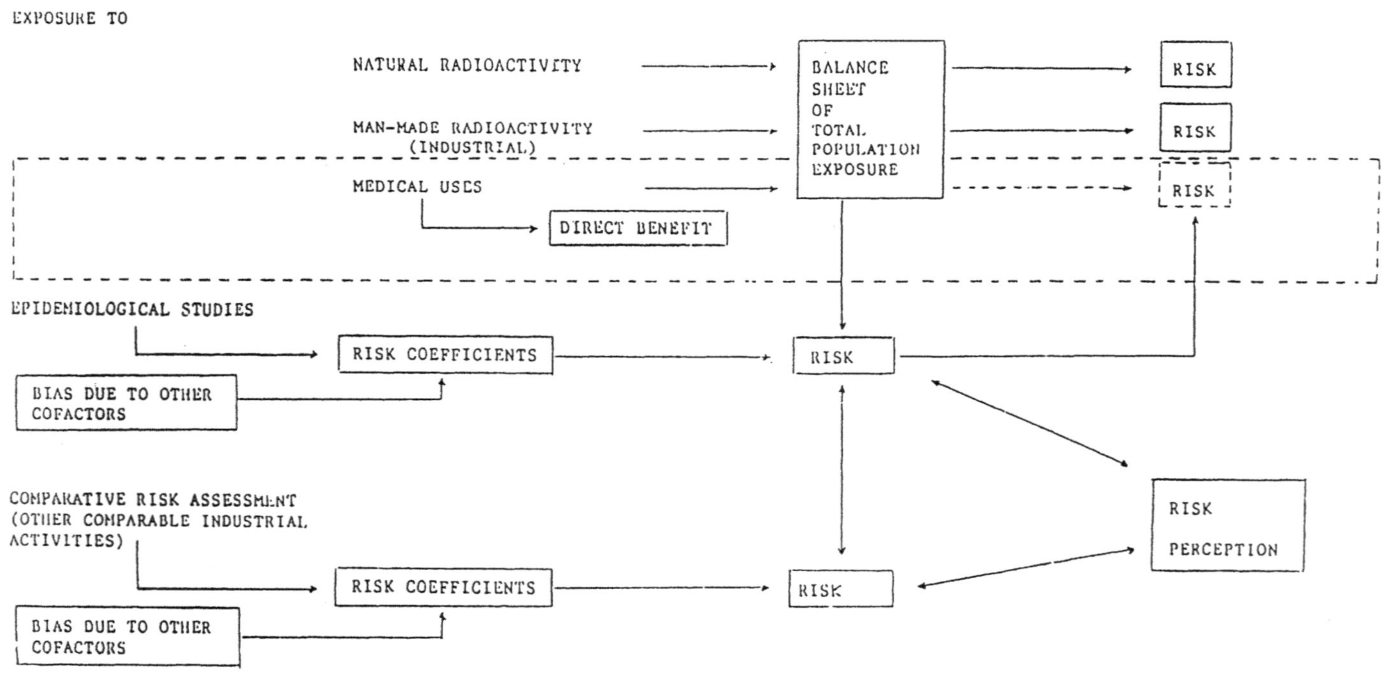

Figure 3. Sector F - Evaluation of radiation risks; broad fields of investigation.

Accident Consequence Assessment
The MARIA-project (**M**ethods for the **A**ssessment of the **R**adiological **I**mpact of **A**ccidents) launched in 1983, and a consequence assessment code (COSYMA - Code System of the Maria-project) is presently being prepared, and will be presented in April 1990 at the second MARIA Workshop. Some post-Chernobyl actions (long distance atmospheric transfer and accident scenarios) complemented the project. This led to a more comprehensive consideration of complex terrain configurations and the impact of large topographical formations, and to the development of computer based decision aiding modules for assessing the positive and negative consequences of the introduction of countermeasures.

Reduction of patient exposure in medical diagnostic radiology
It is hoped that the elaboration of a set of quality criteria for diagnostic radiographic images, based on the evaluation of results on optimization of image quality and patient exposure will stimulate dose reducing efforts in daily work of diagnostic radiology.

Theses quality criteria are at the same time a prerequisite for approaching comparable radiographs, particularly important with a view to 1993 when the large European Internal Market will become reality.

2.2.2. The Nuclear Reactor Accident at Chernobyl and the CEC Post-Chernobyl Actions. After the Chernobyl accident it was immediately recognized that the Radiation Protection Research Programme should reorientate some ongoing work and take up additional research in several areas.

The Radiation Protection Research Programme was consequently revised and additional financial means were given to the Programme in order to establish the necessary data for
- improvement of prediction of immediate and late radiological consequences of accidents;
- increased preparedness for possible future accidents.

The supplementary research work has been grouped under 10 subject areas:

1. Evaluation of the reliability and meaningfulness of long-distance atmospheric transfer models
   Long-distance atmospheric transfer models which describe and predict the spread of radioactivity from the site of an accident must be tested with respect to the data from the Chernobyl accident. Their degree of uncertainty must be assessed. If necessary, the available models must be modified or models better adapted to this accident situation must be developed. It must be investigated experimentally which influence different parameters had on the con-

tamination in the Member States of the European Community, for example the duration of the release or the physico-chemical transformation occurring in the plume.

2. Evaluation of data on the transfer of radionuclides in the foodchain
The large amount of data from the Chernobyl accident must be used to assess dynamic aquatic and terrestrial transfer of radionuclides entering the foodchain under a variety of climatic conditions and agricultural practices. Attention must be given to the intermediary storage and release of radionuclides in natural ecosystems and to the reliability and uncertainty of the models.

3. Feasibility of studies on health effects
The consequences of the Chernobyl accident to the health of the population in Member States are expected to be small and very likely not to be statistically significant. Information available on the health consequences of persons directly affected at Chernobyl will certainly be a valuable source of information for the effects of larger doses of radiation on man. The possible usefulness of epidemiological studies of these consequences must be carefully investigated. The value of observations and data collected must be assured. A report on this problem has been elaborated by a panel of high level scientists well-known in the world of epidemiology(*).

Research related to preparedness for possible future accidents

4. Radiological aspects of nuclear accident scenarios
The study of accident scenarios by probabilistic risk assessment must be extended following the lessons learned from the Chernobyl accident. Emergency management scenarios must be tested. Methodologies must be developed for real-time consequence assessment with adequate provisions for feeding-in data from early environmental measurements.

5. Underlaying data for derived emergency reference levels
The scientific basis for derived emergency reference levels following an accident must be assessed. The uncertainty of the parameters

---

(*) Doll, R., Boice, J.D., Estève, J., Silini, G. and Thiessen, J.W.: "Recommendations for research of an international panel of independent esperts" in "Feasibility of studies on health effects in western Europe due to the reactor accident at Chernobyl and Recommendations for research, Post-Chernobyl action report", EUR 12551, 1990.

relating these levels to the health effects must be evaluated. This could require improved data and models as a function of age for intestinal and inhalation uptake, for metabolism and for dosimetry with regard to the radionuclides likely to be involved in such accidents.

Improvement of practical countermeasures

Countermeasures in case of such accidents must be reviewed and improved with respect to the:

6. Agricultural and aquatic environment
   Studies must be carried out on agricultural practices and on techniques that are most effective in reducing contamination of the food chain. Appropriate techniques must be studied and developed to ensure a safe supply of water for consumption and for agricultural use.

7. Urban environment
   Decontamination research must deal with the physico-chemical mechanisms by which radionuclides interact and are bound to different surfaces in an urban environment.

8. Preventive medication
   Experimental research must include a risk benefit analysis of stable iodine medication, including possibly observations on the health of persons who now had taken such medication and on the risk of cancer in patients given I-131 for thyroid diagnostics.

9. Monitoring and surveillance in accidental situations
   Decision-making following a nuclear accident depends on the correct interpretation of a large number of measurements. The accident at Chernobyl has revealed several short-comings with respect to the practical aspects of the measurement and collection of reliable data, and to their communication to decision-making authorities in a comprehensible form. Studies must be made to define how the multitude of measurements from institutions of different competences can be made more rapidly and more reliably, and how their standardisation and harmonisation can be achieved.

10. Research into treatment methodologies of exposed persons
    The management of exposed persons who might develop radiation syndromes following whole or partial body exposure must be improved. Treatment decisions depend on accident dosimetry including physical and biological methods. Following whole body exposure, better

selection criteria must be elaborated to decide, for example, whether bone marrow transplantation is feasible. In the latter case, more rapid and improved techniques for tissue typing are needed. The treatment of the other patients also requires more study, particularly with respect to situations where many persons are involved, where other trauma present and/or where less than optimal treatment conditions prevail. The situation after localized exposure also requires further development of tests to decide on early optimal treatment schedules.

About 53 contracts have been concluded with the participation of 35 institutes and University laboratories of the European Community.

Due to the late decision on the revision of the Radiation Protection Programme the work started only end 1987. The reports about the studies will be available in 1990.

Two results might illustrate the work underway: Figure 4 and Table 2.

2.2.3. Research Priorities of the 1990-1991 Radiation Protection Research Programme. The 1990-1991 Radiation Protection Research Programme has been adopted on 20 June 1989 (O.J. L 200, 13 July 1989, p. 50).

About 700 research proposal have been submitted to the Commission. They have been evaluated by the Management and Coordination Advisory Committee "Radiation Protection" and the Commission services. About 300 research projects have been selected for contract negotiation for an amount of about 14 MioECU for the two years. The main subject areas to which high priority was given for this period are the following:

a) Human exposure to radiation and radioactivity

Measurement of radiation dose and its interpretation:

- development and implementation of standards and procedures linked to the concepts of dose equivalent quantities for both external and internal exposure;
- radiation measurement and instrumentation for individual and area dosimetry;
- derivation of organ doses and effective dose equivalents;
- assessment of internal exposure.

Transfer and behaviour of radionuclides in the environment:

- environmental behaviour of radionuclides in situations meriting particular attention for long-term behaviour or post-accident conditions;
- natural radioactivity in the environment and its pathways to man;
- influence of speciation, chemical modification, changes in physico-chemical properties and biological conversion, particu-

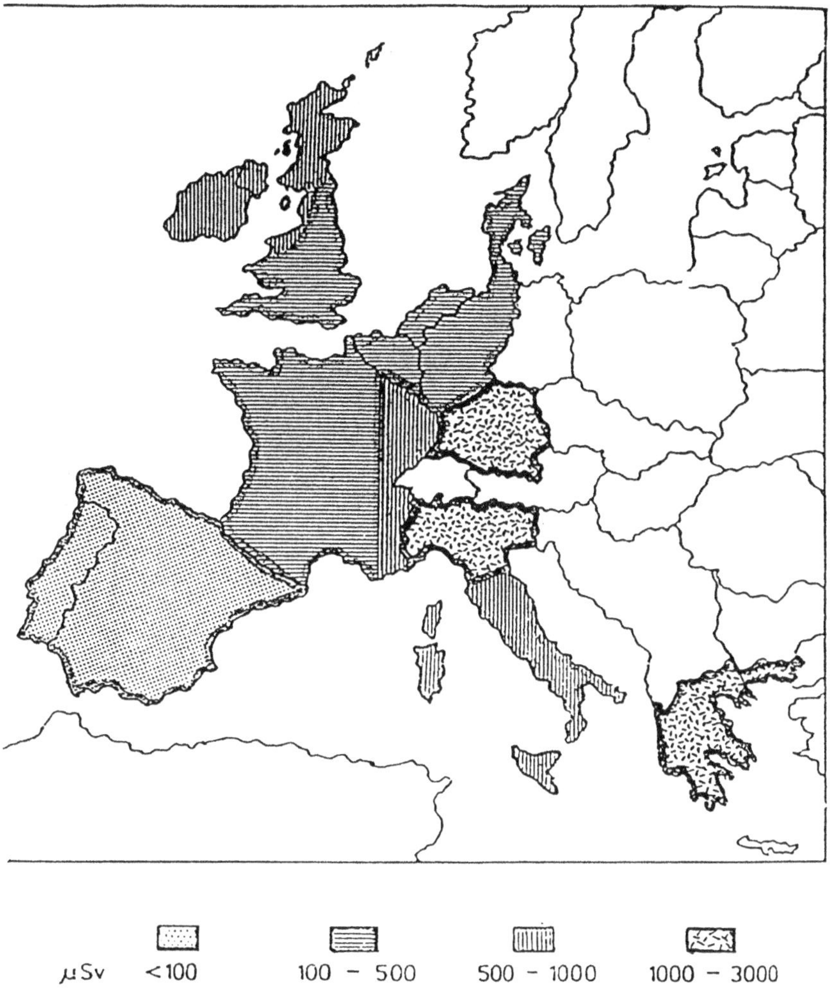

Figure 4. Critical individual effective dose in first year within the EC after the Chernobyl accident.

larly with respect to: long-lived fission and corrosion products, actinides, tritium and effluents from radiopharmaceutics or nuclear medicine;
- the behaviour of accidentally released radionuclides, evaluation of the reliability of transfer parameters and experimental studies;
- the role of retention and release of radionuclides in natural ecosystems such as forests, moorland, swamps, marshlands, water bodies and in marginal agricultural areas;

TABLE 2. The effect of countermeasures on the effective dose to the critical group.

Country (2)	Age Group (3)	Effective Dose to the Critical Individual in the First Year (μSv) (4)		% Dose Saved
		Theoretical Estimate without Countermeasures	With Countermeasures	
France	I	1100	970	12
FRG	I	3000	1900	37
Greece	A	1700	1300 (*)	24
	C	2200	1300 (*)	41
	I	2500 (*)	1000	60
Italy	A	1100	1100	- (5)
	C	1200 (*)	970	19
	I	1200 (*)	590	51
Netherlands	A	440	380 (*)	14
	I	500 (*)	330	34
UK	A/C	1100 (*)	820	25
	I	890	840 (*)	6

(1) The results are given to two significant figures to avoid rounding errors in subsequent calculations, and this degree of accuracy is not implied.

(2) No countermeasures were considered for the critical group for Belgium, Denmark, Ireland, Luxembourg, Portugal and Spain, see Appendices A and B.

(3) For age group I=1 year old infant, C=10 years old child and A=adult.

(4) Where doses for a country are given for more than one age group an asterisk (*) denotes the dose to the critical group.

(5) After implementing countermeasures the reduction in the dose was small.

- development of countermeasures to reduce the contamination in the environment and to impede its transfer to man.

b) Consequences of radiation exposure to man; their assessment, prevention and treatment

Stochastic effects of radiation:

- interpretation of low dose and low dose rate effects with the help of microdosimetry;
- repair and modification of genetic damage and individual radiosensitivity;
- cellular, molecular and animal studies to determine the risk of stochastic somatic effects of radiation with respect to low dose, low dose rate and radiation quality;
- assessment of genetic risks in man;
- action of radionuclides on target cells in relation to radionuclide metabolism and studies on biological models for radionuclide-induced cancer.

Non-stochastic effects of radiation:

- radiation syndromes and their treatment after exposure of large parts of the body;
- irradiation and committed exposure from incorporated radionuclides;
- radiation syndromes and their treatment after local exposure to skin and subcutaneous tissues;
- radiation damage to lens, thyroid and other tissues of relevance in radiation protection.

Radiation effects on the developing organism with emphasis on damage to the central nervous system, hemopoietic system, carcinogenesis and transfer of radionuclides.

c) Risks and management of radiation exposure

Assessment of human exposure and risks:

- evaluation and statistics of the different types of human exposure;
- exposure to natural radioactivity and evaluation of parameters influencing these risks;
- comparative assessment of exposure and risks;
- epidemiological studies in human populations.

Optimisation and management of radiation protection:

- optimisation of radiological protection;
- reduction of patient exposure in medical diagnostic radiology;
- management of radiological protection in normal and accident situations;
- probabilistic risk assessment and real-time models for assessing

the consequences of accidental releases of radioactivity and for evaluating effectiveness and feasibility of countermeasures.

These priorities are the logic follow-up of the preceding programme and the post-Chernobyl actions. They take into account the changing concepts on some radiation quantities, as they are discussed by the International Commission on Radiation Units and Measurements (ICRU) and the International Commission on Radiological Protection (ICRP).

After the selection of proposals the priority areas have shifted in some cases to more specific areas, leaving, however, some subject areas without proposals and consequently open for new initiatives.

3. Conclusions

The research projects will be carried out as shared-cost actions, which means that the CEC participates in research by contributing 25 to 50% of the overall costs, thus having an impact on about 35% of the whole radiation protection research in the Member States.

A special aspect has been introduced into the new programme, concerning more intensive coordination of research teams. It was required that the research proposals had to be coordinated before being submitted to the CEC. The participating scientists had to adjust their research work under common objectives. These preliminary coordination efforts assure that the available, limited financial means and manpower can be used more efficiently. In the new Radiation Protection Programme the 300 selected proposals have been grouped in about 100 coordinated projects.

This exercise is a convincing step towards the CEC objective to create the Europe of research workers. The basic ideas of this particular Community action are the following:

a) Reinforcing the European research community, using available manpower and funds in the *most efficient way*
b) Rationalizing the Community's efforts by linking together, as far as necessary and possible, the research of all Member States into *a coherent network*. This network will result in an intensive and cost-effective interdependence of European research
c) *Stimulating new research approaches* in areas most promising *for the well-being of the individual and the society*
d) Creating a climate of cooperation and collaboration in the Community that will make scientists feel as much at home in another Member State as in their own. Experience has shown that *quality of research can be spectacularly improved* by multiplying the opportunities for international and interdisciplinary contacts between scientists.

EXPERIENCE GAINED WITH OPTIMIZATION OF RADIATION PROTECTION

B. LINDELL
National Institute of Radiation Protection
Box 60204
S - 10401 Stockholm
Sweden

1. Introduction

This lecture is intended to discuss the experience and practical application of the ICRP principle of optimization of radiation protection, a principle which is sometimes believed to be too theoretical to be of practical value. The conclusion is that optimization of protection is, indeed, applicable in practice and that present experience of low exposures of radiation protection personnel and of the public is due to this principle rather than to application of the basic ICRP dose limits. In fact, contrary to common belief, it is the use of constraints based on the optimization principle which, in practice, begin to replace the primary dose limits.

Before discussing the experience, the lecture notes give an introduction to the concept of optimization.

2. What is optimization of protection?

As soon as radiation protection is sufficiently good to prevent all types of deterministic harm, and this should always be the case, protection against stochastic harm, i.e. cancer and hereditary disorders, becomes the immediate problem. Since no dose threshold is assumed for these effects, there would be a certain risk, albeit low, also below the dose limits recommended by ICRP. This remaining risk is not negligible, just as other risks in society are not always negligible in spite of protection efforts. The only justification for continuing practices which still cause some risk is then evidence that all that is reasonable has been done to reduce the risk.

However, risks must not be reduced ad absurdum. For the individual there is some level of incremental risk which is entirely negligible

M. Oberhofer (ed.), Advances in Radiation Protection, 51–63.

because it will not significantly change the total risk picture. Unfortunately, it is not so easy to find individual agreement on what that level is, because individual risk acceptance depends on many things in addition to the reported magnitude of the risk.

What is often not recognized is that even risks which may be insignificant and acceptable to the individual may derive from a source or practice which is likely to cause harm or even death to *some* individuals. W. Rowe, in his book "The Anatomy of Risk", has illustrated this with the drastic example of Russian roulette. The individual risk of participating in a game of Russian roulette is, history shows, acceptable or even enticing to some daring individuals. But would *the game* be equally acceptable when the other participants are one's best friends as it would be if they were one's worst enemies? The answer is that, in addition to the individual risk (probability of harm and possible individual consequence) one must also consider the overall consequences of a practice. Each individual may hope not to get harmed if the probability of individual harm is low but, if many individuals are involved, there will be near certainty of *some* harm.

The concept of source-related consideration of consequences is relatively new. A few decades ago, people responsible for radiation protection had little influence on the way radiation sources were designed and how installations were built. It was then only natural that protection of the individual in an already existing radiation environment was the primary concern, as in old radiological clinics. With increasing concern and extended powers of protection authorities, however, it was recognized that the most effective action against radiation risks was often at the stages of design and construction. Hence radiation protection (against threatening sources) was more and more replaced by requirements on source safety, and protection is now very much source-related.

Ideally, if every radiation source and all operations were subject to the best available technology and planning, there would be nothing further to request and dose limits might seem superfluous. In practice, however, application of the best available technology regardless of cost, may be an impossible requirement on a large scale and would mean an unwise distribution of resources if applied where the gain in reduced safety is small. In theory, one can always reduce exposures somewhat further, at increasing costs and diminishing yields. The question is how far it is reasonable to go.

On this question the ICRP answers with its recommendation that "all doses shall be kept as low as reasonably achievable, economic and social factors being taken into account" (ICRP Publication 26, 1977). From the initial letters of the first part of this phrase, the acronym "ALARA" has been derived, although it is not used by ICRP. To follow this recommendation is, according to ICRP, synonymous with *optimization of protection.*

Optimization of protection is a general requirement, but it is not a guarantee for an acceptable situation. There are two more principles which must apply. One is that the source or practice for which the protection is optimized must be justified, i.e. its introduction should change the overall situation to the better. Another is that no individual shall be exposed to a high risk. The last requirement can be met by making the optimization of protection subject to individual dose limitation.

It may seem odd that this last requirement is still needed if the protection is optimized and the introduction of the source or practice means an improvement. However, the benefit of the practice may not be enjoyed by the same people who run the risk, and under normal condition, the best "overall" situation (e.g. what is collectively best) must not be obtained at the cost of high risks to a few individuals. Therefore the optimization result must be subject to some individual dose constraints. These, which have sometimes been called "upper bounds", must be source-related and chosen by national authorities in such a way that there is no risk that the total of all practices with optimized protection will together cause individual doses which exceed the dose limits recommended by ICRP.

Optimization of protection has sometimes been confused with cost-benefit analysis. This is because CBA has been used to illustrate the principle. However, CBA is only one of several methods which can be used for optimization of protection. These range from sophisticated decision-aiding techniques, including CBA, to the simple use of common sense. Optimization of protection is therefore not necessarily a complicated, abstract concept.

There is a common confusion with regard to the term "optimization". Many things can be optimized, i.e. be subject to an attempt to find the best practical solution under the circumstances. The use of a particular method for diagnostic x-ray examination may be optimized as regards its diagnostic value. However, what ICRP recommends is the more limited concept of optimization of *radiation protection*, no less, no more.

3. Short history

Already in its 1954 recommendations, ICRP stated that "no radiation level higher than the natural background can be regarded as absolutely 'safe'" and recommended "that exposure to radiation be kept at the lowest practicable level in all cases". This recommendation was main tained in later recommendations, although with slight changes in wordings, and in its Publication 9 (1965) the Commission recommended "that any unnecessary exposure be avoided, and that all doses be kept as low as is readily achievable, economic and social considerations being taken into account". This recommendation was interpreted in ICRP Publication 22 (1973), where ICRP Committee 4 (on the Application of the Commission's Recommendations) used CBA to illustrate the optimization principle. The present wording then appeared in ICRP Publication 26 (1978).

In 1982 ICRP published a report prepared by a Task Group under Committee 4. It had the title "Cost-Benefit Analysis in the Optimization of Radiation Protection" (ICRP Publication 37) and in Appendices gave examples of practical applications.

One might say that the period from 1973 to 1982 was characterized by efforts to explain and sell the concept, but that reports on practical applications came first thereafter. The principles were discussed at IAEA seminars on the ICRP system of dose limitation (of which optimization of protection is the major part) in 1979 and 1981, and in the CEC 1st European Scientific Seminar on Radiation Protection Optimization, in Luxembourg in 1979.

Of several meetings which later discussed the practical implication, three are of particular interest:

- 2nd European Scientific Seminar on Radiation Protection Optimization "As Low As Reasonably Achievable", Luxembourg, 1983, (Report EUR, 9173).
- IAEA and NEA (OECD) symposium "Optimization of Radiation Protection", Vienna, 1986, (Proceedings 1986).
- NEA (OECD) ad hoc Meeting on "The Application of Optimisation of Protection in Regulation and Operational Practices" in Paris, 1988, (Proceedings 1988).

Of interest is also a recent report from another Task Group under ICRP Committee 4: "Optimization and Decision-Making in Radiological Protection" (ICRP Publication 55, 1989).

4. Differential cost-benefit analysis

Even though CBA is only one of several methods for optimization of radiation protection, it has attracted much attention because it is the method which has most frequently been used to illustrate the principles.

In short, the basis for the method is the wish of influencing radiation protection - but in theory nothing else - so that the total benefit of a practice is maximised. In the simplest form of the method it may be assumed that the total benefit is the sum of a number of additive components of which the cost of protection (X) and the detrimental effects of radiation (Y) are negative.

Changing protection may mean changing some independent variable (x), e.g. the thickness of shielding or the rate of ventilation, which will influence both the protection cost and the radiation effects. The total benefit will be maximised when its derivative with regard to x is zero, i.e. when a small change of x will neither improve nor reduce the total benefit. If we are restricted to protection only, the only terms which have a derivative other than zero are X and Y, so the condition for maximum benefit by protection alone will be

$$- \Delta X/\Delta x - \Delta Y/\Delta x = 0$$

which also means

$$\Delta X = - \Delta Y$$

When this condition is met, the sum of X and Y has a minimum and therefore the total benefit has the highest value it can have assuming that the other given conditions remain the same. The protection is then optimised, but the result is still subject to the applicable source-related individual dose constraint.

In order to make this calculation possible, X and Y have to have the same dimensions and be expressed in the same units. Since X is a monetary cost, the simplest solution is to express also the harmful radiation effects in monetary terms. The harmful consequences of low doses of radiation to an exposed population would be the mathematical expectation of the number of cases of cancer and severe hereditary effects. On present assumptions this would be proportional to the collective effective dose equivalent commitment S_E^c. To express the harmful biological consequences in monetary terms would therefore be equivalent to expressing collective effective dose equivalents in monetary terms, so that

$$Y = \alpha S_E^c$$

4.1. The value of α

In short, and in more practical terms, improved protection would be worthwile if each dose reduction can be bought for not more than α (dollars, francs, etc.) per unit collective dose, but not if the cost is higher. But, what is the value of α?

Ideally, the value of α should be such that if the same degree of protection was maintained not only with regard to all radiation practices but also with regard to other sources of stochastic risks (e.g. chemical carcinogenes, traffic, etc.), the number of lives saved would be the maximum that could be saved under the circumstances, i.e. for the money that the country could afford to provide for stochastic life saving. "Saving" a life would mean avoiding death of the particular cause which may be eliminated, putting the individual back in the remaining risk situation.

There is sometimes the misunderstanding that α is the amount of money that is actually found to have been paid per unit collective dose avoided by protection efforts. This is not correct if protection efforts have been influenced by considerations which do not relate directly to the collective dose. For example, much higher sums than indicated by the α value may be paid in order not to exceed dose limits or in order to maintain good public relations.

In order to justify higher protection costs under such circumstances some authorities have suggested different values of α under different conditions. Another solution, mentioned although not explicitly recommended by ICRP, is to add a "β term to the "objective" value of $Y = \alpha S_E^c$, to justify higher protection costs when individual doses approach limits or in cases of great public concern. The author of these notes does not believe in this solution which implies that some functional relationship between "Y" and radiation doses can describe the cost which is justified under such circumstances. He feels that the CBA result should be based on one single value for α, but that for the actual decision that result should be seen as only one of several inputs on which to base the decision. Concern about public relations may very well influence the decision but hardly according to an analytic formula.

The question still remains: What should the value of α be? It is easy to give a theoretical answer. The deciding factor is the amount of money (Q) that society wishes to set aside for stochastic life-saving. Is it 0.01 or 0.1 or 1 or 10 per cent of the GNP? It is to this decision any ethical problem relates, the rest is just arithmetics. In theory, it is possible to estimate the number of lives that could possibly be "saved" each year, e.g. those who are wasted in road accidents or from radiation-induced cancer. This number cannot exceed the total annual number of deaths. Furthermore, but only in theory, one might estimate the number of lives (N) which would be saved if a certain value of α was chosen (and the corresponding cost used in optimization of protection in other fields). Finally, the average cost (a) actually paid per unit collective dose eliminated by optimization of protection may be estimated; it must be less than α since some doses may be very inexpensively eliminated and none should be eliminated for a cost higher than α per unit collective dose.

If it is decided (by the government) to give α a very high value, the available sum Q, may be spent to save only a few lives. With a very low value of α, also very few lives would be saved. The maximum number of lives saved would be when Q = N.a, and this would define the "right" value of α.

In practice, values of α have been chosen more or less arbitrarily and even from case to case. This would not provide true optimization of protection, the value of α should be given once and for all, although, since it depends on the value of Q in relation to the GNP, it may differ from country to country.

Values that have been used range from perhaps only a few hundred U.S.$ per man.Sv in medical radiology to 5,000 - 20,000 $ per man.Sv in the nuclear industry. That the latter industry is very public-relation minded is shown by the fact that actual costs are sometimes as high as 1,000,000 $ per man.Sv, which indicates that public relations rather than life-saving has been seen as important.

The low ambition with regard to protection in medicine is subject to criticism and the situation may improve. It is rather the 10,000 $ per man.Sv which is believed to be an appropriate value for α. Whether this is true in relation to other areas of risk may be checked by an assessment of the implied cost per life saved. If we assume that the low dose risk coefficient is 2% per Sv (a likely underestimate in the light of the latest knowledge), the cost per life would be 10,000 $/0.02 = 500,000 $ or less. For societies to be consistent, this should be compared with the willingness to pay to reduce traffic accidents or to protect against chemical carcinogens.

It has sometimes been said that it is unethical to "put a price on the human life" as implied by CBA assessments. However, this is not what is done. It does not help to say that human lives have infinite values if infinite amounts of money are not available for protection, just as it does not mean that I do not value a painting by Rembrandt highly only because I can only afford to pay a small sum for it. Protection optimization by CBA is unethical if the value of Q set aside by society is low. It is not ethical if the value is high but still not infinite. There are no larger sums than the GNP available.

The ethics of CBA was discussed by an expert group called by the Pontificial Academy of Sciences in 1982. Its conclusion was that optimization of protection has the purpose of saving lives and reduce exposures below the dose limits and that "a similar method could be applied to the control of other toxic and mutagenic agents endangering man and his environment".

5. Practical applications of CBA

In the practical application it is rare that there is a continuous protection-related quantity which can be used as the independent variable in a differential cost-benefit analysis. In most practical cases, there is instead a choice between a number of options on how collective and individual doses may be reduced.

In Figure 1 such options are indicated by "O". Each option O_i may be characterized by the cost X_i of going from the initial situation to that option and by the collective dose (not the change) which would be the result if that option is chosen. The current situation is marked O_o. Since it is often difficult to specify the protection cost of the current situation, it is simplest to let X_i be the extra cost of going from the current situation to option O_i.

All options which fall in the shaded region below a line through O_o and with the slope - α will be *cost-effective* in a cost-benefit analysis. However, only one option would give optimized protection. It is the option which is the last to remain in the cost-effective area if this is reduced by moving the line with the slope - α downwards. In the example, therefore, option O_2 is the optimum.

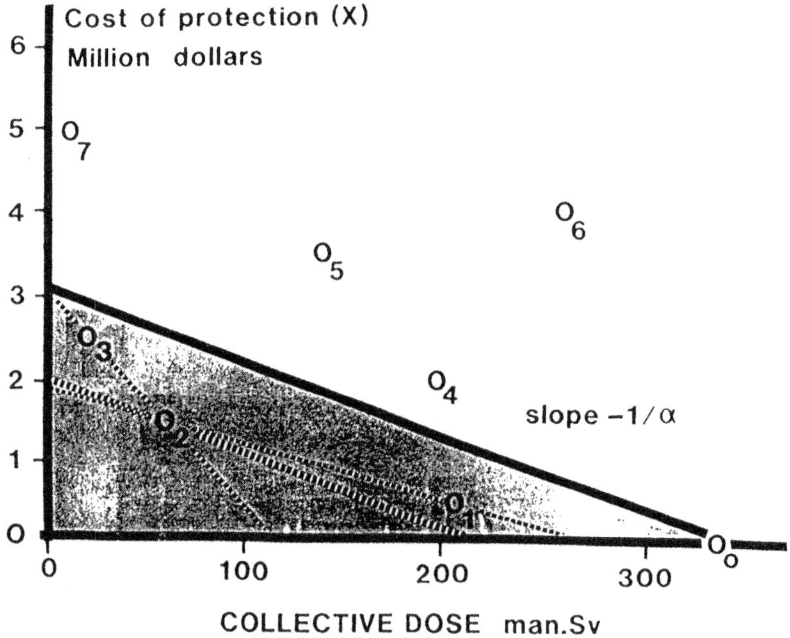

Figure 1.

Again, in practice, there may be much greater differences between the options and the choice may not be very critically dependent on the value of α. In ICRP Publication 37 there is an example with various options of removal of short-lived radioactive gases from a boiling water reactor. The following data were presented (for annual costs and dose commitments):

Option:	O_0	O_1	O_2	O_3
X ($)	0	940,000	2,340,000	7,000,000
S (man.Sv)	120	1.6	0.12	0.8

	0 - 1	1 - 2	2 - 3
$/man.Sv	7,900	946,000	116,000,000
	0 - 2	1 - 3	
	19,500	4,000,000	
	0 - 3		
	58,400		

In this example, option O_0 was actually not the initial stage but a minimum requirement in order to prevent individual dose limits to be exceeded. There was an initial cost of 2.32 M$ for option O_0 (a sand-tank as a delay allowing for some decay) but since this cost was forced by dose limitation it is irrelevant for the optimization study and has been subtracted.

If α is assumed to be around 5,000 - 20,000 U.S.$ per man.Sv, only options O_1 and O_2 are cost-effective. Since the difference in cost per man.Sv is rather small (7,900$ against 19.500$), one might think that option O_2 would be almost as good as option O_1. However, going from O_1 to O_2 involves 946,000 $ per man.Sv and is certainly not cost-effective. Even though it is cost-effective to go directly to option O_2 from the 0 condition, it would in theory be quite cost-effective then to accept the extra 1.48 man.Sv and continue to option O_1. In reality, after one option has been installed, any change may be prohibitively expensive, the optimum option should therefore be chosen directly. Option O_1 is the optimum option for any value of α between 7,900 and 946,000 $ per man.Sv, which again illustrates that there is often a low sensitivity to the value of α.

In addition to the risk of confusing cost-effectiveness with optimization, other mistakes may be made. One would be to compare the annual cost with the annual collective dose. This would be incorrect, since we wish to avoid also doses which each year will be committed to occur in the future. Therefore, the annual collective dose *commitment* must be used. However, in the previous example the difference would be negligible since the options referred to retention of short-lived radionuclides.

In the practical application, the theoretical formalism is less important than the simple rules:

- Never reduce protection if the monetary gain by such action is less than α $ per man.Sv.
- Always improve protection if this can be done for less than α $ per man.Sv avoided.

How much practical application has there yet been of optimization of protection guided by CBA?

First, there is still no universal agreement on the value of α or on how to arrive that value. An exception is the recommendation of a transboundary value of α in IAEA Safety Series No. 67 that, in costing collective doses which fall outside the borders of the source country, no lower value of α should be used than that which is used for doses within the country, and that under no circumstances the transboundary value of α should be less than 3,000 U.S. $ per man.Sv, at 1983 prices.

However, based on various more or less formally agreed values of α within a country or within a practice (e.g. nuclear power production in some countries) a large number of optimization exercises have been reported. Some authors, however, have expanded the theory into what might be seen as pseudo-science, particularly since the scientific and societal basis for costing a unit collective effective dose equivalent is fairly weak, involving great uncertainties. Recent practical applications, however, relate to radiation protection problems of great economic and social importance, e.g. reduction of the radon levels in dwellings and agreements on food restrictions in the case of nuclear catastrophies, like Chernobyl, which may cause contamination of very large quantities of food. In these applications, crude CBA assessments have been of great value, although it has to be recognized that there are factors other than protection cost and collective dose which have decisive influence on decisions, e.g. the public's perception of risks and a number of "intangibles" which are extremely difficult to quantify. One example is the effect that the radioactive contamination of reindeers, game animals and fish had on the living conditions and the culture of the Laplanders in the Nordic countries after the Chernobyl accidents.

Another problem is caused by the communication problem. Lay people understand dose limits and appreciate that authorities want to protect them from high doses. They do not easily understand the rationale for actions against low doses for the purpose of reducing collective doses, and when authorities act for this purpose, the public draws the wrong conclusion that it is individual doses that are high. This is a dilemma which all responsible authorities are familiar with, and the temptation not to act against consequences the existence of which may never be proven is high. Fortunately, however, this problem now also becomes common with regard to carcinogenic substances and there is a hopeful learning process going on.

The nuclear industry is a very special case. Both authorities and managements in many countries have imposed restrictions which are sometimes far beyond "as low as reasonably achievable". However, in many cases where the restrictions have been so stringent that local protection groups argue that there is no room left for "ALARA", it may well be that the restrictions are stringent just because the national authorities had made optimization analyses and found reasons to be restrictive. Whatever the reason, individual doses as well as releases to the environment from commercial nuclear power plants under normal conditions are in many cases very low indeed.

Instead, it is the risk of major accidents with environmental consequences which is the main concern for protection and safety authorities all over the world. Also nuclear safety can be optimized and today there are efforts to combine nuclear safety and radiation protection principles i a unified policy. But to protect not only against collective doses that would appear without protection, but also against collective doses that might appear, is even conceptually a difficult task, and a new art of protection against probabilistic events is developing.

6. Other optimization tools than CBA

It has been said over and over again that cost-benefit analysis is not synonymous with optimization of protection but only one of several available tools, e.g. multiattribute utility analysis and multicriteria outranking analysis and - common sense. The recent ICRP Publication 55 gives a good review of the whole field and is highly recommended for reference. In that report, there is an Appendix illustrating the use of various optimization methods in the case of the choice of the ventilation system in a small uranium mine (the data were taken from a report by Lombard, Oudiz and Zettwoog in 1954, Rapport CEA R5263). The same ICRP publication also gives an annotated bibliography with examples of case studies. The already mentioned 1988 NEA (OECD) ad hoc meeting proceedings are also very useful.

7. Optimization in medicine

It has already been said that, to the extent optimization of protection is applied in the protection of staff and patients in the medical uses of radiation, quite low values of α have been used. This reflects the availability of resources, where protection against stochastic harm has to compete with the obvious and acute demand of life-saving actions on people with severe disorders. The threat against a human life which is not anonymous is usually drawing greater protection resources than the risk of a "stochastic death".

A confusion worth mentioning with regard to optimization of protection of the patient in, for example, diagnostic x-ray examination is that there are more boundary conditions than a high radiation dose. A very low radiation dose might also be objectionable, because it may not provide the required diagnostic result. It is then difficult to optimize radiation protection without also changing conditions which do not relate to protection. Radiologists would prefer to optimize the examination.

This may involve judgments which protection staff might not be competent to question. This problem is quite clear with regard to the taking of an x-ray picture. However, some thought may reveal that similar problems, although not so obvious, make it difficult to talk about "pure" radiation protection also under other circumstances. Food restrictions and the threat against the whole lifestyle of the Laplanders is an example already mentioned.

6. References

Commission of the European Communities. (1979) Proceedings of the 1st European Scientific Seminar on Radiation Protection Optimization: The Practical Problems of Radiation Protection Optimization, Luxembourg, 3-5 October, EUR 700.

Commission of the European Communities. (1983) Proceedings of the 2nd European Scientific Seminar on Radiation Protection Optimization: As Low As Reasonably Achievable, Luxembourg, 8-9 November, EUR 9173.

International Atomic Energy Agency. (1979) Application of the Dose Limitation System for Radiation Protection - Practical Implications. Proceedings of a Seminar, Vienna, 5-9 March 1979, sponsored by IAEA, ILO, WHO, NEA (OECD) and ICRP. IAEA, Vienna.

International Atomic Energy Agency (1982) The Dose Limitation System in the Nuclear Fuel Cycle and in Radiation Protection. Proceedings of a Symposium, Madrid, 19-23 October 1981, sponsored by IAEA, WHO, NEA (OECD) and ICRP. IAEA, Vienna.

International Atomic Energy Agency (1986) Optimization of Radiation Protection. Proceedings of a Symposium, Vienna, 10-14 March 1986, jointly organized by IAEA and NEA (OECD). IAEA, Vienna.

International Commission on Radiological Protection (1955) Recommendations of the ICRP revised December 1, 1954. Brit. J. Radiol. Supplement No. 6.

International Commission on Radiological Protection (1966) Recommendations of the ICRP, adopted 1965. ICRP Publication 9. Pergamon Press, Oxford.

International Commission on Radiological Protection (1973) Implications of Commission Recommendations that Doses be kept as Low as Readily Achievable. ICRP Publication 22. Pergamon Press, Oxford.

International Commission on Radiological Protection (1977) Recommendations of the ICRP. ICRP Publication 26. Annals of the ICRP, Vol. 1, No. 3.

International Commission on Radiological Protection (1983) Cost-Benefit Analysis in the Optimization of Radiation Protection. ICRP Publication 37. Annals of the ICRP Vol. 10, No. 2/3.

International Commission on Radiological Protection. (1989) Optimization and Decision-Making in Radiological Protection. ICRP Publication 55. Annals of the ICRP Vol. 20, No. 1.

Lombard, J., Oudiz, A., and Zettwog, P. (1984) "Contribution a l'optimisation de la protection du personnel dans un mine d'uranium". Rapport CEA R5263.

Nuclear Energy Agency (NEA) of the OECD (1988) Proceedings of an ad hoc Meeting on the Application of Optimisation of Protection in Regulation and Operational Practices, Paris 16-18 March 1988. NEA (OECD), Paris.

Rowe, W. (1977) 'An Anatomy of Risk'. New York, Hohn Wiley & Sons.

NEW OPERATIONAL QUANTITIES IN RADIATION PROTECTION

T.O. MARSHALL
National Radiological Protection Board
Chilton, Didcot, Oxon, OX11 ORQ
United Kingdom

1. Introduction

Over the past 30 years quantities and concepts in radiological protection have undergone substantial changes. Initially radiation doses were measured in terms of the quantity exposure, the ability of photon radiation to produce ionisation in air, which was only indirectly related to effects of radiation on the human body. This shortcoming led to the introduction of the quantity absorbed dose to tissue which gave a measure of the energy deposited in tissue by the radiation concerned. Although an improvement, some ambiguity still remained, concerning the likely effect in a persons body, simply because of the differences in biological effectiveness of some types of radiation relative to others. This led to the quantity dose equivalent in which a modifying factor, Q, is applied to the absorbed dose to allow for this property. Dose equivalent is now the basic radiation quantity and has the advantage that, provided the risk coefficients are known, the risk from radiation of any type or energy can be assessed on a common scale. In Publication 26 the ICRP took this one stage further and recommended that dose limitation for stochastic effects should be based on the quantity effective dose equivalent. This quantity is the sum of the weighted mean dose equivalent in a number of different organs or tissues within the body. The weighting factors are based on epidemiological evidence for the different radiation sensitivities of the organs.

By this quantity the detriment to the individual, as a result of a radiation exposure, can be assessed in terms of the increase in the risk of dying from cancer or there being serious hereditary effects in the first two generations of the decendants. This assessment can be made irrespective of the radiation type and energy and applies equally well to internally or externally received doses. Effective dose equivalent is a primary quantity underlying the Basic Safety Standards of the European Communities.

M. Oberhofer (ed.), Advances in Radiation Protection, 65–85.

Effective dose equivalent is clearly a powerful concept, allowing the comparison of irradiation of different organs or tissues to be made on a uniform basis, but it provides a major difficulty for instrument and dosimeter manufacturers because it is impossible to measure or assess this quantity in practice. ICRU has agonised long and hard over this problem and eventually new operational quantities were recommended in their report number 39 (ICRU (1985)). These quantities are based on the dose equivalent in the surface layers of the human body for individual monitoring and in a 30 cm diameter tissue equivalent sphere for area monitoring. It has been shown that these quantities give a reasonable estimate of the effective dose equivalent and the NRPB and BCRU have recommended their implementation in the U.K. Acceptance of the new quantities elsewhere is generally growing. In particular the CEC has decided to revise the "Technical Recommendations for Monitoring the Exposure of Individuals to External Radiation" (EUR 5287) (CEC (1975)). The new document will recommended the use of the new quantities for individual monitoring.

2. Limiting and Secondary Quantities

2.1. Limiting Quantities

The radiation quantities we should ideally assess stem from the adopted system of dose limitation. For non-stochastic effects ICRP in Publication 26 (ICRP (1977)) places limits on the dose equivalent to individual organs, H_T, while for stochastic effects, the Commission places limits on effective dose equivalent H_E. These concepts have since been adopted by the CEC in their Basic Safety Standards (CEC (1980)) and they are consequently built into the national legislation of the member states of the EEC. Thus, these are the quantities which should, in principle, be measured in radiation monitoring.

The effective dose equivalent is the sum of the weighted mean dose equivalents in six specific organs or tissues, each with its own weighting factor, plus the dose equivalents in a further five of the remaining organs receiving the highest dose equivalents, each with a weighting factor of 0.006. The specified organs and the corresponding weighting factors are given in Table 1.

This quantity provides an estimate of the risk to the individual of serious hereditary defects in the first two generations and of fatal cancers and is formulated to take into account such factors as the variation of dose within the body and the different radiosensitivities of the various organs. It assesses the consequences of radiation expo-

TABLE 1. Risk factors for serious hereditary defects in the first two generations and for fatal cancers from irradiation of various tissues and organs.

Tissue or Organ	Risk Factor		Weighting Factor
	Per Sv	(Sv^{-1})	
Gonads (hereditary)	1 in 250	($4x10^{-3}$)	0.25
Breast	1 in 400	($2.5x10^{-3}$)	0.15
Red bone marrow	1 in 500	($2x10^{-3}$)	0.12
Lung	1 in 500	($2x10^{-3}$)	0.12
Thyroid	1 in 2000	($5x10^{-4}$)	0.03
Bone surfaces	1 in 2000	($5x10^{-4}$)	0.03
Remainder	1 in 200	($5x10^{-3}$)	0.30
Whole body	1 in 60	($1.65x10^{-2}$)	1.00

sure in terms of the detriment to the individual, i.e., his increased risk of suffering the effects mentioned earlier. It is thus more meaningful than the basic quantity, dose equivalent from the point of view of radiation protection in general, in epidemiological studies and cases of litigation concerning claims for compensation.

However, in practice, we cannot assess either H_T or H_E because it requires a knowledge of organ dose. It was to overcome this difficulty that ICRU in Publication 39 (ICRU (1985)) recommended new practical quantities or secondary quantities for operational use which provide an estimate of H_E, avoiding under-estimation and excessive over-estimation, and an upper limit to the value of H_T. Two quantities are recommended for area monitoring: ambient dose equivalent and directional dose equivalent. Similarly two quantities are recommended for individual monitoring: individual dose equivalent, penetrating and individual dose equivalent, superficial.

2.2. Secondary Quantities for Environmental Monitoring

2.2.1. Ambient dose equivalent, H(d).* Ambient dose equivalent, H*(d), at a point in a radiation field is the dose equivalent that would be produced by the corresponding expanded and aligned field in the ICRU (ICRU (1980)) at a depth, d, on the radius opposing the direction of the aligned field. The sphere (ICRU (1980)) is defined as a 30 cm diameter tissue-equivalent sphere with a density of 1 g cm^{-3} and a mass composition of 76.2% oxygen, 11.1% carbon, 10.1% hydrogen and 2.6%

nitrogen.

This quantity is appropriate for *strongly* penetrating radiation and the recommended value for d is 10 mm. It is written as H*(10). The ICRU state, in ICRU Report 39 (ICRU (1985)), that "In the expanded field, the fluence and its angular and energy distributions have the same values throughout the volume of interest as in the actual field at the point of reference. In the expanded and aligned field, the fluence and its energy distribution are the same as in the expanded field, but the fluence is unidirectional". This concept is illustrated in Figure 1.

It is also noted in ICRU Report 39 (ICRU (1985)) that "An instrument that has an isotropic response and is calibrated in terms of H* will measure H* in any radiation fields that are uniform over the dimensions of the instrument". It is implicit in this note that the instrument has the correct energy response.

2.2.2. Directional dose equivalent, H'(d). The directional dose equivalent, H'(d), at a point in a radiation field, is the dose equivalent that would be produced by the corresponding expanded field in the ICRU sphere at a depth, d, on a radius in a specified direction.

This quantity is appropriate for fields containing *weakly* penetrating radiations, and the currently recommended value for d is 0.07 mm. It is written as H'(0.07).

Remarks

1. The above definitions of ambient and directional dose equivalent require neither the presence of a sphere during measurements nor sufficient room at the point of measurement to accommodate the sphere. It merely refers to the dose that would exist in the sphere, if it were placed in an expanded and aligned field in the case of H*(10), or in an expanded field in the case of H'(0.70), corresponding to the field at the point of interest and as such defines the way that the response of instruments should vary with the energy and angle of incidence of the radiation.
2. The ICRU report (ICRU (1985)) defines *weakly* and *strongly* penetrating radiations as follows:

   "If the dose equivalent received by any small area of the sensitive layer of the skin is more than 10 times larger than the effective dose equivalent for a given orientation of the body in a uniform and unidirectional radiation field, the radiation is said to be weakly penetrating. If the dose equivalent received by any small area of the skin is less than 10 times larger than the effective dose equivalent for

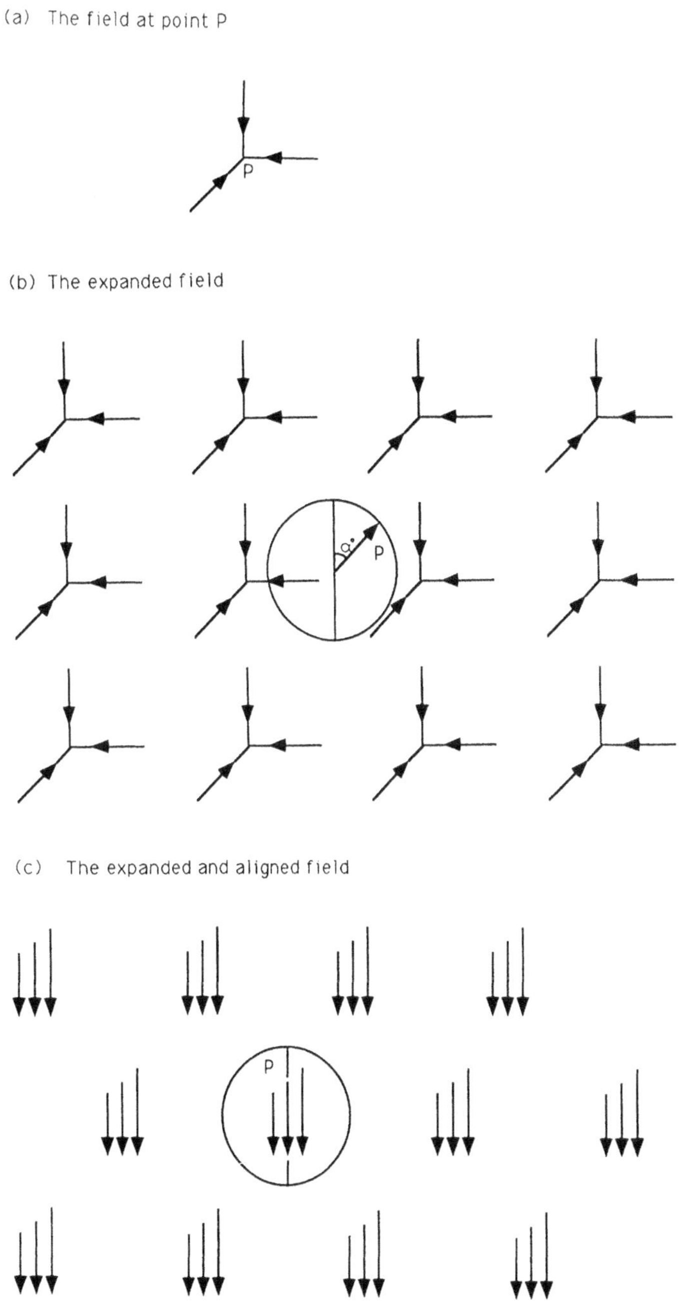

Figure 1. Expanded and aligned fields.

a given uniform and unidirectional field and orientation of the body, the radiation is said to be strongly penetrating".

This definition stems from the relationship between the dose limits for skin (500 mSv) and whole body (50 mSv), and the relative penetration by the radiation to depths of 0.07 mm and 10 mm. If the radiation is attenuated by less than a factor 10 in penetrating to a depth of 10 mm then the whole body is the critical tissue since its dose limit will be reached before that of skin. Hence the radiation is considered to be strongly penetrating. The opposite applies if the radiation is attenuated by more than a factor 10 in a depth of 10 mm and it is in this case classed as weakly penetrating.
By this definition neutrons of all energies are strongly penetrating, but only those beta rays with energies above about 3 MeV and photons with energies greater than 15 keV are considered strongly penetrating.
The ICRU report (ICRU (1985)) notes that "An instrument which determines the dose equivalent at the recommended depth in a plane slab of tissue equivalent material will adequately determine H' for weakly penetrating radiation if the slab surface is perpendicular to the specified direction and the radiation field is uniform over the entrance face of the instrument".

3. The above definitions of ambient and directional dose equivalent require neither the presence of a sphere during measurements nor sufficient room at the point of measurement to accommodate the sphere. It merely refers to the dose that would exist in the sphere, if it were placed in an expanded and aligned field in the case of $H^*(10)$, or in an expanded field in the case of $H'(0.70)$, corresponding to the field at the point of interest and as such defines the way that the response of instruments should vary with the energy and angle of incidence of the radiation.

2.3. Secondary Quantities for Individual Monitoring

2.3.1. Individual dose equivalent, penetrating, $H_p(d)$. Individual dose equivalent, penetrating, $H_p(d)$, is the dose equivalent in soft tissue below a specified point on the body at a depth, d, that is appropriate for strongly penetrating radiations.

The currently recommended depth, d, for monitoring in terms of $H_p(d)$ is 10 mm. It is written as $H_p(10)$.

2.3.2. Individual dose equivalent, superficial, $H_s(d)$. Individual dose equivalent, superficial, $H_s(d)$, is the dose equivalent in soft tissue below a specified point on the body at a depth, d, that is appropriate for weakly penetrating radiations.

The recommended depth for monitoring in terms of $H_s(d)$ is 0.07 mm. It is written as $H_s(0.07)$.

Remarks

The last two definitions require the type testing of individual dosemeters to be carried out on a suitable phantom to represent the body of the wearer and, in particular, to simulate scatter. In Report 39 (ICRU (1985)), ICRU state that, for dosemeters to be worn on the trunk, the ICRU tissue equivalent sphere is suitable. This means that such dosemeters should be type tested for their energy and angular response characteristics on the sphere in terms of the appropriate sphere quantity, i.e., directional dose equivalent. (A more comprehensive discussion of the relationship between the new quantities for environmental and individual monitoring is given in the following chapter on Calibration). Phantoms which represent the human body more closely are conceivable, but bearing in mind the physical differences between one person and another their advantages are not great and, in the interests of uniformity, NRPB (NRPB (1986)) recommend that the ICRU sphere should be used.

3. Requirements for Secondary Quantities in Individual Monitoring

Secondary quantities are essential for legislators working in the field of radiation protection and those responsible for monitoring programmes and dosimetry services. The choice of secondary quantities also has implications for those involved in dosemeter design and development and for those concerned with the type testing of dosemeters. Many of the requirements for secondary quantities are common to all those persons mentioned above but some requirements are specific to particular areas of work.

In the present circumstances discussed above suitable secondary quantities are essential for legislators because it is clearly unacceptable to base legislation on primary quantities which are impossible to measure or assess. Suitable alternatives must be included in Regulations which can be measured or assessed fairly readily with acceptable uncertainties but which nevertheless give a reasonable estimate of the primary limiting quantities. These secondary quantities need to be metrologically sound, to have been extensively and carefully studied, and, preferably have the backing of authoritative bodies such as ICRU and ICRP. Moreover, legislators in their choice of secondary quantities, should not, if possible, place an undue burden on the resources of monitoring programmes and/or dosimetry services.

Those responsible for individual monitoring programmes and those operating individual monitoring services need secondary quantities which are straightforward to use to allow them to work in an efficient and cost effective manner. However, they must also be able to defend their methods of operation, if necessary, to the approval authorities, in the Law Courts and to the satisfaction of their customers; secondary quantities which are metrologically sound with the backing of an international organisation such as the ICRU are clearly essential in this context. At present it is common practice for monitoring services to measure a mixture of radiation quantities. For photon monitoring with film this is the dose equivalent at or near the surface of the body but with TLDs, for the same radiation, it is the dose equivalent at a depth of 0.07 mm and at some specified depth within the range 400-1000 mg cm^{-2}.

For neutrons the maximum of the dose equivalent distribution in a cylindrical phantom (MADE) is assessed and for beta rays the dose equivalent at a depth of 0.07 mm. Thus the quantity chosen not only depends on the type of radiation being measured but also on the technique adopted, which, in principle, poses problems in adding the various components to get the total dose. What is required is a unified system which is additive so that the total dose, from whatever source of radiation, can be computed. Moreover occasionally it is necessary to control the exposure to individuals at, or near to, the dose limits. In these circumstances it is necessary to be able to obtain a reasonably accurate assessment of effective dose equivalent, and this should be possible with the scheme of secondary quantities adopted. Another important aspect in operating monitoring services is quality assurance. In this respect it is important for monitoring services to benefit from taking part in dosemeter intercomparison exercises and scientific discussions. A widely accepted and unified system of dose measurement would help substantially in this respect since it would ensure that in these exercises like would be compared with like, etc. From time to time it is necessary to make large capital investment in individual monitoring services if they are to be kept up to date and reliable in operation. Dosimetry services are likely to make this investment if a unified system of dose measurement exists and the objectives are clear. A good set of secondary quantities could therefore make a contribution towards maintaining good standards of individual monitoring.

Those involved in the design and production of dosemeters and dosimetry systems need clear, and as far as possible, established objectives on which to base their designs, in particular with respect to the variation of response with radiation energy and angle of incidence. Only then can commercial organisations be expected to invest in the production of the necessary equipment and dosemeters.

Finally, those involved in using radiation facilities to type test dosemeters would clearly prefer any new quantities to be formulated in such a way that calibration beams may be established without the costly procedure of producing primary or secondary standards for this purpose.

4. Suitability of the ICRU Quantities

Qualitatively the new ICRU quantities go a long way towards satisfying the requirements of secondary quantities for individual monitoring. They have been studied extensively by ICRU and clearly have the backing of both ICRU and ICRP so their use can readily be defended. It is not difficult to design dosemeters that measure the new quantities so that no excessive burden should be placed on monitoring services in using them.

It is a unified system of dose measurement so there are no conceptual difficulties in adding doses from all energies and types of radiation and moreover, dosemeter intercomparison exercises and the like can be carried out more effectively.

From a quantitative point of view, provided we accept the ICRU (ICRU (1985)) suggestion that for dosemeters to be worn on the trunk the ICRU sphere is a suitable phantom on which to develop and type test personal dosemeters, the definitions do prescribe a required energy and angular response for dosemeters as well as instruments. Moreover, the type testing of dosemeters and instruments is surprisingly straightforward. Conversion coefficients such as those published by ICRP in Publication 51 (ICRP (1987)) allow the conversion of intensities measured in terms of the more conventional quantities, such as air kerma for photons and fluence for neutrons, to the new quantities. There is no need to develop primary or secondary standards for this purpose. The type testing of dosimeters and instruments is dealt with in detail in the next chapter.

Quantitatively the new quantities also go a long way towards satisfying the requirements of secondary quantities in that:

1. Unified quantities are defined for all radiations.
2. These are point quantities. Due to the concept of expansion there is no ambiguity as to where the measurement is made. Questions do not arise as to where or whether the sphere should be there or whether you can measure at a surface, in a hole or in a narrow beam, etc.
3. The quantities are believed to be durable in that they are not likely to change with modifications to the definition of effective dose equivalent or minor changes in dose limits. Had a quantity been designed which was specifically linked to effective dose equivalent it could have been subject to variations if this concept were modified.

4. The quantities are additive in the normal sense.
5. The quantities are compatible with existing instruments.
6. The quantities are relatable to the primary limits.
7. The quantities are calculable by the Monte Carlo method and others, the spherical geometry simplifying some calculations.
8. The definitions are such that the quantities are independently realisable in a standards laboratory.

5. Relationships between the new quantities and effective dose equivalent

In the introduction it was stated that the main aim of the new quantities is to provide an estimate of effective dose equivalent, H_E, avoiding underestimation and excessive overestimation, and an adequate approximation to the dose equivalent to the skin. Computations of dose distributions in an anthropomorphic phantom and in the ICRU sphere allow comparisons to be made between the measurements by instruments designed to measure the new quantities and the corresponding effective dose equivalent rates that would be received by persons at the position of the instruments, if the physical differences between the persons and the phantom are ignored. Two sets of such data are given in Figures 1 and 2 of ICRU Report 39 (ICRU (1985)), which compare $H^*(10)$ with H_E for photons and neutrons under several different geometrical irradiation conditions. Similar and, where appropriate, more recent data (Wittman et al. (1985)) are reproduced below.

Figure 2 shows the ratio of effective dose equivalent, H_E, for different orientations of an anthropomorphic phantom, to the ambient dose equivalent, $H^*(10)$, when exposed to the same photon beam, as a function of photon energy (Williams et al. (1983), Dimbylow et al. (1984), Kramer et al. (1982), Williams et al. (1984), Dimbylow (1985)). The beam, without the phantom in it, is parallel and unidirectional and its fluence is constant over the volume to be occupied by the phantom. The ratios for a stationary anthropomorphic phantom are indicated by diamonds (<>) when the radiation is directed in the anteroposterior direction (sometimes called AP irradiation), by circles (o) when the radiation is directed in the posteroanterior direction (sometimes called PA irradiation), and by crosses (+) when the radiation is directed in the lateral direction. The ratios obtained when the radiation beam is horizontal and the anthropomorphic phantom is rotated about a vertical axis (sometimes called rotational irradiation) are indicated by triangles (Δ).

The ratios obtained when the anthropomorphic phantom is rotated about vertical and horizontal axes so that it receives an equal fluence

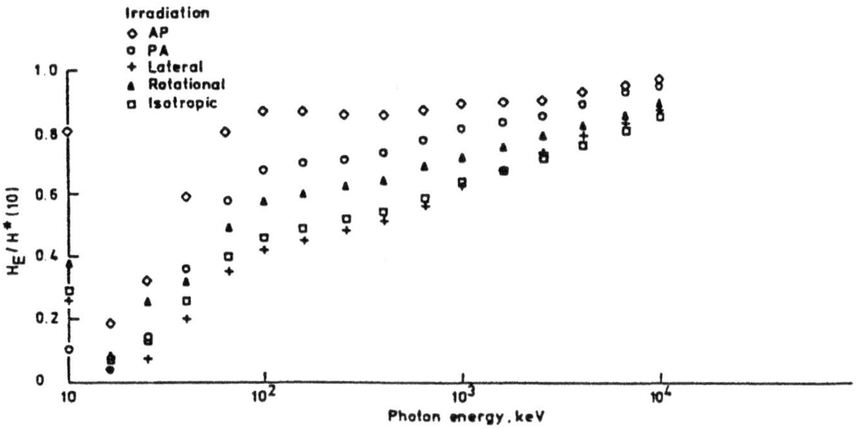

Figure 2. Ratio of $H_E/H^*(10)$ plotted against photon energy for various irradiation geometries.

from all directions (sometimes called isotropic irradiation) are given by squares (□). In Figure 3 similar data (Wittman (1985), Chen (1979), Shiue (1983), Chilton (1981), Lee (1985), Lee (1986)) are given for neutrons using the same symbols for the different geometrical conditions.

It can be seen that, for photons and neutrons, H*(10) always provides an overestimate of the corresponding effective dose equivalent. In the majority of cases the level of overestimation is limited to a

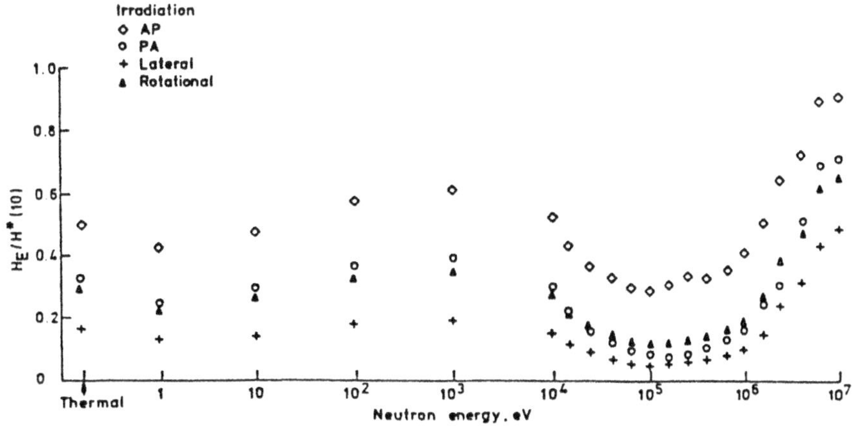

Figure 3. Ratio of $H_E/H^*(10)$ plotted against neutron energy for various irradiation geometries.

factor of two to three, but over limited energy ranges the overestimate may be excessive, i.e., in the region below 40 keV to 1 MeV for PA and rotational irradiation with neutrons, and over practically the whole energy range for lateral irradiation with neutrons. This is of little consequence if the dose rates concerned are very low, but, in cases where dose rates result in doses to personnel that approach the dose limits, and these dose rates can be justified, use of the ambient dose equivalent concept may be over restrictive. In such cases it would be reasonable to characterise the radiation field by obtaining information on the types of radiation present, their energy spectra and their angular distributions. This would allow better estimates of potential effective dose equivalents to be obtained from the measurement of H*(10).

As an indication of the suitability of H_P (10) as a reasonable estimate of H_E to the individual it is instructive to compare the relationship between H'(10) and H_E. It will be shown in the following chapter that for test and calibration purposes personal dosemeters should be tested on the ICRU sphere against the quantity directional dose equivalent, H'(d). Figure 4 (Williams et al. (1983), Dimbylow et al. (1984), Kramer et al. (1982), Williams et al. (1984) Dimbylow (1985)) shows the ratio of H_E to H'(10) for monoenergetic photons under

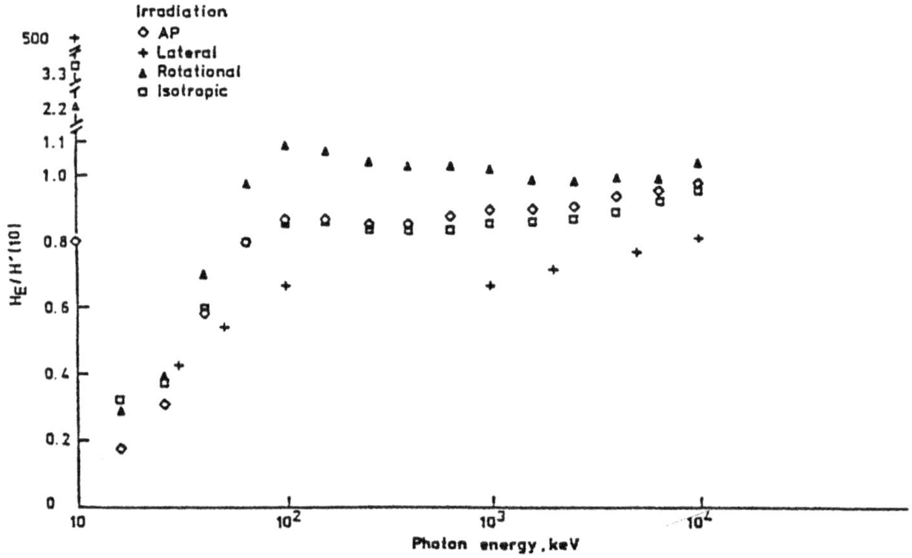

Figure 4. Ratio of H_E/H'(10) plotted photon energy for various irradiation geometries.

several different geometrical irradiation conditions, and Figure 5 (Wittman et al. (1985), Chen et al. (1979), Shiue et al. (1983), Chilton et al. (1981), Lee (1985), Lee et al. (1986)) shows the ratio of H_E to H'(10) for monoenergetic neutrons under several different geometrical irradiation conditions; the same symbols as in Figure 4 apply.

It can be seen from Figure 4 that, for photons, apart from a very large underestimate for lateral, rotational and isotropic irradiations at about 10 keV, H_P (10), insofar as it is approximated by H'(10), would provide an overestimate of effective dose equivalent limited to a factor of three for energies greater than 30 keV.

It can be seen from Figure 5 that, for neutrons, H_P (10), insofar as it is approximated by H'(10), would always provide an overestimate of effective dose equivalent which is limited to a factor of three, other than for lateral irradiation with neutrons in the energy range 20 keV to 2 MeV.

Table 2 shows the ratio of H'(10) to H_E for three practical photon spectra (Bartlett (1985)), for the AP direction, and for rotational and isotropic fields. The values also indicate that H_P (10) would give a very good approximation to H_E.

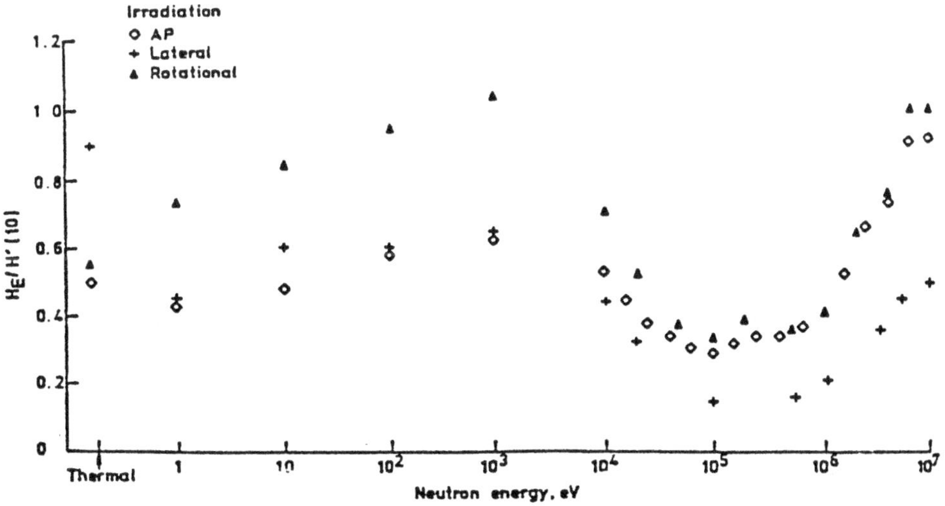

Figure 5. Ratio of H_E/H'(10) plotted against neutron energy for various irradiation geometries.

TABLE 2. Comparison of H' (10) with H_E for three practical photon spectra in AP, rotational (ROT) and isotropic (ISOT) fields.

(a) Details of spectra

Photon Energy Range (MeV)	Photon Fluence (cm^{-2}) I	II	III
0.02 - 0.07	0.32	0.45	0.46
0.07 - 0.5	0.20	0.27	0.28
0.5 - 0.7	0.17	0.22	0.24
0.7 - 1.5	0.05	0.01	0.01
1.5 - 4.5	0.12	0.02	0.01
4.5 - 10	0.14	0.03	0.01

(b) Relationship between H' (10) and H_E

Photon Spectrum	$\frac{H'(10)}{H_E}$ AP	$\frac{H'(10)}{H_E}$ ROT	$\frac{H'(10)}{H_E}$ ISOT
I	1.07	0.977	1.11
II	1.16	1.00	1.16
III	1.22	1.02	1.19

6. ICRU Report 43

ICRU Report No. 43 (ICRU (1985)) is the second of a series of three reports on the new operational quantities. It not only provides the justification for the choice of the new quantities but also provides a great deal of data on the relationship between the new quantities and effective dose equivalent and organ dose. This data can be used to good effect when-ever a better estimate of H_E or H_T is required provided data on the radiation field is available, i.e., its type, energy and angular distribution. Where low doses are concerned the degree of approximation of the new quantities to H_E and H_T given in the preceeding sections is almost certainly good enough but when the control of the exposure of workers at or near the dose limits is necessary and can be justified it would be advisable and less restrictive to obtain the above information on the exposure conditions.

Examples of the data given in ICRU Report 43 are given below. Figure 6 shows the ratio of $H'(10)$ to $\dot{H}_P(10)$ per unit air kerma for photons in the MIRD-5 phantom. The curves are very similar which provides substantial support for the use of the ICRU sphere as a surrogate for the human body. Figure 7 shows the ratio of $H_P(10)$ to H_E for 7 representative irradiation geometries for photons in the energy range 10 keV to 10 meV. These data would assist in providing a better estimate of H_E provided the irradiation conditions are known albeit only approximately. Similar data is shown in Figures 8 and 9 for the ovaries and testes thus allowing a better estimate of the dose to these organs in a practical situation. Similar data is given in ICRU 43 for the female breast, the red bone marrow, the thyroid and the eye lens. All the above data is also given for neutrons. Figures 10, 11 and 12 show these data for neutrons for effective dose equivalent, the ovaries and testes respectively.

7. Summary

It has been shown that, by and large, the new quantities for environmental and individual monitoring do meet the needs of secondary quantities and they do provide a reasonable estimate of H_E and H_T for low doses. Moreover, additional data has been published by ICRU in Report 43 which allows better estimates to be made when this is necessary. As a result these quantities are now becoming generally accepted, in particular by the CEC, the ISO and the IEC.

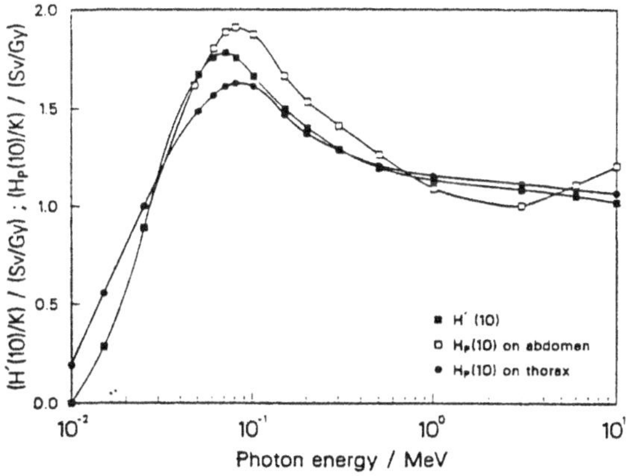

Figure 6. Directional dose equivalent, $H'(10)$, and individual dose equivalents, $H_P(10)$, in the MIRD-5 phantom per unit air kerma (in free air), K, as a function of photon energy; parallel beam from front-to-back (anterior-posterior, AP).

The necessary characteristics of instruments and dosimeters to measure the new quantities, the calibration of these devices and the suitability of current designs for the new quantities will be dealt with in the following chapter.

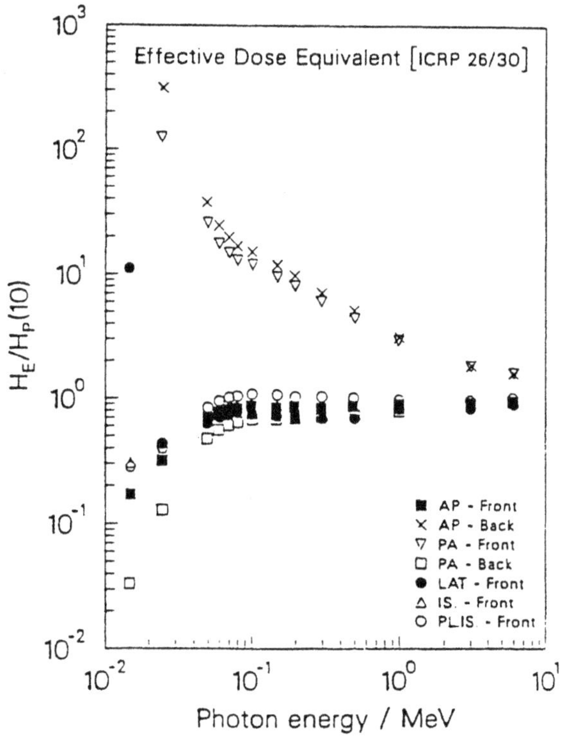

Figure 7. Ratio of the effective dose equivalent, H_E, to the individual dose equivalent, penetrating, $H_P(10)$, as a function of photon energy. Two locations for the personal dosimeter are considered: front of the body (Front) and back of the body (Back). $H_P(10)$ is approximated by the dose equivalent at depth 10 mm along the central axis in the ICRU sphere (see text) (ICRP, 1977, 1979). Five geometries are considered in the calculations: AP, broad parallel beam from front to back (anterior-posterior); PA, broad parallel beam from back to front (posterior-anterior); LAT, broad parallel beam from the side (lateral); IS, isotropic field; PL.IS., planar isotropic field, perpendicular to body axis.

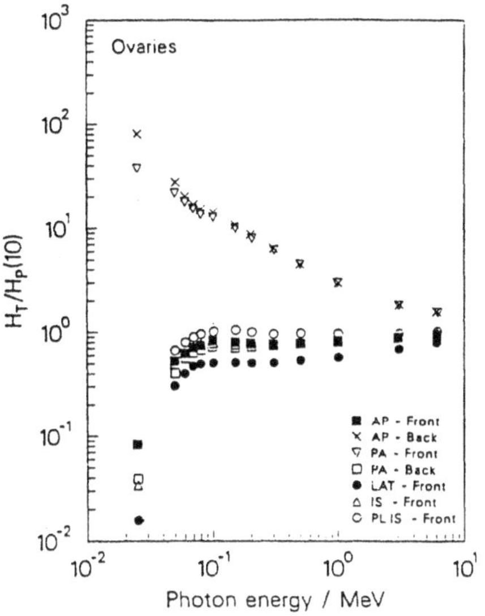

Figure 8. Ratio of the dose equivalent, H_T, to the ovaries (female gonads) to $H_P(10)$ as a function of photon energy; geometries and symbols explained in the caption to Figure 7.

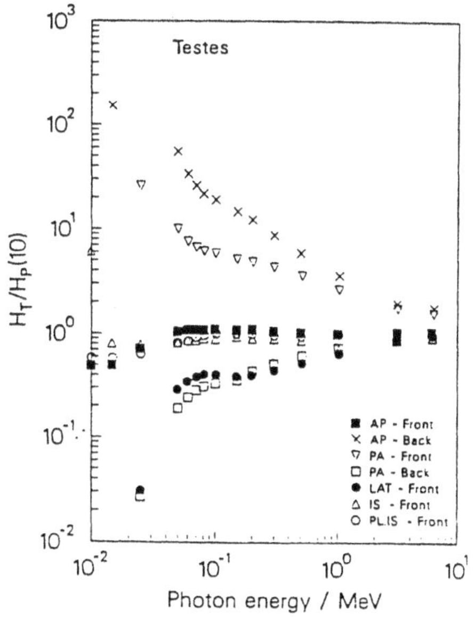

Figure 9. Ratio of the dose equivalent, H_T, to the testes (male gonads) to $H_P(10)$ as a function of photon energy; geometries and symbols explained in the caption to Figure 7.

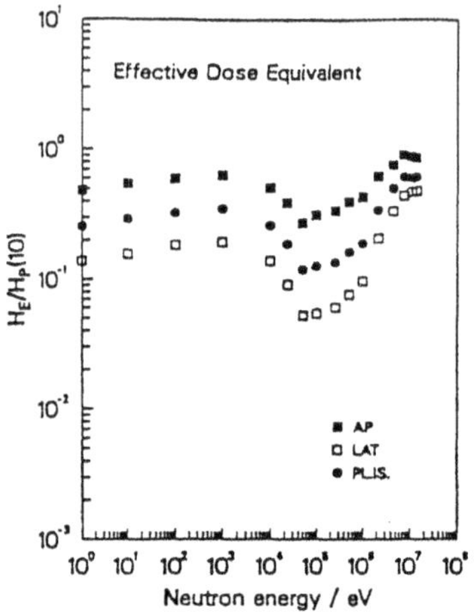

Figure 10. Ratio of the effective dose equivalent, H_E, to the individual dose equivalent, penetrating, $H_P(10)$, as a function of neutron energy. The personal dosemeter is considered to be worn on the front of the body. $H_P(10)$ is approximated by the dose equivalent at depth 10 mm in the ICRU sphere. Three geometries are considered in the calculations: AP, broad parallel beam from front to back (anterior-posterior); LAT, broad parallel beam from the side (lateral); PL.IS., planar isotropic field, perpendicular to body axis.

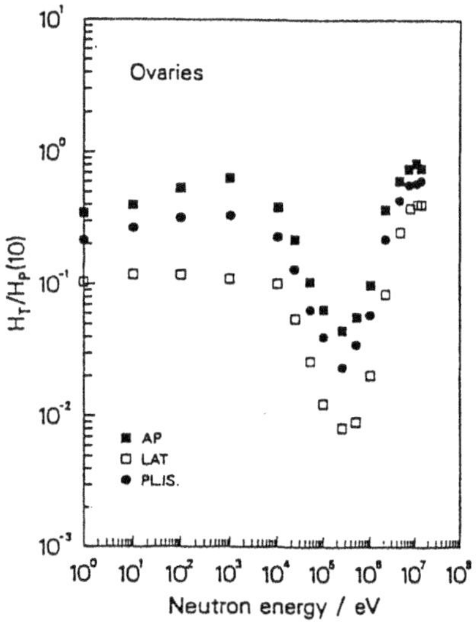

Figure 11. Ratio of the dose equivalent, H_T, to the ovaries (female gonads) to $H_P(10)$ as a function of neutron energy; geometries and symbols explained in the caption to Figure 10.

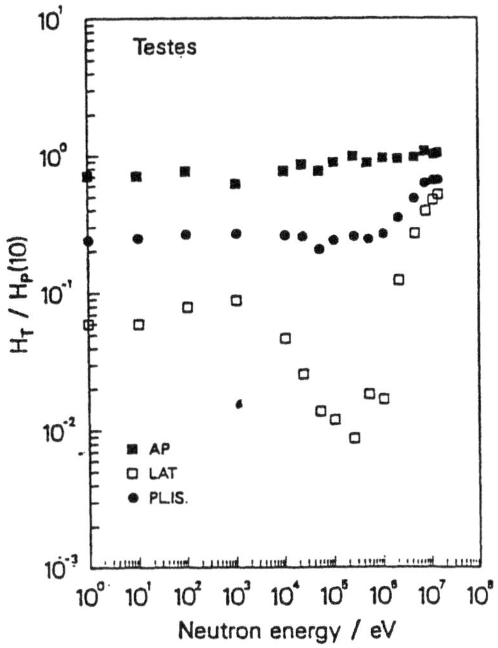

Figure 12. Ratio of the dose equivalent, H_T, to the testes (male gonads) to $H_P(10)$ as a function of neutron energy; geometries and symbols explained in the caption to Figure 10.

5. References

Commission of the European Communities. Technical Recommendations for Monitoring the Exposure of Individuals to External Radiation (1975). (Luxembourg:CEC).

International Commission on Radiological Protection. Recommendations of the International Commission on Radiological Protection (1977). Publication 26. Ann. ICRP 1(3). (Oxford:Pergamon Press).

Commission of European Communities (1980). Council Directive of 15th July 1980 amending the Directives laying down the Basic Safety Standards for the health protection of the General Public and Workers against the Dangers of Ionising Radiation. EUR 7330 (Luxembourg:CEC).

International Commission on Radiation Units and Measurements. Determination of Dose Equivalents Resulting from External Radiation Sources (1985). Report 39 (Bethesda:ICRU Publications).

International Commission on Radiation Units and Measurements, Radiation Quantities and Units (1980). (Bethesda, MD, ICRU Report 33).

National Radiological Protection Board, New Radiation Quantities Recommended by ICRU for Practical Use in Radiological protection: Their Implementation in the United Kingdom, NRPB-GS5 1986.

International Commission on Radiological Protection, Data for Use in Protection Against External Radiation (1987). Publication 51, Ann. ICRP 17 (2/3) (Oxford:Pergamon Press).

Wittman, A., Morhart, A. and Burger, G. (1985) "Organ Doses and Effective Dose Equivalent". Paper presented at the Int. Seminar on Radiation Protection for External Exposure, 19-21 March, P.T.B., Braunschweig, F.R.G.

Williams, G., Swanson, W.P., Kragh, P. and Drexler, G. (1983) "Calculation and Analysis of Photon Dose Equivalent Distributions in the ICRU Sphere". GSF Bericht S-958 (Gesellschaft fur Strahlen- und Umweltforschung mbH, Munich, Germany).

Dimbylow, J.P. and Francis, T.M. (1984) "The Calculation of Dose Equivalent Quantities in the ICRU Sphere with Photon Energies from 0.01 to 10 MeV". Radiat. Prot. Dos. 9, 49-53.

Kramer, R., Zankl, M., Willians, G., Drexler, G. (1982) "The Calculation of Dose from External Photon Exposures Using Reference Human Phantoms and Monte-Carlo Methods". Part 1. The Male (ADAM) and Female (EVA) Adult Mathematical Phantoms. GSF Bericht S-885, December 1982. Munchen, Gesellschaft fur Strahlen-und Umweltforschung mbH.

Willians, G., Zankl, M., Echerl, H., Drexler, G. (1984) "The Calculation of Dose from External Exposures Using Reference Human Phantoms and Monte-Carlo Methods". Part II: Organ Doses from Occupational Exposures. GSF Bericht S-1079, December 1984. Munchen, Gesellschaft fur Strahlen-und Umweltforschung mbH.

Dimbylow, P.J., Private communication, 1985.

Wittman, A., Morhart, A. and Burger, G. (1985) "Organ doses and effective dose equivalent". Paper presented at the Int. Seminar on Radiation Protection for External Exposure, 19-21 March, P.T.B. Braunschweig F.R.G.

Chen, S.-Y. and Chilton, A.B. (1979) "Calculation of Fast Neutron Depth Dose in the ICRU Standard Tissue Phantom and the Derivation of Neutron Fluence-to-Dose-Index Conversion Factors". Rad. Res. 78, 335-370.

Shiue, Y.-L. and Chilton, A.B. (1983) "Calculation of Low Energy Neutron Dose Indices and Depth Doses in the ICRU Tissue Sphere". Rad. Res. 93, 421-443.

Chilton, A.B, Shiue, Y.-L. (1981) "Progress report on Studies Concerning Doses in the ICRU Sphere Resulting from Neutron Beams". In: Proceedings of the Fourth Symposium on Neutron Dosimetry, Munich-Neuherberg 1-5 June 1981, Vol. 1, pp 63-79. Luxembourg, Commission of the European Communities, EUR 7448.

Lee, J.K. (1985) "Assessment of Quantities for Radiation Protection in the ICRU Standard Tissue Sphere for Intermediate Energy Neutrons". Ph.D. Thesis, University of Illinois at Urbana- Champaign.

Lee, J.K., Chilton, A.B. and Ragheb (1986) "Effects of Cross Section Structure on the Dosimetric Response Functions for 0.4-10.0 MeV Neutrons in the ICRU Tissue Sphere (Submitted for Publication)".

Bartlett, D.T. (1985) "Inter-Relationship of the Proposed ICRU Operational Radiological Protection quantities and Effective Dose Equivalent for Practical Radiation Fields. Radiation Protection Dosimetry, Vol. 12, No.2, pp 155-157.

International Commission on Radiation Units and Measurements. Determination of Dose Equivalents Resulting from External Radiation Sources (1985). Part 2, Report 43 (Bethesda:ICRU Publications).

INSTRUMENT AND DOSEMETER CALIBRATION IN TERMS OF THE NEW ICRU QUANTITIES

T.O. MARSHALL
National Radiological Protection Board
Chilton, Didcot, Oxon, OX11 ORQ
United Kingdom

1. Introduction

It is often unclear as to what is meant by the term "calibration". In this chapter it is taken to mean both the "Type Testing" of radiation measuring devices to determine the way their response varies with radiation type, energy and angle, and "Routine Calibration", during which the measurement system as a whole is normalised under specified irradiation conditions with a single radiation source.

The lecture discusses the required characteristics of instruments and dosemeters to measure the new ICRU quantities (ICRU (1985)), the standardisation of calibration radiation beams, the general procedures for instrument and dosemeter calibrations and some results from the calibration of several commonly used instruments and dosemeters. Ways of simplifying the routine calibration of dosemeters and the use of different shaped phantoms are also discussed.

2. Required Characteristics of Instruments and Dosemeters to Measure the New Quantities

2.1. Energy Response Requirements

The required characteristics of instruments and dosemeters stem from the definitions of the new ICRU quantities which were discussed in the previous chapter. For area monitoring instruments the quantities ambient dose equivalent H*(10) and directional dose equivalent H'(0.07) are defined in the ICRU sphere and the required characteristics of these instruments have been derived from the calculation of the dose equivalents at appropriate depths in the sphere when in a radiation field of known type and intensity. This intensity is conventionally expressed in

M. Oberhofer (ed.), Advances in Radiation Protection, 87–106.

terms of air kerma (K_a) for photons and fluence (Φ) for neutrons. These calculations have led to a series of coefficients indicating the required relationship between the reading of instruments designed to measure H*(10) as against air kerma (see Figure 1) or H'(0.07) (see Figure 2) for photons and H*(10) as against neutron fluence (see Figure 3) for neutrons. Note that data are not given for H'(0.07) for neutrons because neutrons of all energies are highly penetrating in terms of the definition given in ICRU Report 39 and hence only the quantity H*(10) for neutrons has relevance for radiation dosimetry.

The new quantities for individual monitoring, in specifying the measurement of dose equivalent within the body of the wearer, require dosemeters to be type tested on an appropriate phantom to simulate attenuation and scatter produced within the person's body. ICRU (ICRU (1985)) state that a suitable phantom for dosemeters intended to be worn on the trunk would be the ICRU 30 cm diameter equivalent sphere (ICRU (1980)). During type testing dosemeters should therefore be exposed on the sphere, or its equivalent, and be tested against the dose equivalent at the appropriate depth in the sphere, i.e., 10 mm for $H_p(10)$ and 0.07 mm for $H_s(0.07)$. The question arises however, as to which sphere quantity individual dosemeters should be tested against. Figure 4 shows the models used for the calculation of the conversion coefficients, 4a for testing the energy response of dosemeters and 4b for their angular response. Conversion coefficients relate the dose equivalent at the required position in the sphere to the field quantity, i.e., air kerma for photons and fluence for neutrons, which would exist in the beam in the absence of the sphere. The coefficients are calculated. Figure 4c shows diagramatically, the experimental arrangement for testing the angular response of dosemeters. In Figure 4a the point D, at which the dose equivalent must be calculated is at distance 10 mm along the principle axis for $H_p(10)$ and 0.07 mm for $H_s(0.07)$. In Figure 4b these distances remain the same but are measured along the radius at angle $\alpha°$ to the principle axis. The conversion coefficient in this case is calculated to provide the dose equivalent at point D against which the reading of the dosemeter is compared in order to determine its response for angle of incidence $\alpha°$.

The coefficients discussed above are numerically equal to those for the quantity directional dose equivalent where the values of d are 10 mm and 0.07 mm for $H_p(10)$ and $H_s(0.07)$ respectively. The energy response of the dosemeter is tested against the conversion coefficients for $\alpha=0°$. Personal dosemeters are therefore in effect tested against the sphere quantity directional dose equivalent. However, this does not mean that personal dosemeters measure directional dose equivalent even ignoring the differences between the sphere and an individual's body, since in a practical situation the radiation field is expanded for this

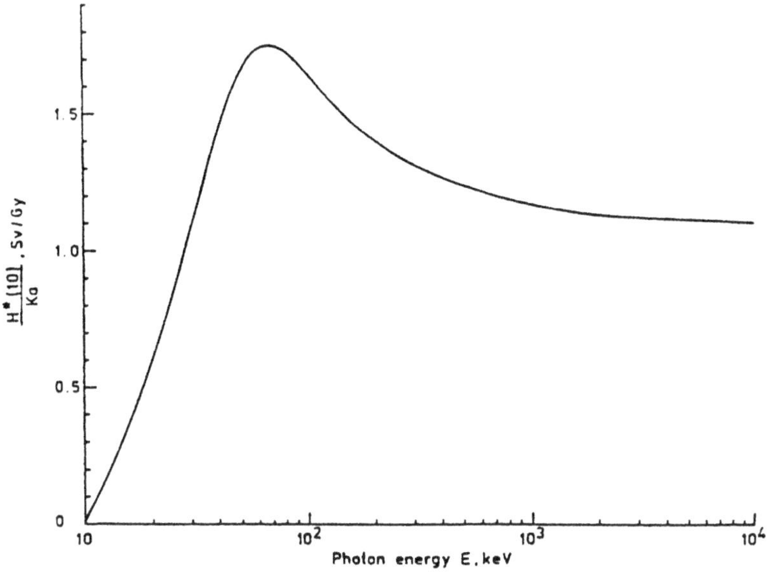

Figure 1. Ratio of H*(10)/K_a plotted against photon energy.

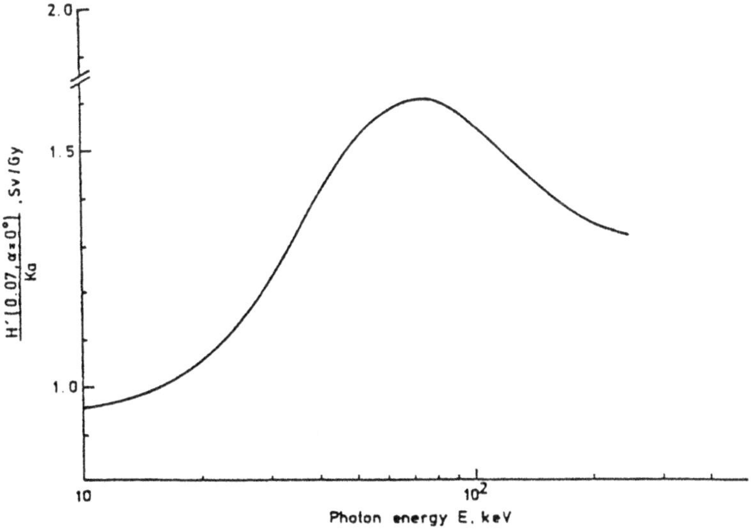

Figure 2. Ratio of H'(0.07, α=0°)/K_a plotted against photon energy.

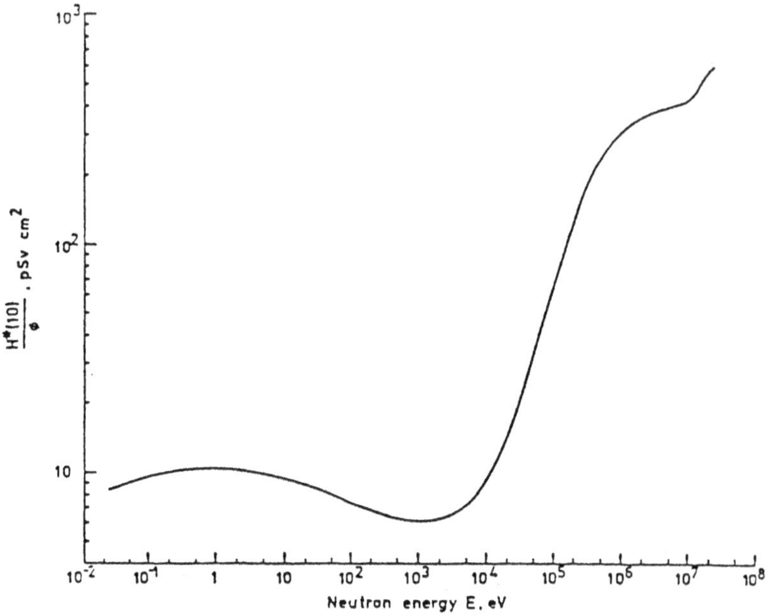

Figure 3. Ratio of $H^*(10)/\Phi$ plotted against neutron energy.

quantity whereas it is not for the quantities for individual monitoring. They are only numerically equal under calibration conditions which normally includes approximate expansion of the field.

2.2. Angular Response Requirements

The definitions of the new quantities also require instruments to have a certain angular response. For $H^*(10)$ an isotropic response in terms of fluence is required because the definition of this quantity assumes alignment of the radiation field. In principle the definition of $H'(0.07)$ requires a response which falls off with angle. However, for photons (excepting energies below 50 keV) the reduction is fairly insignificant for angles less than 90°. A response which is isotropic over the forward 2π Sr of solid angle would therefore be suitable.

The use of computed conversion factors for instrument and dosemeter calibrations is not the approach adopted for beta radiation. The dose rate in calibration beams is either produced by secondary standard sources, or is measured with an extrapolation chamber, in terms of the dose equivalent rate at a depth of 0.07 mm, and at 10 mm for the more energetic beta-ray emitters, in a tissue equivalent medium which provides the same back-scatter and attenuation as soft tissue. The results obtained are practically identical to those which would be obtained in

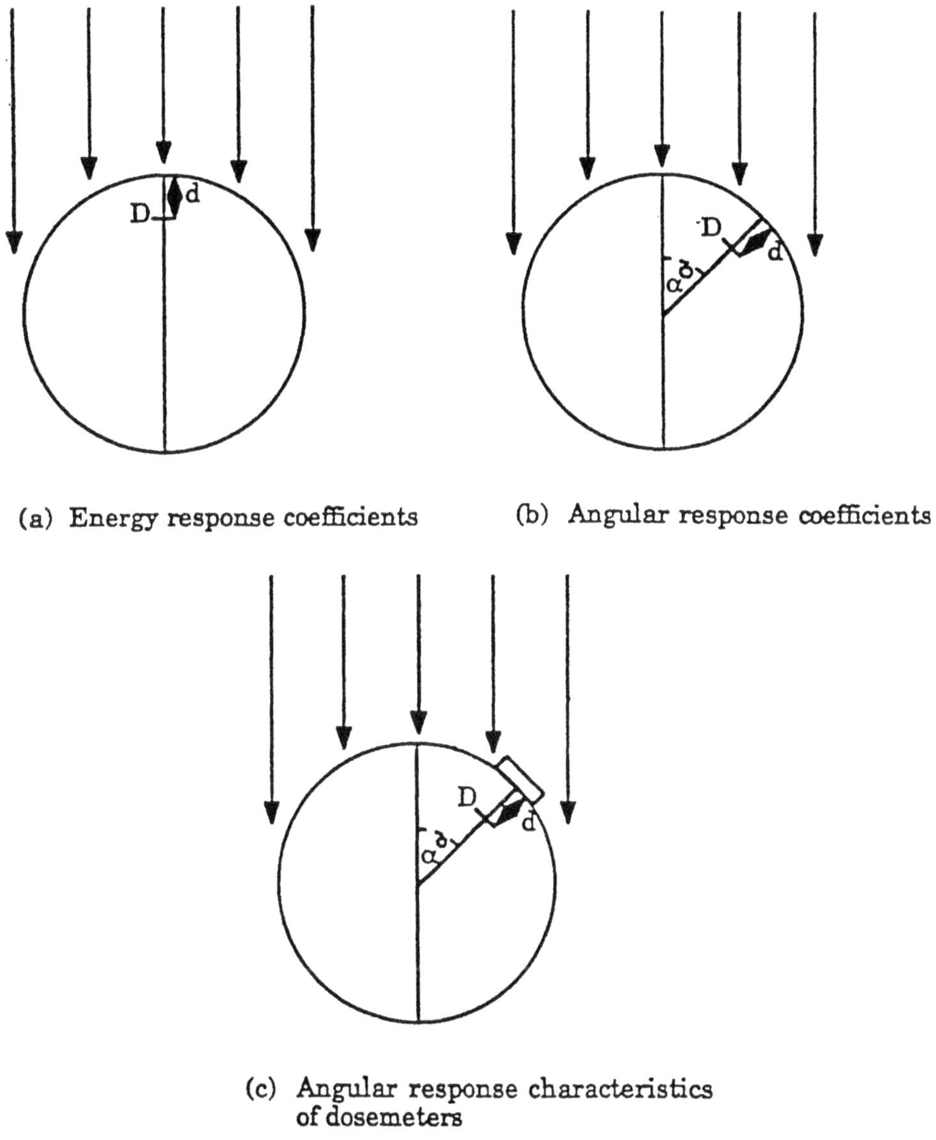

Figure 4. Models for the calculation of conversion coefficients for energy response and angular response compared with the irradiation conditions for type testing individual dosemeters for angular response characteristics.

the ICRU sphere because the range of electrons from common beta-ray emitters is very limited. Hence, the values can be taken as a measurement of H'(0.07) and H*(10). Extrapolation chambers may be used, therefore, as primary and secondary instruments for these quantities in respect of beta radiations.

The required angular response for the measurement of H'(0.07) from beta-rays has been measured. The results are shown in Table 1.

TABLE 1. The angular response of instruments and dosemeters to measure H'(0.07) and $H_s(0.07)$ for beta-rays for two different secondary beta standards.

Nuclide	Distance (cm)	Sources normalised to zero degrees zero°	15°	30°	45°	60°	75°
NRPB-data (NPL standards)							
Strontium-90/ Yttrium-90	20.0	1.000	0.018	0.990	0.057	1.035	0.778
Strontium-90/ Yttrium-90	30.0	1.000	1.002	1.023	1.056	1.019	0.745
Thallium-204	20.0	1.000	0.993	0.974	0.913	0.751	0.480
Thallium-204	30.0	1.000	0.998	0.957	0.890	0.730	0.483
Promethium-147	15.0	1.000	0.923	0.809	0.648	0.484	0.318
PTB-data (PTB standard)							
Strontium-90/ Yttrium-90 (*)							
Type 1	30.0	1.00	1.01	1.05	1.13	1.15	0.89
Type 2	30.0	1.00	1.01	1.05	1.13	1.19	0.95
Thallium-204	30.0	1.00	0.98	0.96	0.90	0.73	0.50
Promethium-147	20.0	1.00	1.00	0.83	0.67		

(*) Type 1: With beam flattening filter.
Type 2: Without beam flattening filter.

The data given in Figures 1, 2 and 3 define the required energy response of personal dosemeters for photons and neutrons respectively when exposed on a phantom. [Data for H*(10) are numerically equal to H'(20, α=0)]. However, as indicated above, H'(10), against which personal dosemeters are type tested, require a prescribed angular response. This is illustrated in Figures 5 and 6 which show a plot of

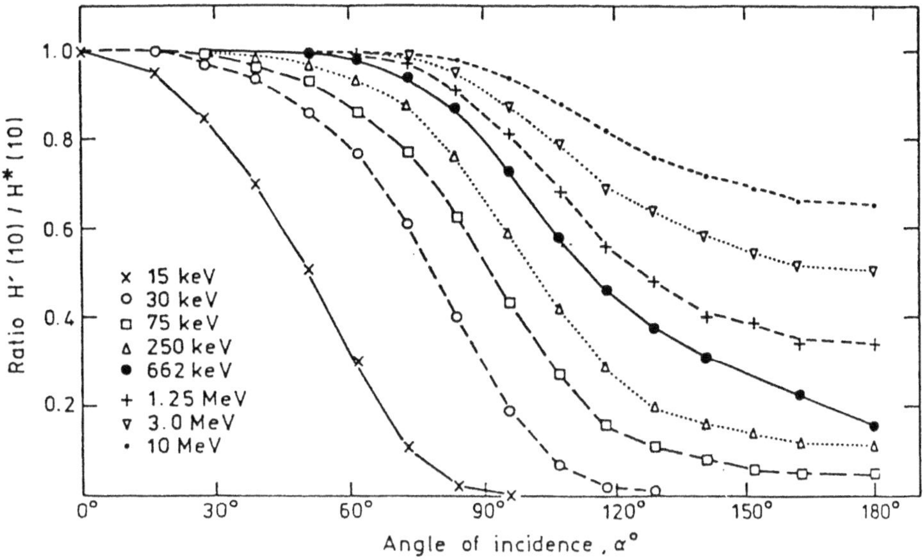

Figure 5. Ratio of H*(10) to H'(10) against angle of incidence (photons).

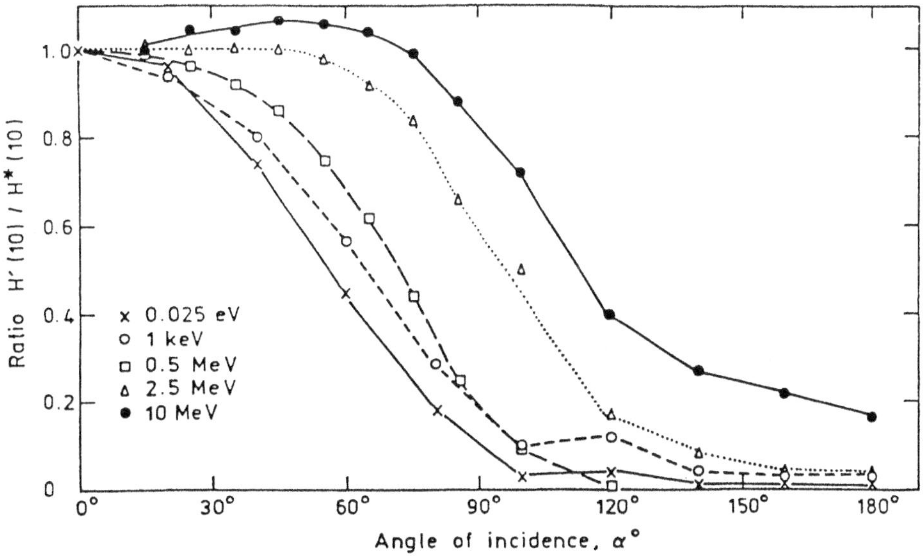

Figure 6. Ratio of H*(10) to H'(10) against angle of incidence (neutrons).

the ratio of H'(10) to H*(10) against the angle of incidence, $\alpha°$, (H*(10) does not vary with angle) for photons and neutrons respectively.

3. Irradiation Beams for Calibration

The ISO series of reference radiations (ISO (1983)) are suitable for instrument and dosemeter calibrations in terms of the new quantities but it is necessary to establish the intensity in the beam in terms of the new quantities. There are two ways in which this can be done. One could produce a set of primary standards for the new quantities or alternatively calculated conversion coefficients may be used to convert the intensity measured in terms of a field quantity such as air kerma for photons or fluence for neutrons to that in terms of the new quantities. These conversion coefficients are readily available. ICRP has published one set in Publication 51 (ICRP to be published) and ICRU another in Report No. 43 (ICRU (1988)). There are some differences between the two sets but these are small and can be ignored for practical purposes. On the other hand to design primary standards for the new quantities would be extremely time consuming and expensive. As a result it is now universally accepted that published conversion coefficients should be used in preference to the design of primary and secondary standards. The conversion coefficients published in ICRP Publication 51 are given in Table 2 for photons and in Table 3 for neutrons.

The coefficients referred above and given in Tables 2 and 3 are calculated for monoenergetic radiations whereas the spectra in practical calibration beams usually have a finite width. However, conversion coefficients for the majority of the ISO series of reference photon radiations are available (Iles (1987)). Values are tabulated in Tables 4.1 and 4.2 for most ISO narrow spectrum photon radiations.

4. The Calibration Procedure

Although the response of area monitoring instruments is defined in the ICRU sphere this is simply to determine the radiation characteristics of these instruments and they must be exposed in air during calibration. On the other hand the definitions of the new quantities for individual monitoring require dosemeters to be calibrated on a phantom, i.e., the ICRU sphere or its equivalent.

TABLE 2. Dose equivalent at various depths on the principal axis of the ICRU sphere per unit absorbed dose to air in free air for photons incident in a plane parallel beam.

Photon Energy (MeV)	Conversion Coefficient, Sv Gy^{-1} 0.07 mm	10 mm
$1.0\ 10^{-2}$	0.930	0.010
$1.5\ 10^{-2}$	0.974	0.271
$2.0\ 10^{-2}$	1.02	0.601
$3.0\ 10^{-2}$	1.10	1.09
$4.0\ 10^{-2}$	1.38	1.43
$5.0\ 10^{-2}$	1.52	1.63
$6.0\ 10^{-2}$	1.58	1.74
$8.0\ 10^{-2}$	1.59	1.73
$1.0\ 10^{-1}$	1.55	1.65
$1.5\ 10^{-1}$	1.42	1.49
$2.0\ 10^{-1}$	1.34	1.38
$3.0\ 10^{-1}$	1.28	1.31
$4.0\ 10^{-1}$	1.24	1.26
$5.0\ 10^{-1}$	1.21	1.21
$6.0\ 10^{-1}$	1.19	1.19
$8.0\ 10^{-1}$	1.18	1.16
$1.0\ 10^{0}$	1.16	1.14
$1.5\ 10^{0}$	1.15	1.13
$2.0\ 10^{0}$	1.14	1.13
$3.0\ 10^{0}$	1.13	1.12
$4.0\ 10^{0}$	1.13	1.11
$5.0\ 10^{0}$	1.12	1.11
$6.0\ 10^{0}$	1.11	1.10
$8.0\ 10^{0}$	1.11	1.09
$1.0\ 10^{1}$	1.11	1.09

TABLE 3. Dose equivalent per unit fluence at depths of 0.07 mm and 10 mm on the principal axis for neutrons incident in a plane parallel beam on the ICRU sphere.

Neutron Energy (MeV)	Conversion Coefficient, 10^{-12} Sv cm^2 0.07 mm	10 mm
$2.5\ 10^{-8}$	7.20	8.00
$1.0\ 10^{-7}$	5.50	10.4
$1.0\ 10^{-6}$	3.7	11.2
$1.0\ 10^{-5}$	2.80	9.20
$1.0\ 10^{-4}$	2.50	7.10
$1.0\ 10^{-3}$	2.80	6.20
$1.0\ 10^{-2}$	8.90	8.60
$1.0\ 10^{-2}$	18.2	14.6
$5.0\ 10^{-2}$	46.6	35.0
$1.0\ 10^{-1}$	95.0	69.0
$2.0\ 10^{-1}$	168	126
$5.0\ 10^{-1}$	219	258
$1.0\ 10^{0}$	292	340
$1.5\ 10^{0}$	292	362
$2.0\ 10^{0}$	283	352
$3.0\ 10^{0}$	305	380
$4.0\ 10^{0}$	329	409
$5.0\ 10^{0}$	301	378
$6.0\ 10^{0}$	302	383
$7.0\ 10^{0}$	312	403
$8.0\ 10^{0}$	341	417
$1.0\ 10^{1}$	368	446
$1.4\ 10^{1}$	359	520
$1.7\ 10^{1}$	421	610
$2.0\ 10^{1}$	516	650

These values of the conversion coefficient relate to the dose equivalent for neutrons as defined before 1985. They should be multiplied by a factor of 2 to obtain the dose equivalent as redefined for neutrons by ICRP in 1985.

TABLE 4.1. Coefficients to convert from air-kerma to directional dose equivalent at 0.07 mm for the ISO narrow (Fluorescence (1) and Bremsstrahlung (2)) series, Cs-137 and Co-60 for various angles of incidence and for rotational and isotropic radiations.

Radiation Type	Mean Energy (keV)	0°	30°	60°	90°	Rot.	Isotr.
1	9.88	0.93	0.92	0.87	0.20	0.41	0.40
1	17.4	0.99	1.00	0.97	0.52	0.46	0.48
1	23.1	1.07	1.08	1.04	0.69	0.54	0.54
1	25.2	1.11	1.11	1.08	0.73	0.56	0.56
1	30.9	1.21	1.21	1.16	0.81	0.62	0.63
2	33	1.24	1.24	1.19	0.84	0.64	0.65
2	48	1.47	1.44	1.39	1.03	0.78	0.80
2	65	1.59	1.55	1.50	1.18	0.87	0.90
2	83	1.58	1.56	1.51	1.22	0.89	0.93
2	100	1.55	1.53	1.49	1.23	0.89	0.93
2	118	1.50	1.48	1.46	1.23	0.88	0.92
2	161	1.39	1.39	1.40	1.23	0.85	0.90
2	205	1.34	1.34	1.36	1.21	0.83	0.88
2	248	1.31	1.31	1.34	1.21	0.83	0.88
Cs-137	662	1.19	1.21	1.23	1.15	0.84	0.90
Co-60	1250 (mean)	1.16	1.17	1.20	1.13	0.88	0.93

The radiation used should be chosen from the appropriate ISO series of references radiations and the instrument or dosemeter should be exposed in a near parallel beam. To achieve this it is recommended that the irradiations should be carried out at a distance of at least 2 m from the source. The intensity to which the instrument or dosemeter is exposed is derived using the appropriate conversion coefficients as discussed in the previous section. These coefficients are applied to the intensity in terms of air kerma for photons or fluence for neutrons as measured at the position of the sensitive volume of the instrument of dosemeter in the absence of such instruments, dosemeters or the sphere.

The conversion coefficients for energy response tests take into account the energy response requirements presented in Figures 1, 2 and 3. The ideal instrument or dosemeter, for the new quantities, tested in this manner, would, after normalisation, have a response equal to 1.0 over the whole of the useful energy range.

In testing personal dosemeters for their angular response they should be placed on the sphere at such a position to achieve the

TABLE 4.2. Coefficients to convert from air-kerma to directional dose equivalent at 10 mm for the ISO narrow (Fluorescence (1) and Bremsstrahlung (2)) series, Cs-137 and Co-60 for various angles of incidence and for rotational and isotropic radiations.

Radiation Type	Mean Energy (keV)	0°	30°	60°	90°	Rot.	Isotr.
1	9.88	0.01	<0.01	<0.01	<0.01	<0.01	<0.01
1	17.4	0.45	0.39	0.21	0.01	0.14	0.11
1	23.1	0.78	0.74	0.53	0.11	0.30	0.26
1	25.2	0.89	0.84	0.64	0.17	0.35	0.31
1	30.9	1.13	1.07	0.88	0.34	0.48	0.45
2	33	1.18	1.12	0.93	0.38	0.51	0.48
2	48	1.56	1.50	1.28	0.68	0.74	0.72
2	65	1.74	1.67	1.46	0.85	0.87	0.87
2	83	1.71	1.65	1.47	0.90	0.89	0.89
2	100	1.64	1.60	1.44	0.92	0.88	0.88
2	118	1.58	1.54	1.41	0.92	0.86	0.87
2	161	1.44	1.43	1.33	0.92	0.82	0.83
2	205	1.37	1.36	1.29	0.92	0.80	0.82
2	248	1.34	1.32	1.26	0.92	0.80	0.82
Cs-137	662	1.18	1.19	1.17	0.96	0.83	0.86
Co-60	1250 (mean)	1.14	1.14	1.13	0.98	0.84	0.87

required angle of incidence and the appropriate conversion coefficient should be chosen from Table 4.1 for the quantity $H_s(0.07)$ and from Table 4.2 for the quantity $H_p(10)$. These conversion coefficients take into account the photon performance requirements for energy and angle presented in Figures 1, 2 and 5. An ideal dosemeter, for the new quantities, tested in this manner, would, after normalisation, have a response equal to 1.0 over the whole of the useful range of energy and angle.

There are some simplifications which may be introduced into the calibration procedure for personal dosemeters. The ICRU sphere is difficult to fabricate and it is impossible to provide its exact chemical composition. It is furthermore not a very convenient shape on which to irradiate dosemeters to measure their energy response because only one dosemeter can be irradiated at a time. Thus at PTB and NRPB an effort is being made to determine a series of correction factors for use with different shape phantoms made of different materials. These factors, when available, will allow type tests to be carried out, effectively using the sphere as the phantom, but actually using a phantom which is much easier to fabricate and use. Unfortunately these correction factors are dosemeter specific and must be determined using the dosemeters themselves.

Similarly, a simplification to routine calibration can be applied. This procedure is carried out regularly with any routine monitoring service and is intended to normalise the response of the system to take into account batch to batch variations in the sensitivity of detectors and changes in the sensitivity of processing equipment, etc. Strictly speaking these irradiations should also be carried out on the sphere but provided the appropriate correction factor is determined and applied other phantoms may be used or the dosemeters may even be exposed free in air.

5. Calibration Results for Some Currently Used Area Monitors and Personal Dosemeters

5.1. Area Monitors

Until recently, virtually all x and γ area survey monitors were designed to measure either exposure or absorbed dose to air in free air and the energy dependence of these instruments varied considerably. In general, ionisation chamber based instruments had a better energy response than Geiger Muller based instruments, although recent designs of energy compensated Geiger Muller tubes are considerably more accurate than their predecessors, both in terms of energy and polar response.

When these instruments are used to measure the new quantities, it is necessary to recalibrate them or, if they have already been calibrated in terms of the old field quantities, use the conversion coefficients to convert these data to the new quantities. Some examples for common instruments are given in Figures 7 to 15 (Marshall et al. (1987)). Table 5 provides more information on these instruments (Marshall et al. (1987)), and in particular indicates the energy range over which each instrument is useful for the new quantities, assuming that a tolerance of ±30% on the energy response is acceptable. For the majority of instruments this will result in an underestimate in the 40 to 200 keV energy region. For instruments with a very accurate response for exposure this will be up to 30%. However, most instruments have a significant over-response in the 60 to 100 keV region for the quantity, exposure, and recalculation in terms of the new quantities will produce only a small or, in some cases, negligible underestimate. See, for example, the results for the Nuclear Enterprises PDR2 family of instruments (Marshall et al. (1987)) (Figure 8). Normally, then, the user can

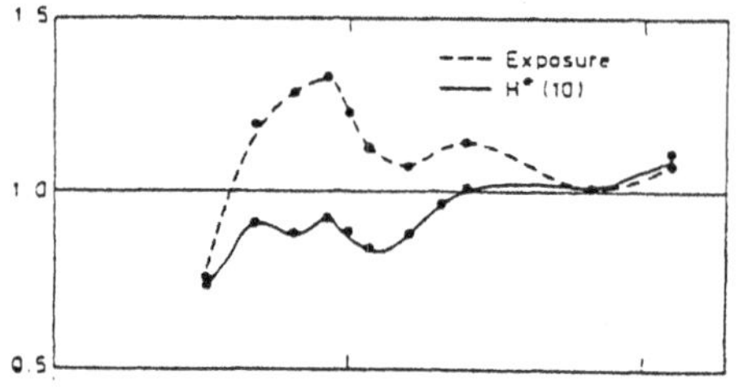

Figure 7. Nuclear Enterprises PDR1, 2601, 1907B (typical).

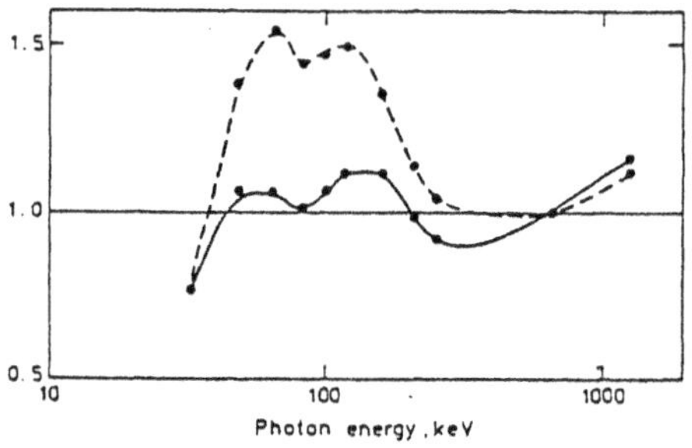

Figure 8. Nuclear Enterprises PDR2, 2602, PDR4 internal detection.

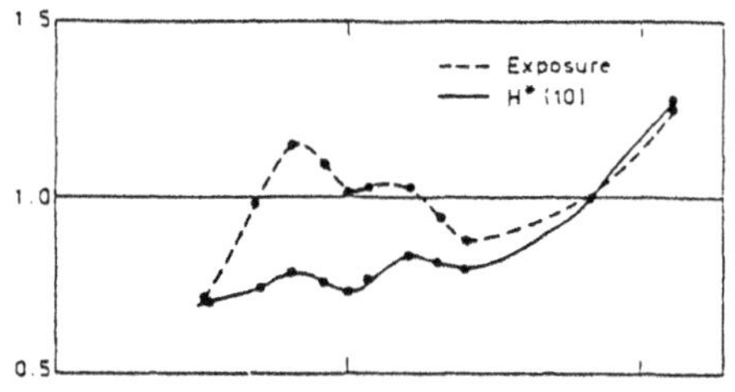

Figure 9. ICI Gammatroll PR190 and other Zp 1201 detector instruments (modern designs).

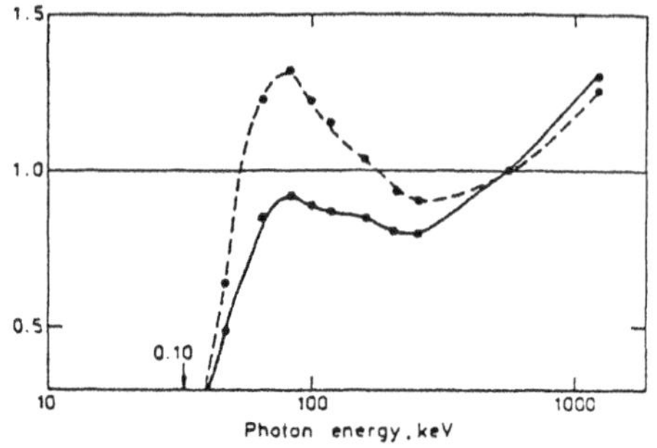

Figure 10. Wallac RD8 3 mSv h^{-1} end on response.

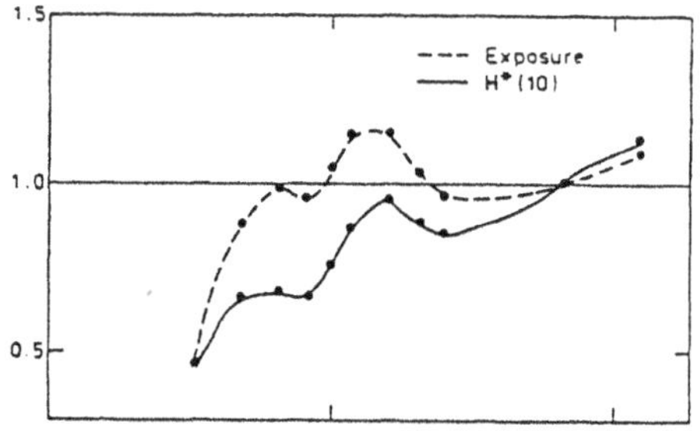

Figure 11. Mini Instruments 900R, 5.10R.

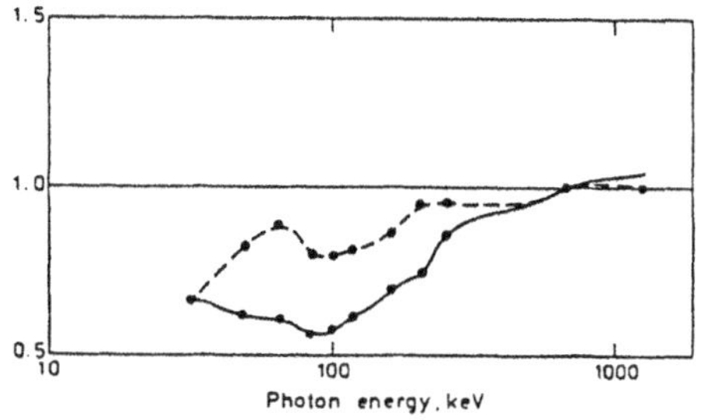

Figure 12. NIS 295.

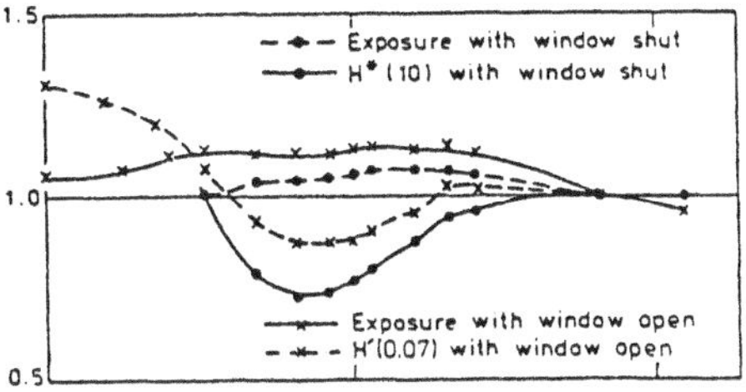

Figure 13. Nuclear Enterprises PDM1 (unmodified).

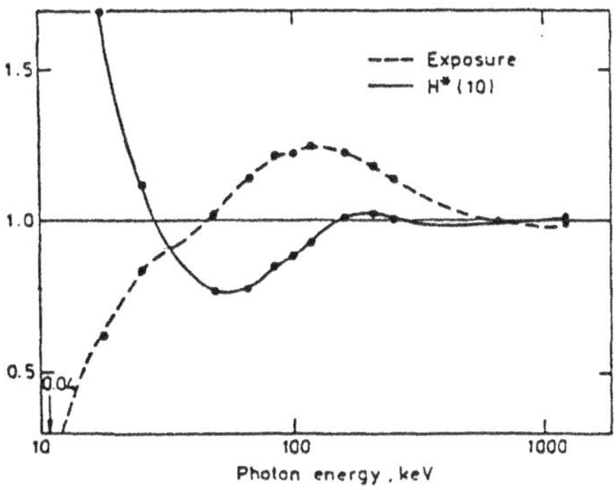

Figure 14. Eberline R02 (unmodified) shutter closed.

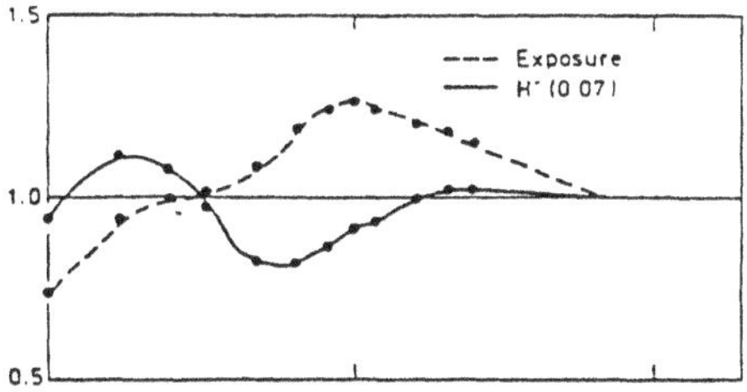

Figure 15. Eberline R02 (unmodified) shutter open.

neglect any error with current instrument designs. However, if the main radiation component is known to be in this energy region it would be wise to assess the under-response and, if necessary, use a multiplication factor to insure that the dose rate is never under-estimated.

Some current designs of instruments and detectors are produced in two forms, either to measure the quantity exposure or the new quantities. These include the Nuclear Enterprises PDM1, the Eberline R02 and R010, and the Mullard and Centronic range of energy compensated Geiger Muller detectors. The photon energy response of the modified Eberline R02 is shown in Figure 16 (Marshall et al. (1987)). It can be seen that, in this form, the instrument is suitable for the measurement of H*(10) or H'(0.07) within the energy range 10 keV to at least 1.3 MeV. Similarly, Figure 17 (Marshall et al. (1987)) shows the energy response of a modified energy compensated Geiger Muller detector for the measurements of H*(10). It is seen to be suitable in this form for the energy range 30 keV to 1.3 MeV.

TABLE 5. Energy range over which some current instrument designs are suitable for the new quantities.

Instrument Type	Manufacturer	Detector	Useful Energy Range for New Quantities
PDR1, 2601, 1907B	Nuclear Enterprises	GM tube	30 keV to at least 1.3 MeV for measurement of H*(10). See Figure 7.
PDR2, 2602, PDR4 (internal), 2604 (internal), 95/0128	Nuclear Enterprises	GM tube	30 keV to at least 1.3 MeV for H*(10). See Figure 8.
Gammatroll PR190 and other instruments using ZP1201 detectors	ICI	GM tube	30 keV to at least 1.3 MeV for H*(10). See Figure 9.
RD8	Wallac	GM tube	60 keV to at least 1.3 MeV for H*(10) end-on response. See Figure 10.
900R. 5.10R	Mini Instruments	GM tube	100 keV to at least 1.3 MeV for H*(10). See Figure 11.
NIS 295		Scintillator	200 keV to at least 1.3 MeV for H*(10). See Figure 12.
PDM1 (unmodified)	Nuclear Enterprises	Ion Chamber	10 keV to at least 1.3 MeV for H*(10) and H'(0.07). See Figure 13.
R02 (unmodified)	Eberline	Ion Chamber (shutter closed)	20 keV to at least 1.3 MeV for H*(10). See Figure 14.
		Ion Chamber (shutter open)	10 keV to at least 1.3 MeV for H'(0.07).

Thus, there will be no difficulties in obtaining instruments with response optimised for the new quantities. It would also be possible, if desired, to convert many instruments to the new quantities by modifying the detector.

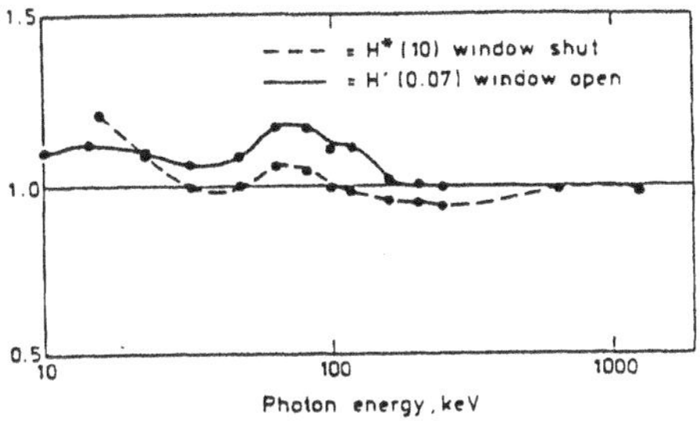

Figure 16. Eberline R02 (modified).

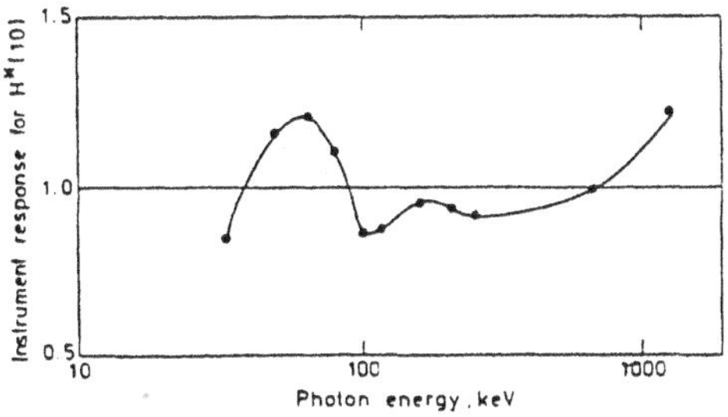

Figure 17. Modified Energy Compensated Geiger Muller detector.

5.2. Personal Dosemeters

The results of the energy dependence type tests on the NRPB thermoluminescent dosemeter (Bartlett et al. (1986)) are shown in Figures 18 and 19 for the quantities $H_p(10)$ and $H_s(0.07)$ assuming these to be numerically equal to H'(10) and H'(0.07) under calibration conditions. It can be seen that the response is close to 1.0 at the higher energies but exceeds 1.0 substantially at the lower energies, especially for the measurement of $H_p(10)$. This occurs because the filter is too thin, i.e., 7 mm instead of 10 m thick.

Note that some of the irradiations were carried out in the ICRU sphere while other were irradiated on a 30 cm cube phantom. It can be seen that with the application of appropriate correction factors the agreement between the two sets of results is good.

The results of tests on the same dosemeter but for its angular response (Bartlett et al. (1986)) are shown in Figures 20 and 21 for the quantities $H_p(10)$ and $H_s(0.07)$ respectively.

In this case note the large overestimate for $H_p(10)$ at 90° incidence. This is because the filter shape is not ideal, i.e., it is a dome rather than slab shaped.

The most commonly used type of dosemeter in the U.K. is the photographic film type and this has been recently type tested for the new quantities. Figures 22 and 23 (Milton et al. (1989)) combine the energy and angle response dependence for $H_p(10)$ and $H_s(0.07)$ respectively. It can be seen that the type of dosemeter measures the new quantities reasonably well for energies above 25 keV.

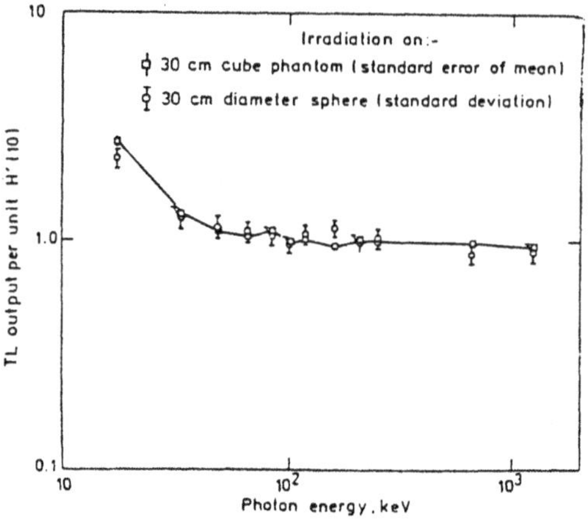

Figure 18. NRPB dosemeter response (*penetrating*).

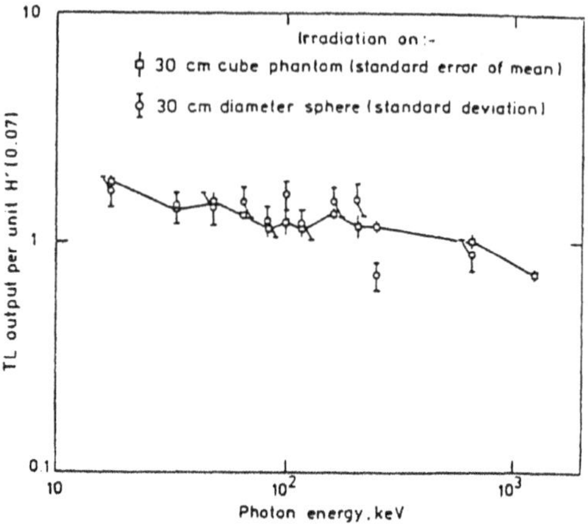

Figure 19. NRPB dosemeter (*non-penetrating*).

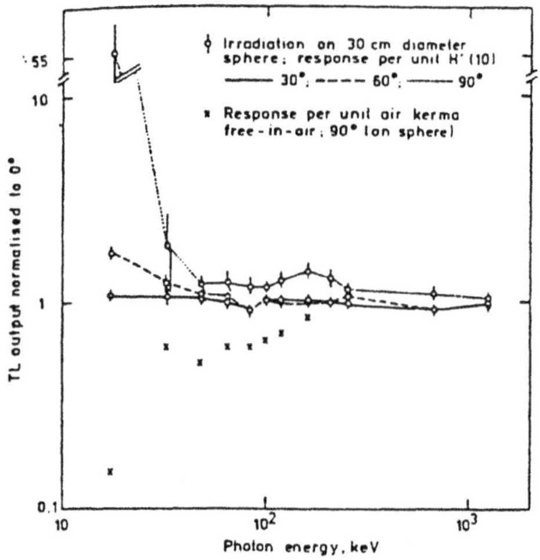

Figure 20. Angle dependence of response of dosemeter normalised at each energy to the response of 0°: *thick* disc.

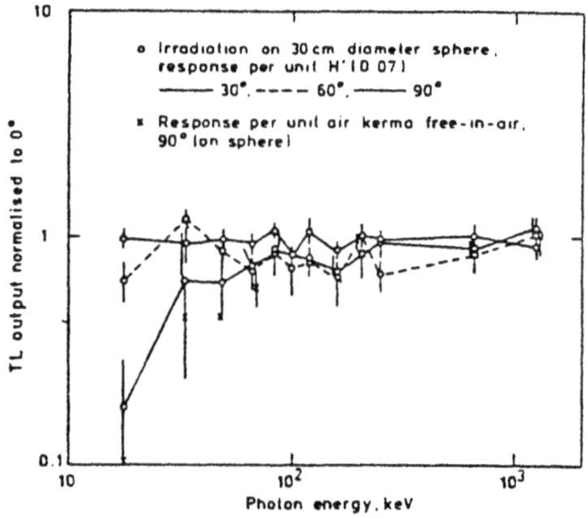

Figure 21. Angle dependence of response of dosemeter normalised at each energy to the response at 0°: *thin* disc.

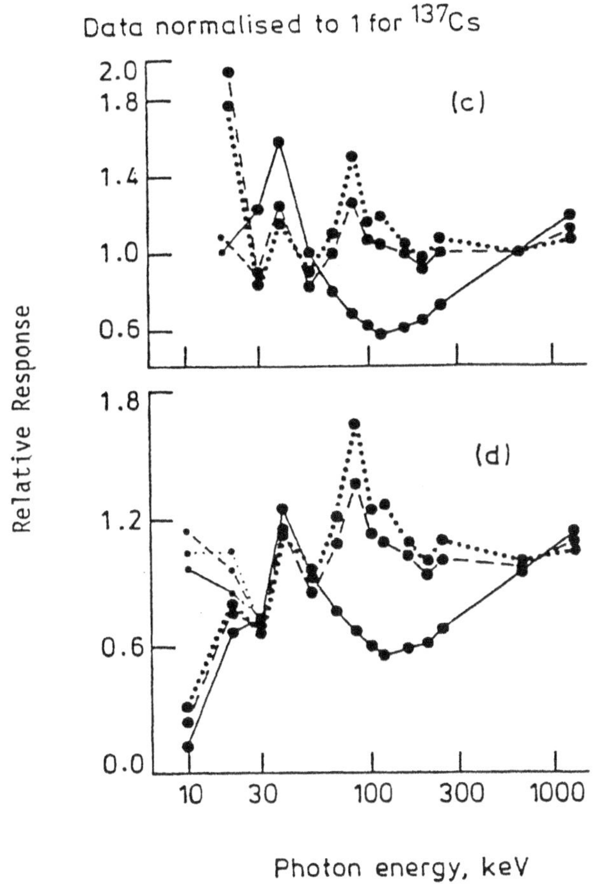

Figure 22 and 23. Type test of NRPB/AERE film dosemeter.

6. References

International Commission on Radiation Units and Measurements, Determination of Dose Equivalents Resulting from External Radiation Sources. Bethesda, M.D., ICRU Reports 39 (1985).

International Commission on Radiation Units and Measurements, Radiation Quantities and Units. Bethesda, M.D., ICRU Report 33 (1980).

International Standards Organisation, X and Y Reference Radiations for Calibrating Dosemeters and Doserate Meters and for Determining their Response as a Function of Photon Energy. International Standards Organisation, International Standard ISO 4037 (1979), and Amendment 1, Low Rate Series of Filtered X-radiations (1983).

International Commission on Radiological Protection. Data for Use in Protection Against External Radiation. Oxford, Pergamon Press, ICRP Publication 51. To be published.

International Commission on Radiation Units and Measurements, Determination of Dose Equivalents from External Radiation Sources - Part 2. Reprint 43 (Bethesda, M.D.: ICRU Publications) (1988).

Iles, W.J. (1987) "Conversion Coefficients from Air Kerma to Ambient Dose Equivalent for the International Standards Organisation's Wide, Narrow and Low Series of Reference Filtered X-radiations". NRPB-R206.

Marshall, T.O., Bartlett, D.T. and Burgess, P.H. (1987) "Current and Future Instrument and Dosemeter Designs to Measure the New ICRU Radiation Quantities". J. Soc. Radiol. Prot. 7(3).

Bartlett, D.T., Burgess, P.H., Francis, T.M., Dutt, J.C. and Dymbylow, P.J. (1986) "The Energy and Angle Dependence of Response to Photons of the NRPB Thermoluminescent Dosemeter in Terms of the New ICRU Quantities, Radiation Protection Dosimetry, 17, No. 1-4.

Milton, M.I.L., Iles, W.J., Bartlett, D.T., Burgess, P.H. and Hill, C.E. (1989) "Type Testing of the NRPB/AERE Film Badge Dosemeter in Terms of the New ICRU Secondary Quantities". Proceedings of 25th Anniversary Symposium of the SRP, Malvern, June 1989, pp. 269-272.

STATUS OF BETA DOSIMETRY

D. F. Regulla
GSF Forschungszentrum für Umwelt und Gesundheit München,
D - 8042 Neuherberg
Federal Republic of Germany

1. Synopsis

The consideration of external beta radiation is important for the risk estimation of essentially the first 10 mm of tissue (skin, eye lenses, eventually testes), which also holds for low-energy photons. However, more penetrating beta radiation can also contribute to whole body doses (Charles (1986), Roesch (1986)). Practical experiences with beta measurements are limited in radiation protection (Piesch et al. 1986)). Indeed, precise and reliable beta dosimetry is not that easy to perform (Swinth et al. (1986)). It was only after the Three Mile Island (TMI) accident that beta radiation did generate attention. At that occasion the individual exposure was assumed to primarily result from beta radiation; appropriate dosemeters however were missing. The experience of TMI has demonstrated the inadequacies of existing techniques for beta dose determinations. Also at some European nuclear facilities protection level beta dosemeters are available only since recently.

In 1983 the outcome of the First International Symposium on Beta Dosimetry, Washington DC, USA, was summarized as follows:

- Radiation protection problems due to beta radiation do exist.
- The state of measurement technology is unsatisfactory.
- Personal and area exposures resulting from beta radiation will essentially not be registered, at present. This holds particularly for gaseous beta emitters (^{85}Kr, ^{133}Xe).

This matter of facts have not experienced principal changes till today (Regulla (1981), DGB (1987), Regulla (1988), Beta (1989)).

Increased beta dose rates may be expected in areas as follows:

- production of fuel elements,
- reprocessing plants,
- disposal of radioactive waste,
- production of uranium from ores,

M. Oberhofer (ed.), Advances in Radiation Protection, 107–117.

- nuclear reactors, and
- nuclear medicine, biology and research using radioactive isotopes.

To demonstrate the possible order of magnitude in beta dose rates, up to 100 Svh^{-1} were reported in areas accessible for personnel (Regulla 1986)).

Beta dosimetry: A stepchild of radiation protection dosimetry (Bö83)? In that article it has been stated:

- a special measuring quantity for personal and area monitoring is missing and
- special standards for beta dosimetry do not exist.

One may add:

- appropriate beta dosemeters were until recently not available and, hence,
- beta radiation was neglected in radiation protection.

2. Quantity of Measurement and Calibration

According to a national standard (DIN (1985)), personal and area doses from electron radiation are represented by the measurement value of a personal or area dosemeter that has been calibrated for the measurement of absorbed dose in soft tissue in a semi-infinite phantom of unity density, 1 gcm^{-3}, in a depth of 0.07 mm, multiplied by the factor 1 Sv/Gy (DIN (1989)).

The measurement quantities recommended by ICRU 39 do reflect the need to determine skin exposure by beta and also low-energy photon radiation, $H_s(0.07)$, and a possible contribution to whole body dose, $H_p(10)$.

Minimum requirements to thermoluminescence (TL) personal and area dosemeters, due to mean electron energies, are 0.2 MeV - 1 MeV, with the reference energy at 0.8 MeV. The latter energy is given by the mean beta energy of a $^{90}Sr/^{90}Y$ source at the measuring target (DIN (1989), ISO (1984)).

Principal criteria for the beta response of a dosemeter are, e.g.,

- the type of beta emitter and, hence, of the beta spectrum under consideration,
- the source geometry (point or plane source), and
- the exposure geometry (source distance, attenuation by absorber layers, angle of radiation incidence).

The calibration of a beta dosemeter should therefore closely approach the realistic field conditions. Since these conditions can be subject of essential fluctuations, depending on the respective place of measurements, the calibration conditions should provide a conservative estimate of the realistic conditions.

3. State of the Beta Dose Assessment

Recommendations for the relevant calculation of body doses from external beta radiation are available in the FRG (SK (1985)). However, existing dose or dose rate survey meters are, with only few exemptions, inadequate for experimental beta dosimetry particularly if medium-energy or low-energy beta radiation is concerned.

Personal film dosemeters as well as standard TL dosemeters are useful for beta detection only above a certain threshold energy. The threshold energy is detector dependent, and ranges approximately between 0.5 MeV $\leq E_{max} \leq$ 0.8 MeV. Below this energy level, the mentioned dosemeters are useful only if appropriate calibration or corrections for the specific radiation circumstances can be applied.

4. Beta Nuclides and Spectra

In case of fuel element production and reprocessing mainly a number of beta emitters must be taken into account, e.g.: ^{238}U, ^{237}U, ^{235}U, ^{234}Pa, ^{208}Tl, ^{241}Pu, ^{239}Pu, ^{236}Pu, ^{241}Am, ^{144}Ce, ^{242}Cm, ^{244}Cm, ^{106}Ru, ^{90}Sr, ^{90}Y.

Computed beta spectra for these radionuclides are partly available from literature (Cross (1983), Hankins (1984)). To some extent, this also holds for experimentally approached beta spectra from nuclear facilities. All of them are characterized by an essential part of the spectrum around and below 0.5 MeV.

Beta radiation dosemeters must be useful for the detection of beta particle energies up to E_{max} = 3.5 MeV (^{106}Rh). Due to valid standards (DIN (1989)) or type test requirements (PTB (1988)) the lower limit of the beta energy range for personal and area dosemeters is stated at 0.2 MeV. Accordingly, external beta radiation of lower energy is not expected to be of biological relevance or must be considered separately.

5. Existing Beta Measurement Techniques and Developments

For survey meters, it is inadequate to discriminate between beta and photon dosemeters, since pure beta radiation fields do hardly exist. Beta dosemeters must adequately be usable in mixed photon/beta radiation fields. As a general requirement, the response of beta dosemeters should be independent of the beta particle energy in the rated beta energy range. The response should similarly be independent of the photon energy in the rated photon energy range. The response for beta and photon radiation must be comparable.

The dimensions of the beta and photon radiation detector should be sufficiently small as compared with changes of the radiation field. This requirement is to allow for an acceptable local resolution also in cases of beta radiation fields with significant gradients, e.g., close to the surface of beta sources (area dosimetry) or at finger tips (partial body dosimetry).

5.1 Area Dosimetry

For use in area dosimetry there are several dosimetry techniques that may principally serve for an adequate detection of beta radiation:

a) thin-walled ionization chambers,
b) thin-layer, tissue-equivalent scintillators for direct skin dose assessment,
c) thick-layer scintillators with subsequent pulse-height discrimination (spectrometer type dosemeter),
d) thin-layer photo elements,
e) simple pulse counters (GM tubes, silicon diodes).

Photon dose and dose rate meters, as commonly found, e.g. at nuclear sites, are of only limited use or no use at all for beta dosimetry (Regulla (1984), Martz et al. 1986)), if sometimes also stated otherwise (KTA (1986)).

5.1.1. Small-volume ionization chamber. At present there is only one beta-gamma dose rate meter commercially available in the FRG, i.e., the type DL1, manufactured by Reaktorwartungsdienst and Apparatebau GmbH, Jülich (Figure 1).

The detector consists of a 15.5 cm^3 ionization chamber with end and lateral windows, air filling under atmospheric pressure, structural material aluminium, wall lining 7 $mgcm^{-2}$. Measurement of gamma component by attached cap of 10 mm thickness tissue equivalent material.

The beta radiation response of the DL1 has been determined at the GSF-SSDL as follows: $^{90}Sr/^{90}Y$ 0.98 (1.07), ^{204}Tl 0.84 (0.85), ^{133}Pm 1.20 (1.00). The data given in brackets are obtained from literature (Heinzelmann and Keller (1986)). The discrepancy still needs clarification.

5.1.2. Large-volume ionization chamber. Also thin-walled ionization chambers of sufficient volume are useful for beta dosimetry, e.g. Babyline 31, Type E433A, manufacturer Nardeux. The beta radiation response of such an instrument has been determined at GSF-SSDL as follows: $^{90}Sr/^{90}Y$ 1.2, ^{204}Tl 0.8, ^{133}Pm 1.0.

If also more sensitive than the DL1, ionization chamber dosemeters

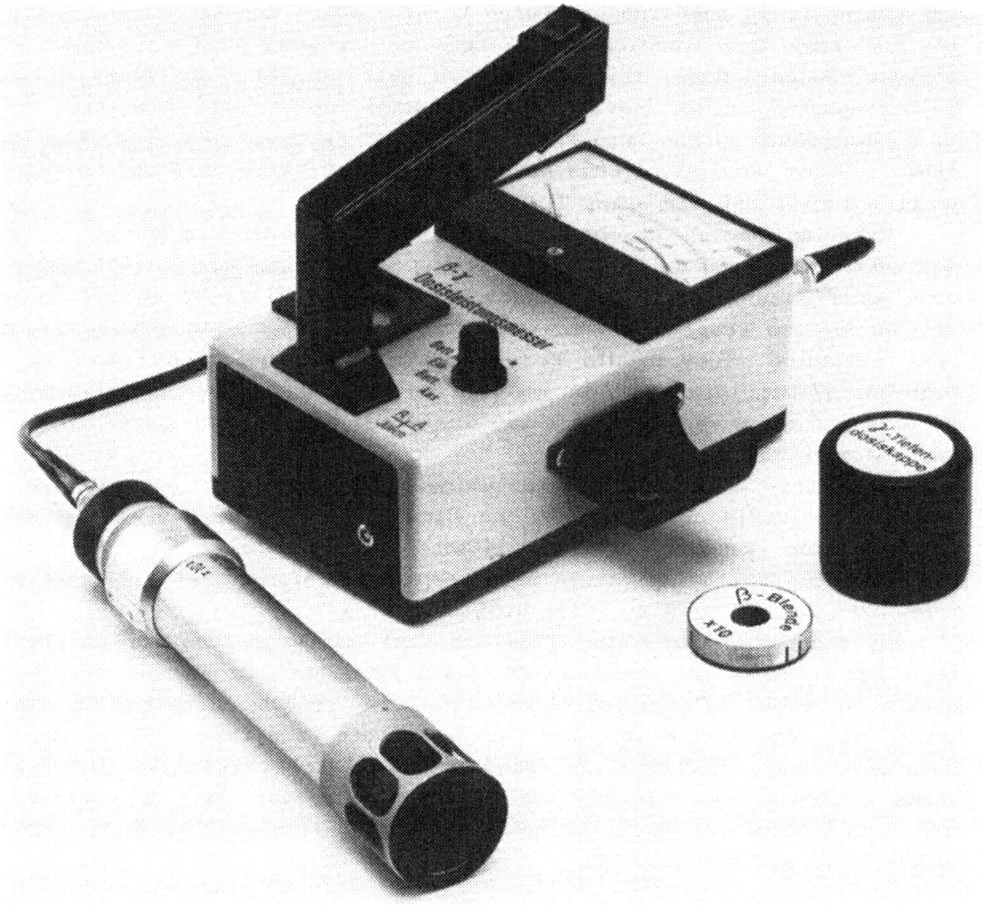

Figure 1. Beta/gamma dose rate meter DL1 (for details see 5.1.1).

of this type are useful for beta dosimetry only in beta fields with small flux gradients, due to the large chamber dimensions. In beta fields with pronounced gradients, such chamber type dosemeters serve more for an indication of beta radiation than for an adequate dosimetry.

5.2 Personal Dosimetry

In view of the lack of adequate beta radiation dosemeters for personal monitoring it is sometimes proposed to use photon dosemeters and multiply the gamma dose reading with a beta-versus-gamma factor in order to achieve the beta dose. This approach of beta radiation dosimetry should be disregarded, since the beta-versus-gamma factor may range from 2.5 to 4, dependent on the beta spectrum and the exposure geometry (Regulla 1984)). Under certain circumstances this factor has been found to range up to a factor 800 (Thompson (1981)).

Existing photon dosemeters with thick detectors such as in TL dosimetry, or detector assemblies using filter-analytical techniques, such as film dosimetry, are characterized by a limited energy range with no or low response to low-energy and sometimes even medium-energy beta radiation. They should be applied to beta radiation fields with high-energy particles only or must be calibrated for the specific type of beta radiation field. Otherwise increased measurement uncertainties must be considered.

From this view also thermoluminescence dosimetry commonly performed with LiF:Mg,Ti chips of the dimensions 3 x 3 x 0.9 mm^3 is not adequate for beta radiation dosimetry in general. For the beta response, the following values have been reported (Dutt et al. (1986)): $^{90}Sr/^{90}Y$ 1.0, ^{204}Tl 0.4, ^{133}Pm 0.05.

By contrast, thin-layer TLDs are much better suited for the purpose for reasons of a decreased field gradient across the detector depth. A reduced TL-output, caused by the reduced detector volume, could be compensated by the use of more sensitive TL phosphor materials, e.g., LiF:Mg,Cu,P (Wang et al. 1986)). This new material shows a 50-fold to probably 100-fold higher sensitivity as compared with traditional LiF:Mg,Ti as manufactured by Harshaw, Cleveland, USA, under the trade name TLD-100.

5.2.1. Thin-layer $CaSO_4$: Tm detectors. The beta radiation response of a commercial $CaSO_4$:Tm thin-layer TL detector (60 µm thickness) has been determined at GSF to be: $^{90}Sr/^{90}Y$ 1.0, ^{204}Tl 1.0, ^{133}Pm 0.6, and hence appears rather valuable for beta radiation dosimetry. As has been evaluated at GSF, this type of detector is good for energy independent beta dose assessment of H'(d) at depths of 0.02 mm upwards. Hence, H'(0.07) could well be determined for needs of radiation protection. However, there is a problem with the over-estimation of low-energy photons in a mixed beta/photon radiation field, since $CaSO_4$:Tm has an elevated effective atomic number and hence enhanced energy dependence of response, as compared with tissue.

5.2.2. Thin-layer LiF detectors. For dosimetry in mixed beta/photon fields an effective atomic number equivalent to tissue is precondition to assure an energy-independent response also for the photon component.

Thin-layer TL detectors may fulfil this requirement to a large extent if prepared out of, e.g., LiF or lithium borate. In case of LiF crystals with a detector thickness of around 0.1 mm the beta response has been determined to be: $^{90}Sr/^{90}Y$: 1.0 (equal to ^{60}Co under CPE conditions), ^{204}Tl: 0.95, ^{133}Pm: 0.5.

5.2.3. Thin-layer LiF/teflon detectors. Even thinner LiF detectors are commercially available, e.g. 0.13 mm, for which LiF powder is imbedded in a teflon matrix. For readout of these LiF/teflon detectors special heater provisions have to be made.

A partial body dosemeter of this type already available routinely is based on LiF/teflon detectors. It has been developed by KFA, Jülich, FRG. Three different detector discs of increasing thickness (0.13 and twice 0.4 mm) allow for the measurement of a kind of depth dose curve (Heinzelmann and Pagenkemper (1987). Evaluation of these TL detectors must be performed manually which is time consuming (Figure 2).

5.2.4. Thin-layer LiF/graphite detectors. A different approach of thin-layer detectors is achieved by an admixture of non-TL graphite to the thermoluminescent crystalline, e.g. LiF, matrix (Christensen (1984)) in which case the TL signal can escape from only utmost thin surface layers. For a LiF/C-type detector with the dimensions 3 x 3 x 0.9 m^3 the following beta response has been reported (Quam (1983)): $^{90}Sr/^{90}Y$ 1.0, ^{204}Tl 0.86, ^{133}Pm 0.23.

Thin-layer LiF detectors are also described with the phosphor powder imbedded in a PTFE matrix with a graphite admixture of 5-10%. As a result, an effective detector thickness of about 0.04 mm was achieved. The spectral response is described as follows (Dutt et al. (1986)): $^{90}Sr/^{90}Y$ 1.0, ^{204}Tl 1.00, ^{133}Pm 0.65.

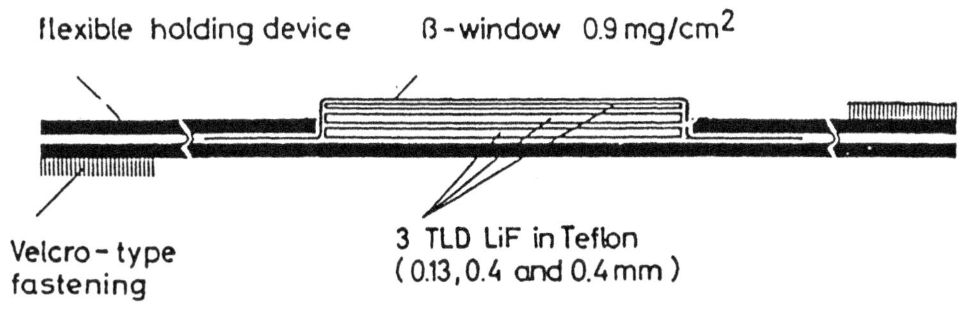

Figure 2. Schematic representation of a skin dose dosemeter based on three TL detectors (for details see 5.2.3).

5.2.5. Laser-heated surface detectors. The TL emission from a detector surface may probably also be achieved by appropriate laser heating in which case a two-dimensional readout from the detector surface could be achieved (Quam (1983)). This technique is so far still at an experimental level; more information is available from the proceedings of the 9th International Conference on Solid State Dosimetry, Vienna, 6-10 November 1989 (Radiat. Prot. Dosim. 33, 1990 and 34, 1990).

5.2.6. Thin-layer LiF/polymid detectors. Another thin-layer TL dosimeter uses LiF powder on a polyimid band by which technique detector dimensions of 6 x 10 x (0.075... 0.11) mm^3 have been achieved (Curry (1986)). The energy dependence of beta response is reported as follows (Dutt et al. (1986)): $^{90}Sr/^{90}Y$ 1.00, ^{204}Tl 0.96, ^{133}Pm 0.54.

5.2.7. TSEE detectors. Most promising for beta radiation dosimetry is, at least in principle, the so-called thermally stimulated exoelectron emission (TSEE) (Kreigseis et al. 1986)). TSEE dosimetry is based on a two-dimensional radiation sensitive layer of less than 1000 Å thickness. Its spectral response is expected to be unity for the whole energy range of beta particles.

This technique has experienced remarkable progress during the last decade. Nevertheless, it has not yet achieved a level that would allow a reliable routine application. (Further details from the 9th International Conference on Solid State Dosimetry, Vienna, 6-10 November 1989).

5.2.8. Phosphate glass detectors. A phosphate glass dosimetry system is recently under laboratory test and may probably provide a possibility for beta radiation dosimetry by a differential scanning of the depth dose profile (Piesch (1989)).

6. Conclusion

Beta exposure can well be avoided by appropriate radiation protection measures. The presence of beta radiation, however, must first be proven qualitatively and quantitatively. Area dose and dose rate meters useful for this task as well as practical experiences are not yet sufficiently available. This holds also for personal dosemeters, particularly partial body dosemeters. The need for beta dosimetry has only recently been recognized: ICRP has established a task group to evaluate on the biological effects of beta radiation and to study on the effective depth of the radiation sensitive basal layer of the epidermis. Similar studies are performed at NCRP, USA.

The ICRP measurement quantities for radiation protection are further in discussion. EURADOS has, at the occasion of the annual meeting in Munich, 1989, declared that research on beta dosimetry deserves a high priority. Beta partial body dosimetry is represented in the new research program of the Commission of the European Communities. ISO/IEC have recognized the need for a partial body dosimetry standard, covering also beta radiation. Relevant research activities can be found in many European research laboratories, last not least at GSF Munich, Neuherberg. Here at GSF, new beta radiation detectors are adopted and equipment for automatic readout is developed.

The earlier mentioned 9th International Conference on Solid State Dosimetry, Vienna, November 6-10, 1989 offered interesting contributions about new developments in the field of beta dosimetry.

7. References

Begriffe und Benennungen in der radiologischen Technik. Dosisgrößen und Dosiseinheiten, DIN 8614 Teil 3 (1985).

Berechnungsgrundlagen für die Ermittlung von Körperdosen bei äußerer Strahlenexposition durch Elektronen, insbesondere durch Beta-Strahlung, Bundesanz. 126a, 21-28 (1985).

Charles, M.W. (1986) "The biological basis of radiological protection criteria for superficial, low penetrating radiation exposure". Radiat. Prot. Dosim. 14, 79-90.

Christensen, P. (1984) "Study of graphite-mixed and boron-diffused TLDs for skin dose assessment". In: Proc. of the International Beta Dosimetry Symposium. USNRC, Washington, D.C., NUREG/CP-0050, 341-350.

Cross, W.G., Ing, N. and Freedman, N. (1983) "A short atlas of beta-ray spectra". Phys. Med. Biol. 28, 1251-1260.

Curry, R.G. and Greenslade, E. (1986) "Extremity dosimeter monitoring service". Radiat. Prot. Dosim. 17, 115.

Dutt, J.C., Greenslade, E. and Marshall, T.O. (1986) "A new approach to the problems of extremity dosimetry". Radiat. Prot. Dosim. 14, 145-150.

Ergebnisse der DGB-Studie (1987).

Francis, T.M., O'Hagan, J.B., Richards, D.J. and Driscoll, C.M.H. (1986) "Responses of thermoluminescent materials to beta radiation and low energy photons". Radiat. Prot. Dosim. 17, 89.

Hankins, D.E. (1984) "Beta dosimetry at LLNL". In: Proc. of the International Beta Dosimetry Symposium. USNRC, Washington, D.C., NUREG/CP-0050 615-633.

Heinzelmann, M. and Keller, M. (1986) "An energy-independent dose rate meter for beta and gamma radiation". Radiat. Prot. Dosim. 14, 179-182.

Heinzelmann, M. and Pagenkemper, M. (1982) "A new extremity dosimeter for beta and gamma radiation". Radiat. Prot. Dosim. 2, 115-118.

Holzapfel, G. and Lesz, J. (1984) "Bonded thin layer detector for integrating solid state dosimeter". Radiat. Prot. Dosim. 6, 200-202.

Koppel, H. (1989) Harshaw Comp., Cleveland, private Mitteilung.

Kreigseis, W., Petel, M. and Scharmann, A. and Wieters, C.U. (1986) "Potentials of TSEE for beta dosimetry". Radiat. Prot. Dosim. 14, 151-155.

Martz, D.E., Johnson, L.O., Rich, B.L. and Daniel, S.H. (1986) "Field tests of a tissue-equivalent beta survey meter". Radiat. Prot. Dosim. 14, 117-121.

Martz, D.E., Rich, B.L. and Johnson, L.O. (1986) "A portable beta spectrometer for tissue dose measurements". Radiat. Prot. Dosim. 14, 183-186.

Messung der Ortsdosisleistung in Sperrbereichen von Kernkraftwerkent. Sicherheitstechnische Regeln des Kerntechnischen Ausschusses, KTA 1506 (1986).

Piesch, E., Burgkhardt, B. and Viglis, M (1989) "Photoluminescence dosimetry-progress and present state of the art". Radiat. Prot. Dosim. 33, 215-226.

Piesch, E., Burkhardt, B. and Anton, R. (1986) "Dose rate measurements in the beta-photon radiation field from UO_2 pellets and glazed ceramics containing uranium". Radiat. Prot. Dosim. 14, 109-112.

PTB-Mitteilungen 96, 2/86, 91-99 (1988).

Quam, W. (1983) "Beta dosimetry using laser heating of hot-pressed TLDs". Health Phys. 44, 75.

Rathbun, L.A., Endres, G.W.R., Fox, R.A., Roberson, P.L. and Scherpelz, R.I. (1984) "Beta particle measurements and dosimetry requirements at NRC licensed facilities", NUREG/CR-3544.

Reference beta radiations for calibrating dosimeters and dose ratemeters and for determining their response as a function of beta radiation energy, ISO 6980 (1984).

Regulla, D.F. (1981) "State of the art in beta dosimetry". In: Proc. 3rd Inform. Seminar on the Radiation Protection Dosimeter Intercomparison Programme, Grenoble, CEC, Luxembourg, EUR 7365, 161-177.

Regulla, D.F. (1984) "Need and practice of beta dosimetry in personnel monitoring". In: Proc. of the International Beta Dosimetry Symposium. USNRC, Washington, D.C., NUREG/CP-0050 263-281.

Regulla, D.F. (1986) "Importance of exposure to beta particles in practical radiation protection". Radiat. Prot. Dosim. 14, 95-100.

Regulla, D.F. (1988) "Beta fields relevant for radiation protection". In: Beta Dosimetry. CEC, Luxembourg, EUR 11363 165-176.

Roesch, W.C. (1986) "Problems related to the critical depth of skin". Radiat. Prot. Dosim. 14, 91-94.

Strahlenschutzdosimetrie, Thermolumineszenzdosimetrie-Systeme, DIN 6818 Teil 6 (1989).

Swinth, K.L., Rathbun, L.A. and Brackenbush, L.W. (1986) "Beta fields and measurement practice at DOE facilities". Radiat. Prot. Dosim. 14, 105-108.

Thompson, I.M.G. (1981) "Future trends in beta dosimetry". In: Proc. 3rd Inform. Seminar on the Radiation Protection Dosimeter Intercomparison Programme, Grenoble, CEC, Luxembourg, EUR 7365, 95-108.

Vorläufige Ergebnisse des AK Beta-Dosimetrie in Kernkraftwerken, Ausschuß Strahlenschutztechnik (S1) der Strahlenschutzkommission (SSK) (1989).

Wang, S., Chen, G., Wu, F, Li, Y., Zha, A. and Zhu, J. (1986) "Newly developed highly sensitive LiF:Mg,Cu,P TL-chips with high signal to noise ratio". Radiat. Prot. Dosim. 14, 223.

STATUS OF NEUTRON DOSIMETRY

M. OBERHOFER
Commission of the European Communities
DG XII - Joint Research Centre Ispra
I - 21020 Ispra (VA)
Italy

1. Introductory General Remarks

When dealing with various subjects in radiation protection, neutron dosimetry, a topic of particular actuality, cannot be omitted.

Neutron dosimetry has always been a subject of much concern to radiation protectionists, because of the complexity of this matter.

Neutrons behave so differently from charged particles and photons when passing through matter, because of their exclusive interaction processes with atomic nuclei.

Neutrons do not loose energy either by deceleration as a sequence of direct ionisation or by deceleration through radiation emission. Their interaction with orbital atomic electrons is so weak, that it can be neglected.

From our nuclear physics courses we know that for dosimetric purposes the following elementary nuclear processes are important:

- elastic scatter at nuclei;
- inelastic scatter at nuclei;
- capture;
- exchange-reactions;
- nuclear fission.

Let us look in some more detail into these different processes.

In the first, forth and last case energetic charged particles are produced (recoil protons, protons, alpha-particles, etc.) which most of their kinetic energy lose by ionisation processes. In the case of inelastic scatter and capture, photons are produced, which lose their energy by way of secondary electrons, mostly distant from their place of origin. With some delay energy also can be transferred by radiation emitted from activated nuclei.

The distribution of the various nuclear processes strongly depends on the neutron-energy with often a wide distribution (spectrum) and

M. Oberhofer (ed.), Advances in Radiation Protection, 119–133.

also on the composition of the irradiated substance. The energy-transfer is described in terms of kerma (*k*inetic *e*nergy *r*eleased in *ma*terial), K, which mathematically is expressed by:

$$K = \frac{\eta(E_n)}{\rho} \phi$$

with:
ϕ = particle fluence of monoenergetic neutrons;
E_n = neutron energy;
ρ = density of the neutron exposed medium.

The energy-dependent material-constant $\eta(E_n)/\rho$ is called fluence to kerma conversion-factor.

$\eta(E_n)/\rho$ again can be mathematically expressed by:

$$\frac{1}{\rho} \sum_i \sum_j n_i \sigma_{ij} (E_n) \overline{W}_{ij}$$

with:
n_i = number of atoms of type i of the exposed material;
σ_{ij} = cross-section of nuclide i for reaction of type j;
$\overline{W}_{ij}$ = sum of average initial kinetic energy of charged particles produced in reaction j with nuclide type i.

In case of a neutron-spectrum, ϕ is replaced by the spectral neutron fluence $d\phi(E_n)/dE_n$ and the whole expression integrated over E_n.

Because of the low range of the charged secondary particles compared to the mean free path length of the primary neutrons in a homogeneous neutron-absorber, secondary particle equilibrium may prevail. Neglecting charged particle breaking-radiation (bremsstrahlung) in the case of secondary particle equilibrium, kerma is equal to absorbed dose. This dose can be calculated with the equation presented above only if the indirect energy transfer by photons, produced in capture and inelastic scatter-processes can be neglected. If photon-radiation is produced, its contribution to the absorbed dose needs to be considered, if the point of interest lies within a large volume, the human body, for example.

Usually the absorbed dose in some neutron exposed material, $D_{material}$ has to be deducted from the absorbed dose in some equilibrium-detector, $D_{detector}$ by a relationship, which mathematically is given by:

$$D_{material}/D_{detector} = \frac{(\eta(E_n))}{\rho}_{material} / \frac{(\eta(E_n))}{\rho}_{detector}$$

Because of the strong neutron energy dependence of this ratio, the knowledge of the neutron spectrum at the point of interest needs to be known the more, the broader the spectrum is.

The calculation of the sensitivity of the detector is usually difficult, because it generally depends on the kind and the energy of the charged secondary particles produced within the detector. Mostly the response of the detectors is not sufficiently sensitive to heavy recoils.

In the case of detectors, which are based on the measurement of electric charges one complication is the fact that heavy particles at some energy-threshold do not ionise anymore (protons below 10 keV, for example). This is why ionisation chambers with tissue equivalent walls and tissue-equivalent filling-gases, the historic Rossi-Failla chamber (Rossi and Failla (1956)), Figure 1, for example, can be made use of only at rather high neutron-energies for the energy-independent measurement of the absorbed dose in tissue. Also the application of the

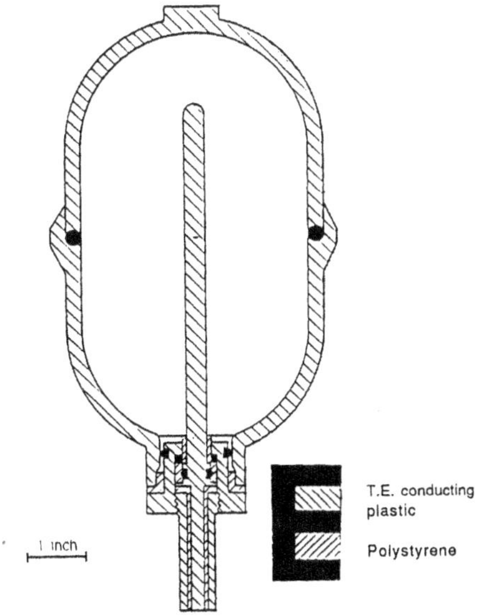

Figure 1. Tissue-equivalent Rossi-Failla chamber filled with tissue-equivalent gas:
hydrogen : -38.1%;
methane : -22.2%;
oxygen : -37.6%;
nitrogen : -2.1%.

Bragg-Gray principle is limited, because of the short range of the charged secondary particles. Very small detector probes are needed, available now in form of tiny TL-dosemeters, though with often insufficient sensitivity.

In radiation protection dosimetry, one needs to consider quality factors, Q, for arriving at the quantity dose-equivalent, H, the corresponding formula being:

$$H = D \times Q$$

Up to recently it was and still is common practice to obtain the **ma**ximum **d**ose **e**quivalent in a phantom (usually a slab or cylinder), called MADE, by multiplying the maximum absorbed dose (maximum of the depth distribution in the phantom and/or human body) with a quality-factor averaged over the whole neutron-spectrum, the effective quality factor, $\overline{Q}$:

$$H_{MADE} = D_{max} \times \overline{Q}$$

$\overline{Q}$ is calculated according to the expression:

$$\overline{Q} = \frac{1}{D} \int_{0}^{\infty} Q \frac{dD}{dL_{\infty}} dL_{\infty}$$

with the dose absorbed delivered by particles with different L_{∞} (L = linear energy transfer).

The effective quality factors have been calculated for the different energies of undirectional broad beams of monoenergetic neutrons at normal incidence and are listed up in Table 1 in column 4 from thermal neutrons up to 3 x 10^3 MeV neutrons and shown graphically in Figure 2.

For the same neutron energy range in columns 2 and 3 in Table 1 conversion factors are listed up, which allow the transformation of neutron fluence-rates in units of cm^{-2} s^{-1} into dose equivalent rates at the point where the dose equivalent rate is maximum in units of μSv h^{-1} and/or mrem h^{-1}. With the help of these factors and the neutron fluence-rates known the MADE can be obtained for the exposure conditions mentioned above. Conversions are now also available for the effective dose equivalent for defined geometrical relationships between a uniform field of neutron fluence and a realistic body phantom.

In practical circumstances the situation is complicated by the fact that neutrons are present with all kinds of energies, that is with spectral distribution, which changes very much mostly from place to place. Often in practice efforts are made to obtain an average or

TABLE 1. Conversion factors (neutron fluence rate, cm^{-2} s^{-1}, corresponding to a dose equivalent rate of 1 μSv h^{-1} and 1 mrem h^{-1}) and effective quality factor $\overline{Q}$ as functions of neutron energy. (The factors may also be used to relate neutron fluence rate and dose equivalent index rate). From (CECRP (1980)).

Neutron energy MeV	Conversion factor		Effective quality factor $\overline{Q}$
	(cm^{-2} s^{-1}) per (μSv h^{-1})	(cm^{-2} s^{-1}) per (mrem h^{-1})	
$2{\cdot}5 \cdot 10^{-8}$ (thermal)	26	260	2·3
$1{\cdot}10^{-7}$	24	240	2
$1{\cdot}10^{-6}$	22	220	2
$1{\cdot}10^{-5}$	23	230	2
$1{\cdot}10^{-4}$	24	240	2
$1{\cdot}10^{-3}$	27	270	2
$1{\cdot}10^{-2}$	28	280	2
$2{\cdot}10^{-2}$	17	170	3·3
$5{\cdot}10^{-2}$	8·5	85	5·7
$1{\cdot}10^{-1}$	4·8	48	7·4
$5{\cdot}10^{-1}$	1·4	14	11
1	0·85	8·5	10·6
2	0·70	7·0	9·3
5	0·68	6·8	7·8
10	0·68	6·8	6·8
20	0·65	6·5	6·0
50	0·61	6·1	5·0
$1{\cdot}10^{2}$	0·56	5·6	4·4
$2{\cdot}10^{2}$	0·51	5·1	3·8
$5{\cdot}10^{2}$	0·36	3·6	3·2
$1{\cdot}10^{3}$	0·22	2·2	2·8
$2{\cdot}10^{3}$	0·16	1·6	2·6
$3{\cdot}10^{3}$	0·14	1·4	2·5

effective neutron-energy with application of the corresponding conversion-factor.

Prerequisite to the transformations is the knowledge of the fluence-rate or the fluence, if the dose-equivalent has to be determined apart from the neutron-energy or the neutron-spectrum.

Various methods are available for this purpose, special neutron counting systems, spectrometers, threshold detectors, nuclear-track detectors, earlier also the nuclear emulsion. Elaborate calibration tech-

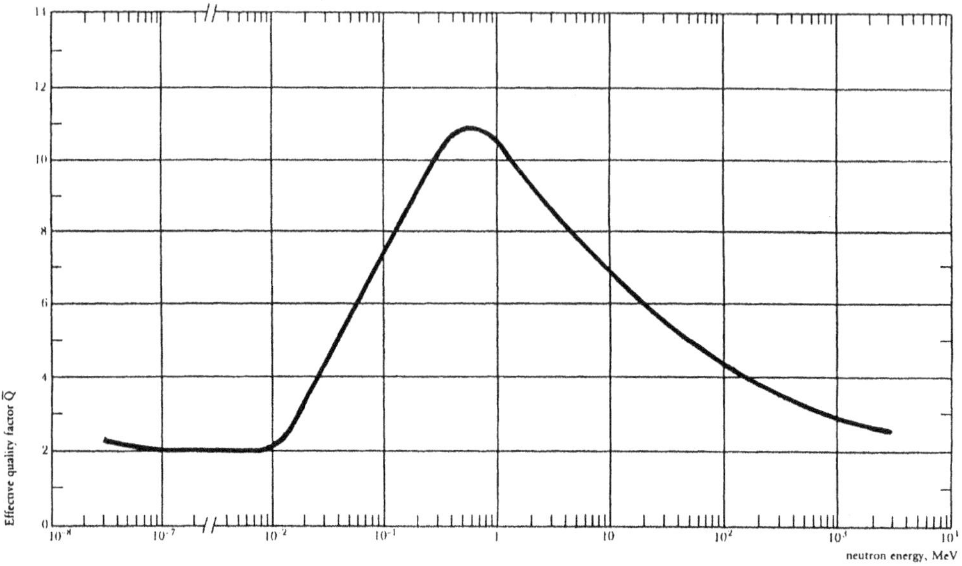

Figure 2. Effective quality factor for neutrons, from (CECRP (1980)).

niques are made use of in this connection.

Parallel to the method outlined attempts have always been made to "measure" neutron dose-equivalents with energy-independent devices. It is evident that the probe method cannot be used because dose-equivalent is not a physical quantity and is no measure of the absorbed energy per mass-unit.

Therefore, by proper choice of detector-substances, geometrical form of the detectors and neutron-absorbers of different configurations one tried to obtain the ratio of some quantity M and the dose-equivalent, which is as much independent of neutron-energy as possible.

It is very difficult to achieve this objective with one detector, often several detectors are used and the quantity M calculated by applying some mathematical algorithm on the single measuring values M_1, M_2, ... M_n from the different detectors:

$$M = f(M_1, M_2, \ldots M_n)$$

A typical example is the multisphere- or BONNER method (Bramblett et al. (1960)).

This method makes use of several spheres produced from polythene with different diameters in the centre of which thermal neutron detectors (like cylindrical or spherical BF_3 or He-3 counters, thermoluminescence or other solid-state detectors) are placed. For the different sphere-diameters different measuring-values (readings) M_1 up to M_n are obtained, depending on their dependence on neutron energy or the spectral distribution of the neutrons.

With the help of suitably chosen weighing factors by linear combination of M_1 to M_n M is obtained more or less independent of neutron-energy (for constant neutron-fluence-rate and/or fluence). Within the range of thermal neutron energies up to 10 MeV it is possible to achieve an accuracy in the neutron dose-equivalent assessment in the order of ±30%. The method is evidently used for area dosimetry, as well as instruments, which use only one moderator-combination like the Anderson and Braun neutron rem-counter (Anderson and Braun (1964)) and similar ones produced on the same principle, ultimately the Dineutron dose-equivalent meter equipped with two polythene-spheres of different diameters containing helium counters (see Figure 3). In this latter example the dose-equivalent results from a combination of two count-rates, the ratio of which allows to characterize the neutron-field and as a result the instrument internal computer calculation of the dose-equivalent rate and/or the dose with quality-factor indication.

Another method used to arrive at personal neutron dose-equivalents in μSv, for example, is the so-called albedo-dosemeter system.

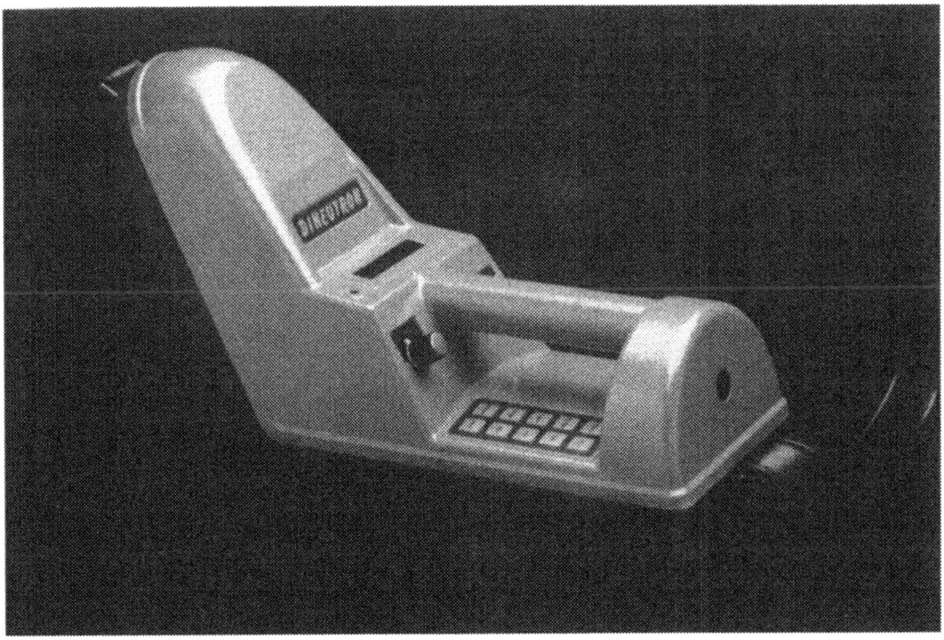

Figure 3. The DINEUTRON dose-equivalent meter.

Some explanation is needed here.

The human body contains hydrogenous material which moderates and scatters neutrons. The back-scattered neutrons will re-emerge from the body with lower energy. They are called albedo-neutrons. Albedo personnel dosemeters - there are many types - are based on the detection of the thermal and epithermal albedo neutrons. The problem consists here in the fact that both, the fraction of neutrons re-emerging with thermal energy and also the neutron fluence to dose-equivalent conversion factor vary with energy (see Figure 4) in such a way that the overall-response of simple albedo dosemeters is highly energy dependent. Figure 5 shows how in principle an albedo dosemeter system works. Usually the thermal neutron detectors are LiF or $Li_2B_4O_7$ TL phosphorous with the Li enriched in Li-6 and B enriched in B-10 for obtaining maximum thermal neutron sensitivity. Because the TLDs are also sensitive to γ -radiation, the γ-contribution to the TL-signal must be corrected for, the reason why usually couples of LiF and $Li_2B_4O_7$ TL-detectors are used with one detector in the pair having Li and/or B depleted in Li6 and/or B-10, responding practically only to photon radiation. Figure 6 shows schematically some types of LiF albedo-dosemeters where the number 6 means LiF with Li enriched in the Li-6 isotope and the number 7 LiF depleted in Li-6.

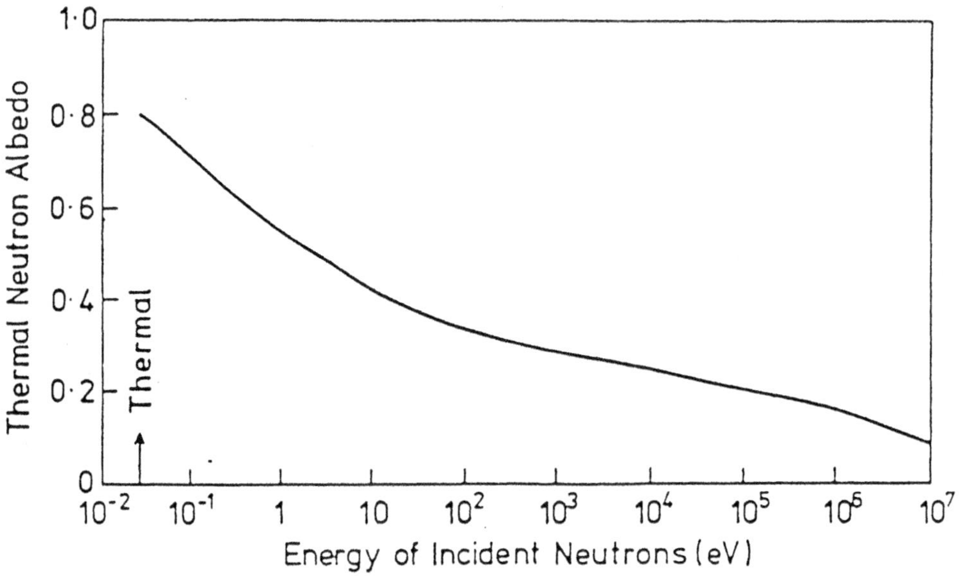

Figure 4. Thermal neutron albedo from the body as a function of incident neutron energy (from Oberhofer and Scharmann (1981)).

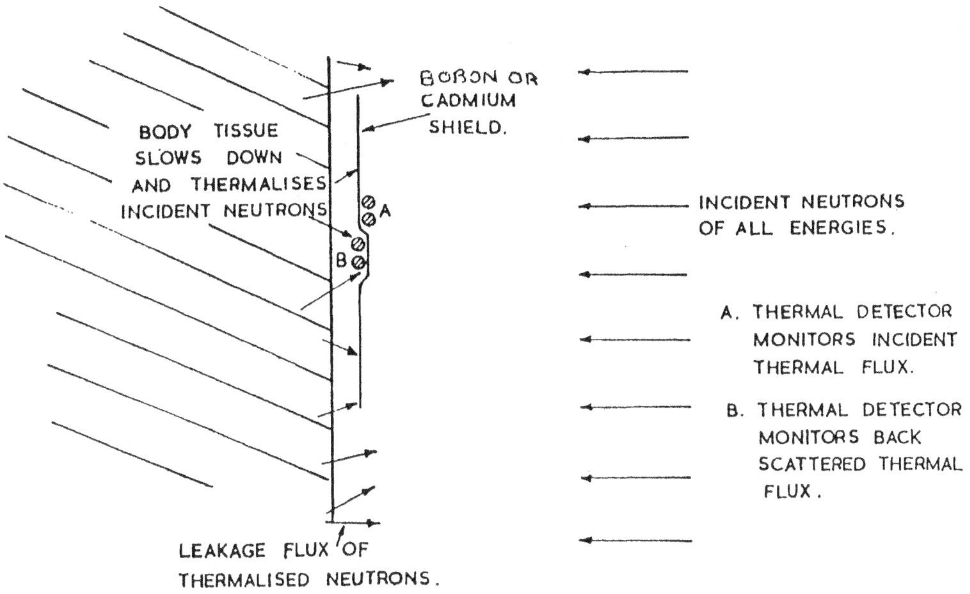

Figure 5. Schematic diagram of prototype albedo dosimetry system.

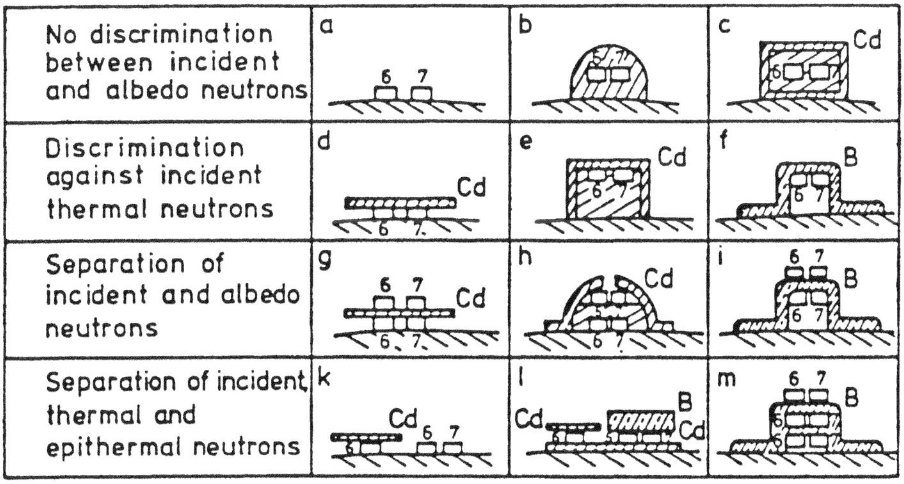

Figure 6. Schematic diagram of types of albedo dosemeters (from Oberhofer and Scharmann (1981)).

Complicated mathematical algorithms yield the neutron-dose equivalent. The method is also used for area-monitoring.

The overall neutron energy dependence of the systems is such that they overrespond to thermal and epithermal energies and are best suited for environments where these neutron-energies prevail.

Calibration in the proper energy ranges is mandatory for best dose-equivalent results.

The calibration procedures have to be reviewed with regards to the recent introduction of new (ICRU) quantities for operational radiation protection purposes out of which in this context two are of interest, the:

- ambient dose-equivalent, $H^*(10)$, and the
- individual dose-equivalent, $H_p(10)$.

Much discussion is going on also with regards to what extent a proposed increase of Q for neutrons by a factor 2 (ICRP (1985)) will effect calibration of neutron instruments in terms of MADE or the new ICRU quantities.

2. Recent Developments

Much work has been done in the past decades to find satisfactory solutions to the neutron measurement and dosimetry problem and many efforts were made in this direction particularly in the past years, as was evidenced at the Symposium on Neutron Dosimetry at Neuherberg in the Federal Republic of Germany in 1987. New basic physical data were obtained and models developed which are required for the understanding of the interactions of neutrons with matter. This holds also for the physical, biological and epidemiological basis of the quality factor for neutrons. Apart from conceptional considerations the last years were characterized by numerous contributions in the instrumental field with new types of proportional counter instruments and improved solid state nuclear track detectors having been the most outstanding achievements.

2.1. Area Monitoring

Besides the continuation of the moderator type instruments, for which the newest commercial representative is the Dineutron, mentioned already earlier, and which is widely used now in the nuclear industry because of its sound design and low weight (2.5 kgs only, compared to heavier instruments of the older generation), physicists and instrument-designers have been working on dose-equivalent meters based on microdosimetric techniques employing low pressure tissue-equivalent proportional counters (TEPC). These instruments measure absorbed dose on cavity chamber principles. Due to the low gas pressure and single-event registration with pulse height analysis, additional information on radiation quality can be obtained. Such instruments are therefore able to determine dose-equivalent by a method which is close to the definition of this quantity and differs from all other methods used in the realisation of radiation protection instruments for neutrons. The development of a number of such devices is so advanced, that in the meantime they could be intercompared with the result that the instruments are able to measure dose-equivalent of mixed-neutron photon radiation and to evaluate a mean quality factor. In one special case (Saion and Watt (1988)) a TEPC was designed and developed which can be operated to discriminate in favour of intermediate neutrons in mixed fields, certainly of interest to many nuclear installations with degraded neutrons. Hopefully TEPC instruments will be commercialized for an achievable price in not too distant future.

2.2. Individual Monitoring

One can differentiate between two classes of individual neutron dosemeters:
- alarm dosemeters, and
- dosemeters for record purposes.

2.2.1. Alarm dosemeters. Whereas the design, and production of γ- pocket and alarm dosemeters never was a serious problem, personal active dosemeters and alarm devices for neutrons have proved to be difficult and expensive to produce.

Several attempts have been made and various products presented, based upon proportional counters and silicon surface barrier detectors. One proportional counter device (Quam et al. (1982)) uses microdosimetric principles to measure the stopping power of recoil protons and deduces the quality factor for application to the measured absorbed dose, thus obtaining the dose-equivalent. Originally the device has been designed for use in space by astronauts for measuring doses due to cosmic radiation.

One problem of such devices is battery-life and miniaturisation.

Only little efforts are made to obtain a neutron-chirper, though such a device would be very useful for a limited number of persons, reactor-control personnel and safeguard-inspectors, for example.

At the Joint Research Centre Ispra a prototype of such a pocketable device has been produced, which makes use of the historic Hornyak-button (ZnS (Ag)-powder in lucite).

2.2.2. Dosemeters for record purposes. One type of passive neutron detector for dose record has already been mentioned, the albedo-dosemeter, normally based upon TL phosphors. Together with this personal dosemeter often solid state nuclear track detectors are made use of, often for compensation of the decreased sensitivity of the first one at higher neutron energies.

The principle is described by J.A.B. Gibson in (Gibson (1988)) as follows:

"Non-photographic registration of tracks of ionising particles in insulating solids, such as mica, glass or organic polymers, is based on the creation of radiation damage along the tracks which may be made by fission fragments: heavy recoil nuclei, alpha particles or protons. The latent tracks are made visible by applying a chemical etchant which preferentially dissolves the damage volume along the track. By using thin material it is possible to etch right through the material and then use a spark counter to count the tracks automatically. Alternatively, an alternating voltage can be applied across thicker types of material (CR-39) in the etching solution, producing breakdown "trees" at the tip of each track which can then be counted easily by microscopes equipped with image recognition facilities and a microcomputer. This is the technique of electrochemical etching (ECE). The original tracks can be from fissile targets or other radiators (e.g. B) adjacent to the track-registering material or may be produced as proton recoils in the plastic material itself or adjacent radiators."

Much experience has been gained in the past years with the recoil proton registration by polycarbonate-foils, which are increasing made use of for personal neutron dosimetry purposes. In particular CR-39 plastic with ECE proved to be superior to many other plastics following extensive work for developing a stable, reproduceable material with a low neutron energy threshold (150 keV) and good energy response.

This response is shown in Figure 7 for the NRPB-PADC dosemeter (Gilvin et al. (1987)), which apparently represents the latest state of the art.

Remains to draw the reader's attention to a new development, which reminds on the classic quartz fibre electroscope or pen-dosemeter for gamma and/or thermal neutron radiation, the bubble or superheated drop detector (SDD) for fast neutrons, a promising Chalk River development (Ing and Birnboim (1984)).

In the invention superheated liquid drops are incorporated in a fluid or semi-solid gel. Charged particles (neutron produced recoil-protons, for example) interact with the drop and cause it to expand. In the gel, it expands to a readily visible dimension so that the number of bubbles can be counted and related to the incident neutrons fluence (see Figure 8).

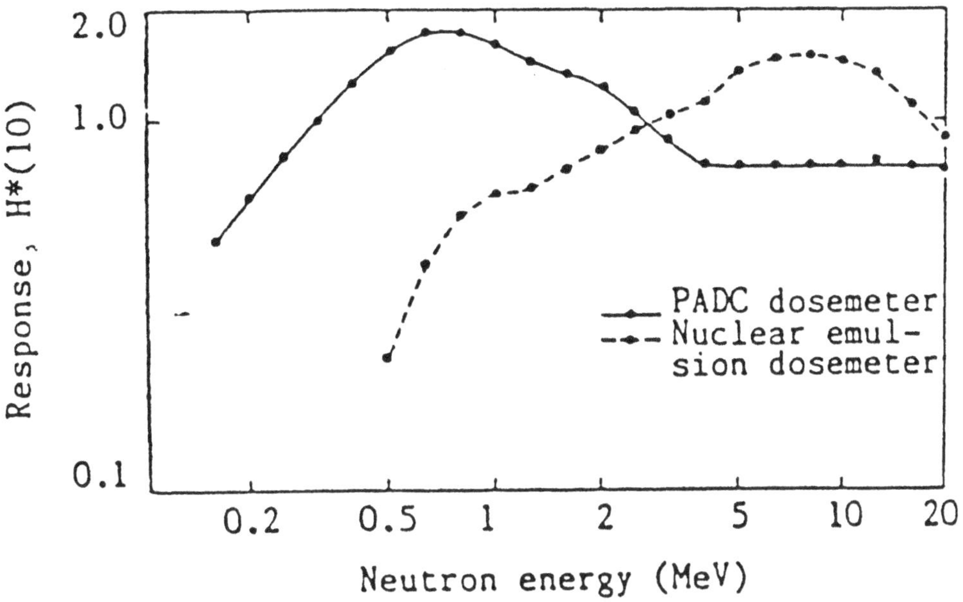

Figure 7. PADC CR-39 and nuclear emulsion dosemeter responses.

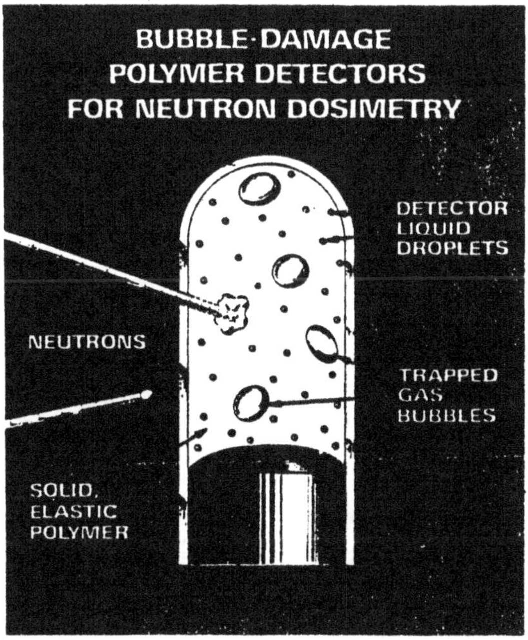

Figure 8. Schematic diagram of the bubble-damage polymer detector (from Ing and Birnboim (1984)).

The neutron sensitivity is reported to be high, up to one bubble per μSv with zero-background and negligible sensitivity to gamma radiation.

If some problems, like loss of neutron sensitivity with storage prior to irradiation, temperature sensitivity during irradiation and bubble stability after irradiation can be overcome, the bubble pen dosemeter has real changes for both immediate read-out and possibly for recording the neutron dose-equivalent over longer periods.

3. Concluding Remarks

Neutron dosimetry always was a challenge and still is because of the many unsolved problems in spite of the increasing research-work performed in the field. New concepts and the previsted quality-factor increase for all secondary particles produced by neutrons will complicate the actual situation and require more research and development to satisfy above all the practical needs.

A completely new approach to the subject is the detection of upsets of "soft errors" in dynamic random access memories (DRAMs), a technique which is interesting because of the analogy between an array of microelectronic elements and a volume of human cells (Bradford (1983)). With the advance of microelectronics the method may find application in the field of neutron dosimetry and lead to a completely new generation of personal neutron dosemeters.

4. References

Anderson, I.O. and Braun, J. (1964) "Nucleonics". 6, 5, 237-241.

Bradford, J.N. (1983) "EUR 8395" pp. 935-942.

Bramblett, R.L., Ewing, R.I. and Bonner, T.W. (1960) "Nuclear Instr. and Meth.". 9, 1.

Commission of the European Communities Radiological Protection - 21 (1980) "Council Directive of 15 July 1980 amending the Directives laying down the basic safety standards for the health protection of the general public and workers against the dangers of ionising radiation EUR 7330.

Gibson, J.A.B. (1988) "Rad. Prot. Dosim." 23, 1/4, 109-115.

Gilvin, P.J. et al. (1987) "Rad. Prot. Dosim." 20, (1/2) 99-102.

ICRP (1985), "Radiat. Prot. Dosim.", 11, 135.

Ing, H. and Birnboim, H.C. "Fifth Symp. on Neutron Dosimetry Neuherberg, FRG.

Ing, H. and Birnboim, H.C. (1984) "Nucl. Tracks Radiat. Meas." 8, 285-288.

Oberhofer, M. and Scharmann, A. (editors) (1981) "Applied Thermoluminescence Dosimetry". Adam Hilger Ltd., Bristol.

Quam, W. et al. (1982) "CONF-811012-63". 637.

Rossi, H. and Failla, G. (1956) "Nucleonics". 14, 32-37.

Saion, E.B. and Watt, D.E. (1988) "Rad. Prot. Dosimetry". 23, 1/4 pp. 265-268.

HIGH-DOSE MEASUREMENT TECHNIQUES

D.F. REGULLA
GSF Forschungszentrum für Umwelt und Gesundheit München
D - 8042 Neuherberg
Federal Republic of Germany

1. Introduction

Radiation plants for high-level industrial radiation processing are increasing in number. The objectives of processing are many-fold, e.g., sterilization of medical supplies and pharmaceuticals, preservation of foods, radiation treatment of agricultural products, disinfection and pre-sowing-irradiation of seeds, polymer modifications, radiation vulcanization, curing of paint coating, modification of wood and concrete, radiation treatment of waste products, improving stability of integrated circuits during fabrication in microelectronics and material testing due to radiation damage. The irradiated objects can be gaseous, liquid or solid, homogeneous or heterogeneous.

The radiation processing plants are based mainly on the gamma radiation of radioactive nuclides, Co60 and Cs137. Source loading with Co60 is between 0.1 MCi and approximately 15 MCi, for one processing plant. Recently, beta radiation of radioactive Sr90/Y90 has been applied for treatment of particularly semiconductor materials. Apart, electron radiation from accelerators with energies from 0.1-10 MeV is of great significance and use. Also the possibility of applying gamma radiation or bremsstrahlung from short-lived radionuclides formed at nuclear reactors is under consideration.

During the development of the irradiation process, it is necessary to perform an accurate and reliable quantification of radiation in terms of absorbed dose, dose rate, mean energy of particles, spatial dose distribution within the product and other parameters. This is for the safe and optimum control of the irradiation plant and for high-quality treatment of the products.

As for dosimetry, appropriate high-level measurement techniques are increasingly requested that are capable to integrate the overall dose of the product during the radiation treatment. The measurements

M. Oberhofer (ed.), Advances in Radiation Protection, 135–157.

should provide high precision, in some cases as low as some 5%. Hence dosimetry in radiation processing has a decisive position in controlling many complex processes.

The magnitude of absorbed doses used in radiation processing ranges from around 10 Gy to above 100 kGy, at dose rates from 0.1 Gys^{-1} to 10 $MGys^{-1}$.

It is not a simple matter to meet these requirements by an appropriate dosimetry technique and to standardize the radiation measurements in industrial radiation processing quantitatively in terms of absorbed dose. There is a number of complicating factors:

1) the large variety of irradiation conditions, source-product geometries, environmental parameters, etc.;
2) the problem of correct and meaningful calibration, that is, making the dose values relevant to the actual processing effect of interest, i.e., killing of micro-organisms, sprouting inhibition, cross-linking or degradations of polymers, etc.;
3) localization of dose determination, especially when the size of the dosimeter and its container may be a factor;
4) the lack of a completely reliable dosemeter, one that is reproducible, easily evaluated, and free from systematic error;
5) the tendency of some radiation users to oversimplify dosimetry problems, in spite of the many intricacies and technical problems.

The significance of these difficulties has apparently not been universally understood and not always been respected by plant operators. Adequate instrumentation for dosimetry is up to now not considered during the planning phase of the radiation processing plant and, hence, not available during plant operation. This is the more conceivable, since dosimetry is obviously the only and generally accepted overall method for quality control in the plant.

2. Implementation and Use of Dosemeters

Accurate and reliable dosimetry as a tool of quality control particularly in food irradiation is worth stressing since public health is or may be concerned. The United Nations Development Programme (UNDP) together with FAO and IAEA, upon request, are in the position to provide technical assistance and guidance in dosimetry, apart form plant realization.

> "..... the appropriate dose level is the key to the technologically and economically proper application of food irradiation (Codex Alimentarius (1984))."
>
> "..... doses delivered to products are of vital importance in industrial development of ionizing radiation and national and

international exchanges of irradiated products (Vidal (1986))."

Dosimetry also allows for the setting of the plant parameters; time setting only is often not sufficient for reasons of different products, densities and geometries, as well as non-stable source parameters at accelerators.

The dose-effect relationship is transferable from one plant to another; this principle is already known from radiation therapy. For such purposes accurate and reliable dosimetry is considered mandatory and, in case of food irradiation, serving for different purposes, e.g.:

a) Transfer of dose recommendations between plants for safe and economical achievement of the desired biological effect in the product.
b) International acceptance between producers resp. irradiators and users, i.e., harmonization of international trade.
c) Balance between safe achievement of the intended effect and an economic treatment. Safe treatment also means exclusion of radiation influence to food consistency and taste; economy also means exclusion of unnecessary overdosage by reducing the dosimetric uncertainties.
d) Reproducibility of the dose-effect relationship, also in long-term view, for reasons of any probably necessary re-evaluation of dose, e.g., in case of juridical claims for compensation.
e) Optimization of safe product treatment aiming to achieve and secure health or wholesomeness to the benefit of mankind, considering irradiated food as part of the human environment.

3. State of the Art in High-Level Dosimetry

High-level dosimetry is not easy to perform; conventional dosimetry procedures as known from radiotherapy and radiation protection can not be used.

The selection of a suitable dosimetry method provides matching its performance with the specific operational requirements, e.g., a long-lived radiation induced effect which changes steadily and reproducibly as a function of dose within the dose range of interest. Further the influence of systematic errors affecting the system response should be sufficiently small or correctable: this also holds for the effects of differences in the spectral distribution between calibration and product irradiation fields.

Another criterion is the traceability of the system calibration to the national standards. Up to now, there is no national laboratory providing a primary standard in the field of high-level dosimetry.

The dosimetric methods as developed in the past for the high-level dose range, showed different approaches of the above criteria with respect to accuracy. It was therefore appropriate to introduce a hierarchy of dosimetry with corresponding requirements to the tolerated overall uncertainty. From the Table 1 it becomes evident that 4 classes of dosemeters have been settled followed by a class for the indication of irradiation.

TABLE 1. Hierarchy of dosemeters and overall uncertainties tolerated on a 95% confidence level.

Primary standard dosemeters	±1,5%
Reference standard dosemeters	± 3%
Transfer standard dosemeters	± 5%
Routine dosemeters	± 10%
Go/no go (dose) indicators	without

The decisive role in high-level dosimetry has therefore been taken over by the reference standard dosemeters. A survey on reference standard dosemeters is given in Table 2; it is shown that these types of standard dosemeters have increased significantly in number, since 1977. Most of them may equally serve as transfer standard dosemeters which truly holds for the alanine/ESR system (see also chapter: "IAEA/GSF International Dose Assurance Service"). Readout for the different methods is made, e.g., by spectrophotometry, electron spin resonance (ESR) spectroscopy, titrometry or oscillometry.

For routine dosimetry, different methods are in use, e.g., change of absorbance due to irradiation of polymethylmethacrylate (PMMA) in red, amber and clear solid forms (PMMA is produced under different trade, names, such as perspex, acrylic, plexiglass, lucite and radix).

There are other radiochromic types of dosemeters, such as solid dye film, liquid dye in optical waveguide, polymer label, and triacetate film; further aqueous ferrous-cupric sulfate solution; thermoluminescence and lyoluminescence. The access to the dose information is by spectrophotometry, either in the ultraviolet or visible range, and by equipment capable to detect luminescence (Table 3).

4. Survey on High-Level Dosemeters

High-level dosimetry systems require calibration traceable to national standards. Accepted reference or secondary standard dosimetry methods for this purpose are calorimetry, ionization chamber dosimetry or Fricke dosimetry.

TABLE 2. Examples of reference standard dosemeters available in 1977 and 1986.

Dosemeter Type (1977)	Dose Range in Gy	Dosemeter Type (1986)	Dose Range in Gy
Fricke (1)	20-20x10^2 (photon radiation)	Cont'd existing	
Calorimetry (2)	(electron radiation)	Cont'd existing	
		Alanine solids/ESR (*) (Gamma and electron radiation) (3)	1-10^5
		Ceric-Cerous sulfate solution (3)	10^3-10^5
		Ethanol-chlorobenzene solution (4)	10^2-10^5 0
		Potassium dichromate solution (5)	10^3-10^5

(*) Electron spin resonance spectroscopy
(1) Ellis (1977).
(2) Miller (1985).
(3) Matthews (1982).
(4) Kovacs et al. (1985).
(5) Sharpe et al. (1985).

For a comprehensive discussion of the various dosimetry methods and materials applicable to the radiation types and energies under consideration, see also ICRU Reports 14, 17, 34, 35 and 44.

The measuring quantity is the one of the calibration dosemeter, usually exposure or absorbed dose in water. Absorbed dose in materials other than water may be determined by applying the appropriate conversion factors.

During calibration and use of the high-level dosemeters, possible effects of ambient parameters to the dosemeter response, mainly temperature and humidity during irradiation and storage, have to be taken into account.

4.1. Alanine Dosimetry

The alanine dosimetry system provides a means of determining absorbed dose in materials in terms of dose in water. Under the influence of ionizing radiation, radiolysis takes place in the alanine, generating free radicals. The identification and assessment of the concentration of the free radicals is based on ESR spectroscopy.

TABLE 3. Examples of Routine Dosemeters.

Dosemeter Type	Readout System	Approximate Dose Range (Gy)
Dyed polymethyl-methacrylate (1)	Visible spectrophotometer	10^3-$4x10^4$
Clear polymethyl-methacrylate (2)	UV spectrophotometer	10^3-10^5
Cellulose triacetate (3)	UV spectrophotometer	10^4-$4x10^5$
Lithium borate, lithium fluoride, etc. (4)	Thermoluminescence reader	10^{-4}-10^6
Radiochromic dye film, solution, paper, optical waveguide (5)	Visible spectrophotometer or densitometer	1-10^6
Ferrous-cupric solution (6)	UV spectrophotometer	10^3-$3x10^5$
Amino acid, saccharide, etc. (7)	Lyoluminescence reader	10-$5x10^4$

(1) Barrett (1982), Whittaker et al. (1985), Glover et al. (1985), Whittaker (1987).
(2) Barrett (1982), Chadwick (1977).
(3) Tanaka et al. (1984).
(4) Cameron (1970), McLaughlin et al. (1979).
(5) Miller and McLaughlin (1981), Liu et al. (1985), McLaughlin et al. (1985).
(6) McLaughlin et al. (1981).
(7) Thwaites et al. (1976).

ESR can be defined as resonant absorption of electromagnetic energy in paramagnetic substances by transition of the spin of an unpaired electron between different energy levels, in the presence of a magnetic field. Usually the first derivative of the absorption curve is registered by an XY-recorder as a function of the magnetic field, yielding the typical ESR spectrum.

Alanine dosimetry is a relative method; absorbed dose is determined by using an experimentally derived calibration curve. The calibration curve is determined by measuring one or several alanine dosemeters irradiated to known absorbed doses that adequately span the range of utilization of the system.

The evaluation of free radicals by ESR spectroscopy is non-destructive. Alanine dosemeters can be read out repeatedly and hence serve for documentation purposes.

Alanine is an organic solid state material that for dosimetry purposes serves best in polycrystalline form: CH_3-$CH(NH_2)$-COOH. Alanine dosimetry uses pure basic materials, no additives or dopants are required. The purity should be analytical grade (99.5%) or better. Alanine of appropriate purity is commercially available. Most frequently L-type alanine is selected for dosimetry for reasons of high yield of free radicals and linear dose-effect relationship over a wide range. Grain sizes of 100 µm to some 100 µm are appropriate for good dosimetric characteristics. The physical form of the alanine dosemeters typically includes powder, compressed solids or cast solids and films. Identification codes can be applied directly on the sample surface, e.g. bare code, stamping, permanent ink.

4.2. Calorimetry

Calorimetry is a convenient way to measure doses at industrial electron accelerators, where high absorbed doses (1-100 kGy) are delivered at dose rates of 10^2-10^5 Gys^{-1} or even higher.

Calorimetry has already been applied as routine dosimetry at industrial electron accelerators with energies between 6 and 10 MeV. The high rate of energy deposition of an electron accelerator is an advantage here, making the thermal isolation less critical. Temperature measurements can be made either on-line or off-line at very short times before and after irradiation, thereby minimizing the effects of temperature drift. Calorimeters for industrial routine dosimetry use water as the absorbing medium. Recently, water calorimeters have been criticized because of suspected endo-thermal or exo-thermal reactions (heat defect), and though this phenomenon has mainly been reported at lower doses, the need for a closer investigation of the water calorimeter is obvious.

The effects of possible sources of error, such as lack of thermal equilibrium and heat defect, may be checked by comparison of the calorimeter response with a reference, or by incorporation of an electrical heater for calibration. A built-in heater will, however, deposit its energy very inhomogeneously as compared to the deposition of energy by radiation, and a comparison with other monitors was therefore preferred. Graphite calorimeters were constructed for this purpose. The precision of individual calorimeter measurements is considered to be ± 2%.

4.3. Fricke Dosimetry

Classic ferrosulphate (Fricke) dosimetry is widely used at comparatively low dose rates. It is based on the oxydation of an aerated ferrous sulphate solution. It has been calibrated by calorimetry and other physical standards. Achievable precision is within 1%-2%. Its principal draw-backs are the rather limited dose range and a dependence of yield on LET, and high dose rate.

The analysis of the ferric ion is mostly achieved by spectroscopy at 304 nm; measurements are also possible at the more intense 224 nm ferric ion peak. Precision depends on the purity of the solution and cleanliness of ampule walls; cleanliness of the ambient working conditions are of major importance, which makes the method time consuming.

The excellent dosimetric qualities of the standard ferrosulphate system and the large body of experimental data available make it the reference system for other dosemeters. There are, however, areas that require further attention. Surprisingly, the molar extinction coefficient for ferric ions is still a matter of some controversy, the most recent determinations being about 2% smaller than the widely published values. The effect of temperature on $G(Fe^{3+})$ cannot be regarded as constant with temperature.

It is seen that the yield of Fe^{3+} is increased with increasing the temperature. At temperatures up to 100°C there is a linear dependence with a coefficient of 0.06%°C^{-1}.

4.4. *Acidic Aqueous Dichromate Solution*

A dosemeter system is based on the radiation-induced reduction of silver and potassium dichromate in perchloric acid. As for the mechanism the reduction of Cr(IV) to Cr(III) is used. Measurement of radiation yield is done by a spectrophotometry, i.e., for the solution composition 0.5 mM $Ag_2Cr_2O_7$, 0.1 M $HClO_4$ at 350 nm (2 to 10 kGy) and for 0.5 mM $Ag_2Cr_2O_7$, 2.0 mM $K_2Cr_2O_7$, 0.1 M $HClO_4$ at 440 nm (10 to 50 kGy). The G-value for dichromate reduction when the dosemeter is exposed to gamma radiation was found to be 0.397.

Advantages of the dichromate dosimetry system are high precision (c.v. = 0.5%), stable measurement signal before and after irradiation, low temperature coefficient (-0.2% °C^{-1}). The stability, accuracy and reproducibility of response are such that this system may be considered for secondary standard dosimetry. So far only few drawbacks have been found with this dosemeter system. Used as a routine dosemeter it may in some cases be inconvenient with a system in the form of a liquid. Breakage of the ampoules followed by spillage of the solution on to expensive goods should be considered a potential risk.

We have found that commercially available silver dichromate does not have a reliable stochiometric composition. This means that solutions of exactly known concentrations cannot be made by simply weighing out the corresponding amount of the salt. However, the absorbance response of the dosemeter solution has a high, long stability, both for unirradiated and irradiated samples.

4.5. Ethanol-Monochlorobenzene Solution

The ethanol-monochlorobenzene (ECB) solution dosimetry uses the concentration of HCl, produced during irradiation as a measure of absorbed dose. The concentration of HCl can be determined by different methods; this makes the use of this system suitable over a wide dose range, i.e., 10 Gy-1 MGy. This dosemeter is applicable within limits to both electron and gamma irradiation.

A dosemeter solution contains, e.g., 24 vol.% monochlorobenzene, 4 vol.% water, 0.04 vol.% acetone and benzene and 71.92 vol.% ethanol. Chemicals used for preparation of solution and for titration are of analytical or reagent grade.

Readout is performed by either spectrophotometry, amperometric titration (coulometry), conductivity, pH evaluation or oscillometry. The oscillometric evaluation has been worked out as a quick routine method. Since the conductivity of a solutions is changed by altering the amount of ions in the solution, the conductivity can be followed by high frequency oscillometry using a high frequency oscillator circuit. Because there is no galvanic contact between the solution and the electrodes, the measurements can be carried out in a closed system. The ampoule, containing the solution under test, is placed between the plates of the capacitive cell. The oscillometric measurements keep track of changes in the Q factor of the oscillator circuit. The method which requires calibration, is applicable in the 1 kGy-100 kGy dose-range. As a routine method, reproducibility can be made better than ± 5%.

4.6. Ceric-Cerous Solution

The ceric-cerous dosemeter is useful for the measurement of absorbed doses in the range 1 kGy-40 kGy. The addition of cerous ions to the solution or the generation of cerous ions by pre-irradiation inhibits the effects of impurities. The presence of additional cerous ions facilitates potentiometry as an analytical method since the unirradiated solution may be used as the reference electrode. The difference in electrochemical potential between the irradiated and unirradiated solutions is proportional to absorbed dose.

The main disadvantage of the ceric-cerous dosemeter is that solutions containing high concentrations of cerium are liable to absorb energy significantly greater than light elements if the radiation spectrum is degraded. At cerium concentrations of 0.01 M and less, the mismatch between the cerium solution and water is not very good. A correction can be applied if necessary by measuring the absorbed dose in solutions containing different concentrations of cerium.

4.7. Radiochromic Dyes

Colorless derivatives of the family of aminotriphenylmethane dyes can be made radiochromic; they change from colorless to a deeply colored state as a function of radiation absorbed dose. A dye derivative which has found widespread use in the radiation processing industry is hexahydroxyethyl aminotriphenyl aceto-nitrile. Readout of radiochromic dosimetry materials is by simple photometric means.

Radiochromic materials can be fabricated in many different forms including liquid solutions, solid state solutions (plastic films), gel coatings, impregnated paper, and optical waveguides. Two forms are currently available commercially, plastic film and optical waveguides.

A solid-state solution is a free-standing film of 40-50 μm thickness. It is radiochromic nylon film available in bulk form as 15 x 15 cm^2 sheets or cut into detectors, 1 x 1 cm^2, with the thickness measured and sorted into bins which have an average thickness ± 2%. They are packaged 1000 pieces per box, sealed in an aluminium laminate pouch, and have a shelf life of several years.

Radiochromic detectors change color when exposed to any electromagnetic radiation with wavelengths shorter than 350 nm. Sun-light and light of fluorescent lamps have such UV components that will color the detectors. To minimize this effect the user may:

a) shield lights and windows with a UV inhibited plastic cover,
b) use incandescent lights or 'gold' fluorescent lights which are essentially above the UV region, or
c) expose detectors in light tight envelopes. The radiochromic detectors may suffer from humidity, dose rate and insufficient build-up time.

Organic solvents with a high refractive index are used to fill a fluorinated plastic tube having a low refractive index, thus resulting in a liquid core optical waveguide. If a radiochromic dye is dissolved in the liquid the result will be a radiation sensitive optical waveguide, a so-called opti-chromic dosemeter.

The absorbed dose range for radiochromic nylon films is 1 kGy-200 kGy, for opti-chromic waveguides 0.01 kGy-2 kGy or 0.1 kGy-20 kGy.

4.8. Dyed and Undyed Acrylic Plastic

Polymethylmethacrylate (PMMA) in the form of thin strips is currently in common use as a dosemeter in the 5 kGy-50 kGy absorbed dose range, which is of particular interest in the sterilization of goods by ionizing radiation. Two distinct types of systems have been developed, both of which depend on a radiation induced change in optical absorption as the means of assessing absorbed dose.

Clear perspex HX is in principle the simpler of the systems. Polymerization of monomers without additives is initiated by benzoyl peroxide and is continued until a specified (low) amount of residual monomer remains. On irradiation, degradation products of the polymer build up in the matrix, and optical density measurement in the 300-320 nm region allows their concentration to be determined.

Red Perspex 4034 is the second type of system, although other formulations have been developed. Dye molecules are incorporated into the PMMA polymer during manufacture, and on irradiation polymer radicals react with dye molecules to produce optical absorption in the visible region of the spectrum (600-650 nm in the case of Red Perspex 4043). The manufacturing process is again closely controlled, and in addition the polymer is conditioned at a specified temperature and humidity before being sealed in metal foil packets.

Both systems exhibit variations in radiation response with changes in environmental conditions and both also show pronounced time-dependent changes in optical density after irradiation. In addition there are inevitable batch variations in response. Calibration curves available with the dosemeters are relevant to one set of conditions only, and users are recommended to recalibrate using their own measurement system under their own conditions only, and users are recommended to recalibrate using their own measurement system under their own conditions, A correction for the thickness of the dosemeter must be applied to absorbance measurements, and both types of dosemeter show an absorption before irradiation which must be taken into account.

4.9. Thermoluminescence

The term thermoluminescence (TL) refers to the thermally-stimulated emission of light from a crystal that has undergone some means of electronic excitation. TLD materials typically used for radiation dosimetry are crystals in which imperfections are introduced by doping with impurities. The most common of these phosphors are LiF:Mg,Ti, CaF_2:Mn, CaF_2:Dy, $Li_2B_4O_7$:Mn, Al_2O_3, and $CaSO_4$:Dy. Many others are available. Physical forms include single crystals, sintered pellets, powder, and powder imbedded in a teflon matrix. The selection of a particular TLD phosphor depends on the dose range of interest, the irradiation environment and the material for which absorbed-dose measurements are needed. For example, LiF is more suitable than CaF_2 for measurements of absorbed dose in water. The physical form of the TLD is largely a matter of convenience.

Depending on the particular phosphor and instrumentation, TLDs can be used successfully for radiation absorbed doses as low as 2×10^{-5} Gy and as high as 10^4 Gy. The upper detection threshold for a given system is determined either by saturation in the instrumentation or by saturation effects in the phosphor.

Elevated temperatures result in loss of TL signal, and are precluded during irradiation and post-irradiation storage. Many TL materials exhibit enhanced fading or increased background when exposed to ultraviolet light, and precautions must be taken in packaging and handling to minimize this exposure. TLD can be read only once for a given irradiation. Moreover, high-level doses may significantly and irreversibly change the TL response of a phosphor.

4.10. Organic Lyoluminescence

When lyoluminescent material is placed in contact with a proper solvent light is emitted, the so-called lyoluminescence (LL). Its intensity as a function of time follows a glow curve, named so because of analogy with the thermoluminescent glow curve. The actual shape of the curve reflects the rate at which dissolution takes place. The measured LL yield is obtained by integrating the glow curve from the beginning to an arbitrarily chosen point in its tail. The usual practise is to choose the cut-off point in the region where only a weak "slow" component is present. The yield is usually expressed as a ratio of LL intensity integral versus mass of the dissolved sample. The constancy of LL yield per unit mass has been investigated by many laboratories, and the results indicate that if the mass of dissolved sample is kept within certain narrow limits then it is possible to attain 0.5% precision. Requirements to the volume of solvent appear less stringent.

Lyoluminescence has been observed in a large family of mono- and oligosaccharides, actually in every investigated commercially available sugar. It has also been found in certain polysaccharides, like starches and dextrans and in deoxy sugars. Similarly, it has been observed in all investigated water-soluble amino acids and amino sugars.

LL dosimetry bridges the gap between the dose ranges of ferrosulfate (Fricke) dosimetry and perspex systems. The discovery of a method of stabilization against fading by a suitable heat treatment made it possible to use with confidence mailed glutamine dosemeters.

4.11. Quartz Paramagnetic Resonance

ESR spectroscopy has successfully been settled in the field of solid state dosimetry in the last decade. An extremely wide dose range can be covered by selection of suitable dosimetric materials. With the well studied alanine proper dosimetry is feasible from the medical therapy level, i.e., around 1 Gy, to the sterilization level, i.e., around 0.5 MGy. For the subsequent ultra-high dose range as applied in, e.g., material testing or core dosimetry, no approved dosimetry method does exist, on a first sight. Presently the dosimetric properties of different quartz samples of the crystalline and glassy state are studied by ESR spectroscopy, on the basis of the paramagnetic E' centre (Wieser and Regulla). So far a dose range of 1 kGy-100 MGy can be covered with high precision. The dose range can be adopted by proper selection of a proper quartz material. The system has successfully been tested for irradiation temperatures up to 300°C.

4.12. Ionization Chamber Technique.

For completeness it should be mentioned that also the ionization chamber method is available commercially for high-level dose rate measurements. With a chamber volume of 0.125 cm^3, dose rates up to 60 Gys^{-1} can be determined on a high confidence level, i.e., more than 0.2 $MGyh^{-1}$ (Pychlau).

5. High-Level Absorbed Dose Calibration.

Since most methods described are good for relative dosimetry only, their response must be calibrated. As mentioned already, there is no primary standard available in high-level dosimetry. Instead it is common practice at some national laboratories that the national dosimetric standard is realized indirectly by a calibrated radiation field, e.g., at the National Bureau Of Standards (NBS), Washington D.C., USA, through characterizing a ^{60}Co source array by calorimetry.

A variant of the calibrated radiation field practice is based on a multi-step procedure (Figure 1). At first, the reference type high-level dosemeters are calibrated against secondary standard ion chamber dosimetry in the reference beam of usually a ^{60}Co therapy type source. The high-level reference dosemeters are then brought into the high-level calibration source, frequently a Gammacell type source of AECL Canada, to assess the absorbed dose rate up to the 100 kGy range. For control, a cross-check in dose rate is made with a second high-level system; most widely used is the ferrous sulfate aqueous solution (Fricke dosemeter). This procedure is practised at GSF, Neuherberg, and comparably at the National Physical Laboratory (NPL), Teddington, United Kingdom.

Either reference or transfer standard dosemeters are used for the

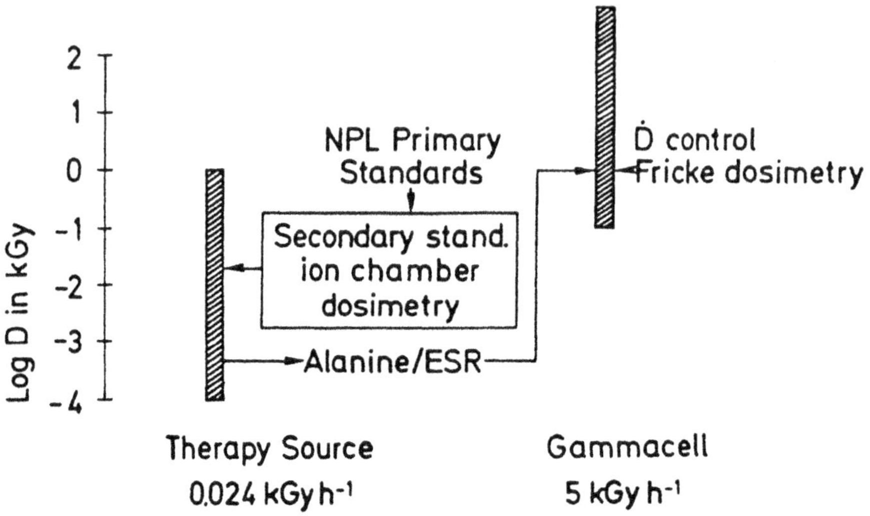

Figure 1. Schematic representation of a multi-step calibration procedure in high-level dosimetry.

calibration of radiation fields.

The calibration of dosemeters by a national laboratory (NBS, NPL) or the high-dose reference laboratory (GSF) for purposes of application in food irradiation is usually provided in terms of absorbed dose in water.

6. IAEA/GSF International Dose Assurance Service (IDAS).

Harmonization and reliability in dosimetry are pre-requisits for the acceptance of food irradiation by the public and for international trade affairs. This goal has for long time been met by national activities based, e.g., on radiochromic dye (NBS for the USA) and Fricke solution (NPL for the UK and upon request also for other countries). However, there was no concerted international effort to achieve measurement standardization of dosimetry and dose assurance for large radiation sources for long time. It was the IAEA, that has first been establishing a programme on high-level photon dosimetry for the development of quality control measures by standardizing dosimetry in radiation processing. One of the programme components is the International Dose Assurance Service (IDAS) which on joint project basis is operated by both the GSF Munich, Neuherberg and the IAEA, Vienna. IDAS is addressed to the high-level radioprocessing facilities of the more than 120 Member States of the IAEA.

The operational scheme of IDAS is layed down in Figure 2. Member States may address their requests for assistance administratively to IAEA which assistance is metrologically provided by GSF. GSF informs the IAEA about the dose results. IAEA evaluates the ratio between the dose as given by the participant and the dose as reported by GSF. All data are kept confidential.

While IAEA triggered the high-dose standardization programme, GSF has developed an appropriate dosemeter system lending itself, by accuracy, for reference dosimetry and, by reliability, for transfer dosimetry (Regulla and Deffner (1982)). Throughout a series of intercomparison studies including field tests together with several candidate transfer dosimetry systems, the alanine/ESR metrology as developed by GSF was found to be the most appropriate system for the particular task. It fulfills all dosimetric criteria which were requested, provides a highly stable (free radical) signal up to 70°C irradiation and storage temperature, a dynamic dose range from 1 Gy to 50 kGy and allows for repeated readout, i.e., verification and documentation of the dose assessment. IDAS represents for Member States that have agreed

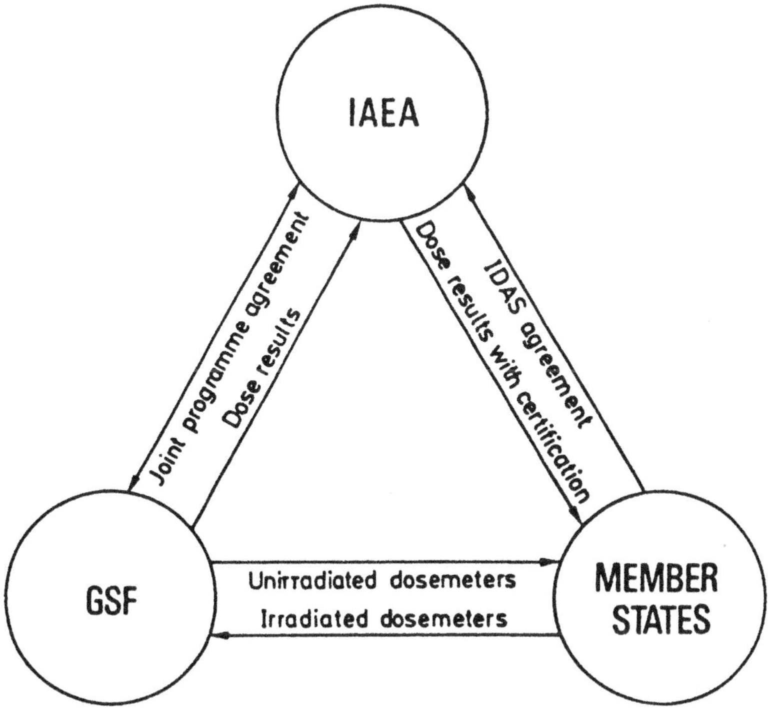

Figure 2. Operational scheme of the IAEA/GSF International Dose Assurance Service (IDAS).

officially to participate, an internationally accepted dosimetry standard for the high-level dose assurance.

The GSF part of the IDAS programme is performed with a dosemeter set-up as shown in Figure 3. Apart from the alanine samples there is a buildup cap to provide charged particle equilibrium as well as protec-

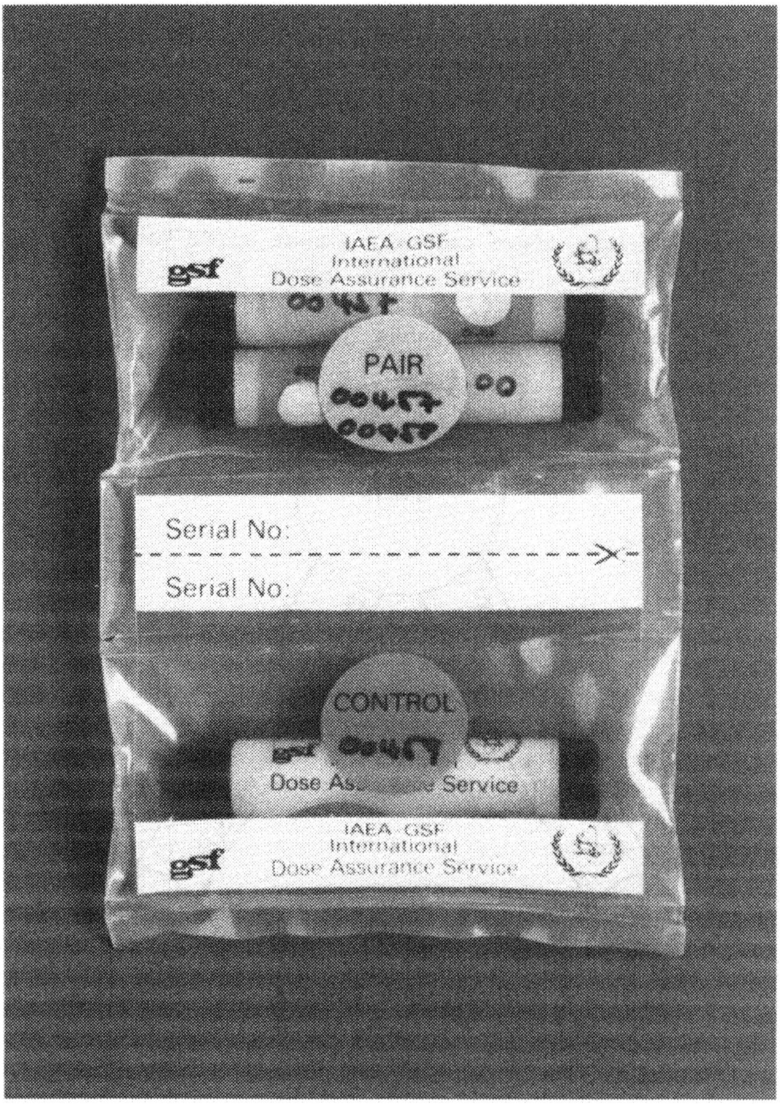

Figure 3. A mailing set of the GSF alanine/ESR transfer dosemeter.

tion against weather influences and dust. The dose curve of the alanine system is given in Figure 4. It is envisaged that IDAS may also serve to national authorities for licensing, commissioning, quality control and inspection of radioprocessing plants.

Currently, IDAS deals with photons. The service for electrons from about 0.5 MeV to 10 MeV will be implemented, in the near future. Moreover, the range of processing temperatures will be extended down to -60°C. Both corresponding research projects are under investigation at GSF, within IAEA research agreements. ESR tabletop readout instruments are being developed by industry in cooperation with GSF.

The radioprocessing dosimetry standardized in this manner should in future be an acceptable basis of justification for the regulatory approval and international clearance for trade of irradiated food. In this connection there are still many IAEA member states from which application for IDAS participation has not yet been registered, so far. Particularly the USA, Canada, Latin America, Japan, Australia, but also European countries are concerned (Figure 5). It appears desirable that the authorities of those countries which have already agreed to the

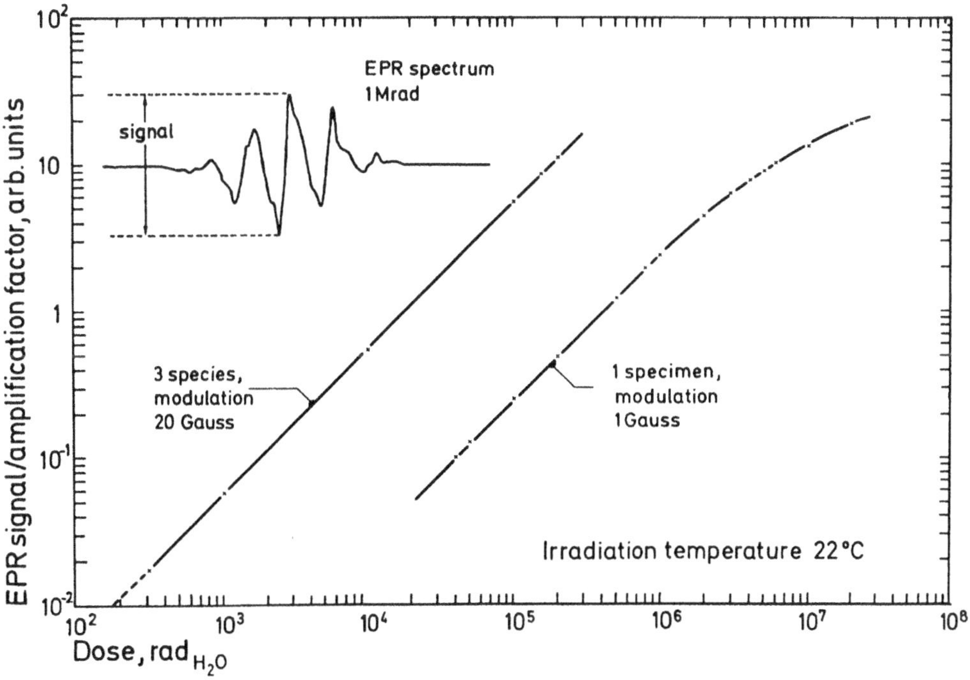

Figure 4. Example of dose response of an established reference/transfer dosimeter (Alanine/ESR).

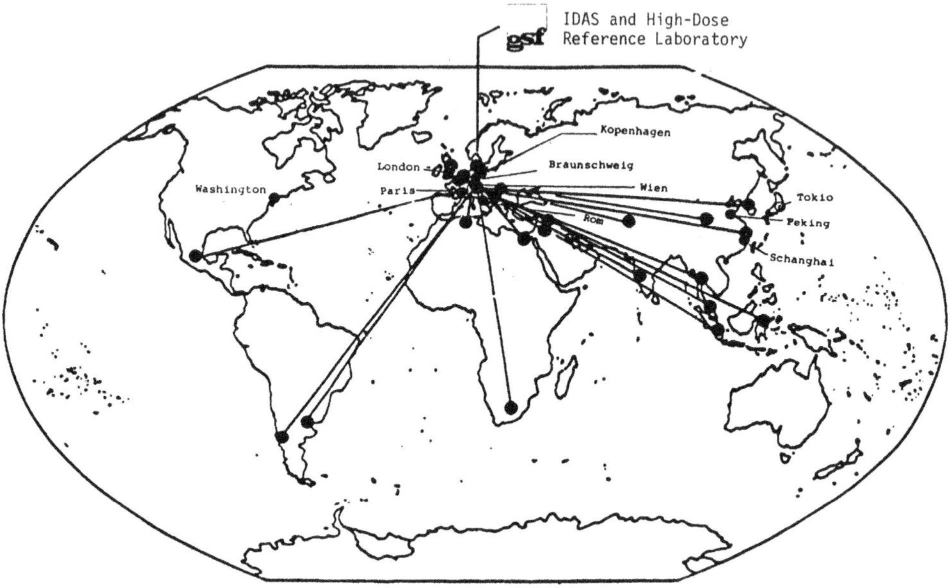

Figure 5. Global origin of present IDAS participants.

operation of radiation processing plants prepare the necessary legislative prerequisites to allow processor plants participate in the IDAS programme; existing national dose assurance programmes should not represent a principal hindrance since IDAS, in this case, could serve as a cross-check system for international link.

7. International Dose Calibration Network.

Standardization and harmonization in high-level dose application, and that in the absence of primary standards dosemeters, would definitely take profit from a centralized dose calibration. This could best be provided by a world-wide high-level dose calibration network of established dosimetric facilities and a carrier of dose information which can easily be distributed. Alanine/ESR dosemeters are prepared to permanently carry dose information from Gy to MGy, at common environmental conditions, and allow for repeated readout. A network equipped with ESR spectrometers would be able to realize calibration and hence dosimetry in this dose range at any site of the world. The existing IAEA/WHO Secondary Standard Dosimetry Laboratories (SSDLs) appear suitable to take over the role of network nuclei. First results with Italy (Regulla et al.) and the People's Republic of China (private communication), showing a local dose verification within around 1%, have proven the applicability of the basic idea.

8. Problems and Needs in High-Level Dosimetry.

The problem in high-level dosimetry are many-fold; several national and international organizations and committees are involved in improving the situation (Table 4). As mentioned above, primary standards are missing. The "Comité Consultatif pour les Etalons de Mesure des Rayonnements Ionisants" of the Bureau International des Poids et Mesures (BIPM), did in 1985 not feel the necessity for radiation fields of very high dose rates to be installed at BIPM at present. The traceability of working standards to the international system for absorbed dose should, according to the committee's opinion, be achieved by linking through the use of dosemeters with a suitable wide working range. It was agreed that BIPM would participate, as best it could without a high-output source, in establishing a coherent high-dose measurement system. One of the principal conclusions recommended that national laboratories and international organisations that operate reference services using passive dosemeters should send a sample from each batch for irradiation at BIPM.

Also missing is a reliable and cheap routine dosimetry technique, particularly in the "low-dose" range from 1 to 10^3 Gy. Recent dose results of simultaneously exposed dosemeters at an industrial radiation processor under operational conditions revealed excellent agreement between alanine/ESR and ceric-cerous dosimetry within around 1%; by contrast, polymethylmetacrylate dosemeters at the same time underestimated the dose rate by up to 30% causing an equally high over-exposure, as a result of compensation by extended time setting (Regulla (1980)).

TABLE 4. National and international organizations active in elaborating recommendations, standards, guide-lines, manuals or codes of good practice in high-level dosimetry.

International	ICRU	BIPM			
	IAEA	WHO	FAO	CEC	ESNA
National	ASTM				

While in the field of photon dosimetry remarkable progress has been achieved within the passed decade, more efforts need to be done in the field of electron dosimetry, particularly for low and medium energies from some 100 keV to 4 MeV.

Difficulties arise from mapping the spatial dose distributions in the product particularly in defining D_{max} and D_{min}. Limited experiences are available on the dose distribution close to interfaces or discontinuities between materials of different atomic number and density, which may occur in non-homogeneous products or between products and the packaging material.

Other dosimetric problems may arise from the dose quantity not clearly defined in high-level dosimetry. While exposure measurements are still practised, the results are often just stated in units of absorbed dose, i.e., in Gray, which probably refers to absorbed dose in water although not being stated. If conversions from exposure to absorbed dose are made, then the conversion factor is often not applied uniformly at different laboratories. There are almost no interpretations made in terms of dose in the irradiated products themselves.

Other dosimetric problems are due to the spectral distribution in a large industrial irradiator facility which may differ considerably from that under which the dosemeter system was calibrated, or due to the dose rate that ranges in routine from 0.01 Gys^{-1} to 1 Gys^{-1} in case of gamma-rays up to 10^{10} Gys^{-1} in case of electrons, from linear accelerators.

Apart from the dose rate it is the irradiation temperature during the irradiation process that may particularly affect the response of available dosemeter systems; from literature it is known that the temperature coefficients of response may significantly change in value and sign.

While these subjects are of significance already in technologically advanced countries and this under moderate climatic conditions, they particularly hold for developing countries with their additional relevant difficulties, particularly due to tropical or subtropical climates with emphasis to humidity.

One should, in this context, remember that even the relevant technical recommendations of the International Standard Organization (ISO) and International Electrotechnical Commission (IEC) for dosimetry in therapy and radiation protection are mostly prepared by experts from technologically advanced countries respecting their own climatic conditions and omitting to consider the more stressful environment in tropical zones.

Since developing countries are most interested in producing longerlived fruits and shelfware by using radiation processing, they should have most interest in reliably meeting the dose recommendations of international recommendations and customers. Dosimetry thus is an essential part and pre-requisite of international trade; nonqualified or non-reliable dosimetry may be of direct influence to the economic situation of a country. As a result, there is a strong need for reference and routine standard dosemeters apart from appropriate calibration facilities, in these countries.

Attention should also be paid to effects on the dosemeter response due to the gaseous environment. The dosemeter response can already be affected by an enhanced portion of oxygen or nitrogen as compared with the normal atmospheric composition. Very little is known about the effect on the detector response or structure through aggressive chemical agents that are present in a high-level radiation area, e.g., ozone and various nitrous oxide gases produced by radiation in air, and hydrogenated halogens or other halogen-containing volatile compounds produced by irradiation of halogen-containing materials, e.g., PVC packaging.

Progress should particularly be achieved for suitable dosimetry to be applied to the processing of free-flowing food and bulk solids. Apart, standards and guidelines are needed for the dose harmonization in trans-boundary trade and acceptancy. Attention must also be paid to the appropriate training for the personnel of processing facilities, public health services and regulatory authorities.

9. References

"Codex Alimentarius", (1984) Volume XV, Codex General Standard for Irradiated Foods and Recommended International Code of Practice for the Operation of Radiation Facilities Used for the Treatment of Foods, FAO/WHO, Rome, CAC/VOL XV-Ed. 1.

Barrett, J.H. (1982) "Dosimetry with dyed and undyed acrylic plastic", Int. J. Appl. Radiat. Isot., 33, pp. 1177-1187.

Cameron, J. (1970) "Lithium fluoride thermoluminescence dosimetry", in: Manual on Radiation Dosimetry. Holm, N.W., Berry, R.J., Eds., Marcel Dekker, NY, p. 405.

Chadwick, K.H. (1977) "Solid state dosimetry at high-doses", in: Ionizing Radiation Metrology. Casnati, E., Ed., Editrice Compositori, Bologna, pp. 195-211.

Ellis, S.C. (1977) "The dissemination of absorbed dose standards by chemical dosimetry, mechanism and use of the Fricke dosemeter", Ionizing Radiation Metrology, Casnati, E., Editrice Compositori, Bologna, pp. 163-180.

Glover, K.M., King, M., and Watts, M.F. (1985) "Calibration and intercomparison of red 4034 perspex dosemeters", in: High-Dose Dosimetry, IAEA, Vienna, pp. 373-395.

Kovacs, A., Stenger, V., Foldiak, G., and Legeza, L. (1985) "Evaluation of irradiated ethanol-chlorobenzene by conductivity method", in: High-Dose Dosimetry, IAEA, Vienna, pp. 135-142.

Liu, Zhan-Jun, Radak, B.B., and McLaughlin, W.L. (1985) "Food irradiation dosimetry by opti-chromic technique", Radiat. Phys. Chem., 25, pp. 125-134

Matthews, R.W. (1982) "Aqueous chemical dosimetry", Int. J. Appl. Radiat. Isot., 33, pp. 1159-1170.

McLaughlin, W.L., Jarrett, R.D., and Olejnik T.A. (1981) in: Preservation of Food by Ionizing Radiation, Vol. 1, Josephson, E.S., Peterson, M.S., Eds., CRC Press, Boca Raton, Florida, pp. 189-245.

McLaughlin, W.L., Lucas, A.C., Kapsar, B.M., and Miller A. (1979) "Electron and gamma-ray dosimetry using radiation-induced color centers" in: LiF. Radiat. Phys. Chem., 14, pp. 467-480.

McLaughlin, W.L., Radak, B.B., and Farahani, M. (1985) "Measurement quality assurance in the radiation treatment of food", Transactions of the American Nuclear Society, 49, pp. 8-9.

Miller, A., and McLaughlin, W.L. (1981) "Evaluation of radiochromic dye films and other plastic dose meters under radiation processing conditions" in: High-Dose Measurements in Industrial Radiation Processing, Technical Reports Series 205, IAEA, Vienna, pp. 119-138.

Miller, A. (1985) "Calorimetry at industrial electron accelerators", Nucl. Instr. Meth. in Phys. Res. B10/11, pp. 994/997.

Pychlau, P.: private communication (PTW Freiburg, FRG).

Regulla, D.F. (1980) Private communication.

Regulla, D.F, Deffner, U., Wieser, A., Bartolotta, A., Onori, S.: to be published.

Regulla, D.F., and Deffner, U. (1982) "Progress in alanine/ESR transfer dosimetry" in: High-Dose Dosimetry, IAEA, Vienna, pp. 221-235.

Regulla, D.F., Deffner, U. (1982) "Dosimetry by ESR spectroscopy of alanine", Int. J. Appl. Radiat. Isot., 33, p. 1101.

Sharpe, P.H.G., Barrett J.H., and Berkley, A.M. (1985) "Use of dichromate solutions as reference dosemeters in the 10-40 kGy range", Int. J. Appl. Radiat. Isot., 36, pp. 647-652.

Tanaka, R., Mitomo, S., and Tamara N. (1984) "Factors influencing the sensitivity of cellulose triacetate dosemeter upon irradiation with electron beams and gamma-rays", Radiat. Phys. Chem., p. 23.

Thwaites, D.E., Buchan, G., Ettinger K.V., Mallard J.R., and Takavar, A. (1976) "A new sensitive technique for study of radiation effects in amino acids", Int. J. Appl. Radiat. Isot., 27, pp. 663-664.

Vidal, P. (1986) "Alanine/ESR transfer dosimetry", Newsletter 16, Association of International Industrial Irradiation, pp. 162-163.

Whittaker, B., Watts, M.F., Mellor, S., and Heneghan, M. (1985) "Some parameters affecting the radiation response and post-irradiation stability of red 4034 perspex dosemeters", in: High-Dose Dosimetry, IAEA, Vienna, pp. 293-305.

Whittaker, B. (8 April 1987) "The gammachrome YR system. UK Panel Meeting of Gamma and Electron Irradiation Dosimetry and Control of Radiation Processing", National Physical Laboratory, Teddington, Middlesex, England.

Wieser, A., and Regulla, D.F.

RECENT PROGRESS IN SOLID STATE DOSIMETRY AND TRENDS FOR THE FUTURE

D.F. REGULLA
GSF Forschungszentrum für Umwelt und Gesundbeit München
D - 8042 Neuherberg
Federal Republic of Germany

1. TLD Equipment

The appearance of TLD equipment has significantly changed during the recent past. Most impressive is the PC assisted TL analysis and dose evaluation. A commercial version of such PC program is offered by the name "TL Glow Curve Deconvolution" (Harshaw). This is a software-implemented analytical process which enables the well known glow curves to be segregated into their individual glow peaks. By analytical manipulating the resulting peaks and determining relationships between areas under the peaks, it expands the dosimetric information of the presently available integral data. Most important is a reduction of the minimum measurable dose by as much as a factor of 10; further inaccuracies due to fading are reduced and last not least, this program offers a possibility to determine the time elapsed between a major radiation accident and readout.

Common to almost all new TLD systems is a screen-display of the glow curves, area quantification within segments, and linear combinations of the component data.

Most TLD systems offer a software for data management. One of them (Harshaw) appears quite sophisticated but already elaborated for routine. It performs other data transmission, storage, and logical functions required in most TLD systems. These include: Mass storage of TLD records, individual TL chip calibration, real-time data display/printout, menu-driven TLD readout operation, correlation with user's data base, incorporation of dose algorithms, transmission of data to other computers, documentation and recall of archival data.

The above mentioned TLD glow curve deconvolution requires an adequate linear time-temperature profile for the readout cycle. Efforts of the industry to make improvements in this field are remarkable. While most older systems still operate with heater planchets and the related

M. Oberhofer (ed.), Advances in Radiation Protection, 159–169.

drawbacks, a new gas heating is worth mentioning which allows a linearly controlled heater ramp up to 400°C (Harshaw). Hitherto TLD readers with gas heating operated with a constant gas temperature for readout. Another recent approach to go new paths in TLD heating systems is by direct laser heating of the detector crystals. Details are given below.

It is essential for the present state of TL dosimetry, that important accessories are available on an adequate quality level. This concerns particularly the oven for sample pre- and post-annealing and desk top radioactive check sources.

1.1. Laser-Heated Desk Top TLD System.

The TLD reader identifies, evaluates, and anneals sealed TLD badges containing up to eight dosemeter spots. Their thickness can be chosen from 3 mg/cm^2 to 25 mg/cm^2. Standard 7 mg/cm^2 thick $CaSO_4$:Tm and 10 mg/cm^2 thick MgB_4O_7:Tm elements are supplied.

Configurations available are 3 x 3 mm spots of:

a) phosphor plus polymer binder on a polymer substrate,

b) phosphor plus inorganic binder on a thin sheetglass substrate.

Configuration b) was developed to eliminate any non-radiation induced signals for high sensitivity applications.

The badge can be equipped with a new direct fast neutron dosemeter that is presently under development. The TLD reader is designed to read and anneal this element once it becomes commercially available. The badge contains all necessary filters for operation with the standard dosemeter elements but will accept other energy correction or thermal neutron absorption filters ranging in thickness from 1 mg/cm^2 to 1000 mg/cm^2. Thus it may be used in a routine mode or for research purposes.

The reader opens and reseals the badge and performs automatically desired pre-programmed pre- and post-anneal treatments. The other badge housing is equipped with a proprietary "smart chip" which is interrogated and updated during the readout cycle. It contains information on the type of badge, identification, history of its use, calibration constants, background signals and optimized heating programs of each of its up to eight dosemeter elements. The seal is tamper proof and water tight to 10 meters of water. The badge will not be affected by a normal dishwasher or washer cycle. A special snapon back cover prevents dirt collection and provides convenient means for fixing the badge to clothing by a belt loop and a clip.

The laser heating power delivery is "time-shaped" for optimal heating and annealing of each individual dosemeter spot.

1.2. TLD Array for 2-D Mapping of Dose Distributions.

A laser-heated TLD system has been developed for mapping radiation fields encountered in radiation processing, radiotherapy and radiation hardness testing of electronic devices. Thin (8 mg/cm^2) TL dosemeters of 1.5 mm diameter are printed on a 3 mm x 3 mm grid over a 30 cm x 30 cm area (total array size 101 x 101 dosemeter spots). Using precision X-Y translation, each spot is heated in less than 800 msec (depending on the user-specified TL phosphor material and layer thickness) by a CO_2 laser, and the TL emission is collected by optical components and measured by a PMT. Readout rate is up to 4 dosemeters per second. If lower resolution is desired for faster evaluation, the reader can be set to read every fourth or every sixteenth spot (in a square pattern) or the operator can set the resolution at 1, 4, or 16 over specified areas of a single sheet.

After the dosemeters are evaluated, the maximum dose is obtained and dose contours are computed and plotted on the high-resolution colour graphic monitor. A screen dump feature quickly transfers the contour plot to a large format digital colour plotter. The default contour intervals are in 5% increments for 75-95% and 10% for 35-75% of maximum dose. The user can specify any desired contour intervals not included in the default set.

The system also has a zoom feature. The arbitrary magnification capability allows the user to enlarge areas where the dose contours are crowded together in the normal size plot. Using the cursor, the user simply selects the area to be enlarged, and this function is repeatable to arbitrary magnification.

Advantages of the system over conventional 3-D water phantom scans or film densitometry are:

1) the system requires only a few minutes of beam time for exposure of any number of sheets in a plastic phantom
2) the dosemeter sheets, only .006 inches thick, are annealable and reusable.

The system is also ideally suited for dose mapping of high intensity radiation fields, providing a dynamic range of at least 4 orders of magnitude and permitting exposure times from nanoseconds to several months.

1.3. TL Dose Sampling by Fiber Optics.

A radiation dosimetry system has been developed based on laser heating of microgram quantities of TLD phosphors via a fused silica optical fiber. The probes, at 1 mm overall diameter, can be inserted into small spaces and remotely evaluated for realtime in vivo dosimetry. The target application is radiotherapy patient dosimetry.

The system consists of a rack mount size laser/control unit, containing the low power laser for TL element stimulation, a microprocessor which operates the system and calculates radiation dose from the recorded glow curve, and all power supplies. Twelve fiber-optic probes are connected to the front panel and each can be evaluated independently. An external microcomputer is used for graphic display and communication between the user and the system.

Probes of arbitrary length (exceeding ten meters) can be placed in any position for remote dosimetry. Because the communication between the dosemeter and reader is optical, the system can be operated in high EMI environments. The twelve probes are evaluated sequentially through an optical multiplexer, and the readout procedure consumes approximately 10 seconds per probe. After readout, the probes are instantly available for additional dose measurements. The operator may choose to address a small number of probes frequently, or may simply allow the reader to sequentially evaluate the entire set.

Sterilizable, reusable probes for gamma, electron and fast neutron radiations, with sensitivities of 100 mrad have been developed.

2. TLD Materials

TL detection threshold at less than 1 μGy? Today the most common and adequate TLD material is lithium fluoride doped with magnesium and titanium, LiF:Mg,Ti (TLD-100). A number of users of this material demand for higher sensitivity to practise low-level dosimetry at some 10 μGy or less. This demand holds particularly for patient dosimetry in X-ray diagnostics and short-term environmental monitoring. It was only recently that such a LiF became available. Produced in the People's Republic of China it is based on LiF doped with magnesium, copper and phosphorus, LiF:Mg,Cu,P. We tried to verify particularly the published figures on the detection threshold and reproducibility. All measurements were made on the basis of the special annealing procedure as recommended by the original publication. Using the GSF precision TLD equipment and the adopted Chinese annealing procedure a lower detection threshold of about 0.1 μGy and, for individual reproducibility, a coefficient of variation of about 1% at and above a 10-fold detection threshold were found. Both figures appear rather revolutionary in TL dosimetry and worth being verified also at other laboratories. For this, our measuring procedure is carefully described together with other dosimetric findings, i.e. interspecimen scattering, batch-to-batch response, energy dependence of response and fading.

3. Impact of TLD Operators on TLD Precision

Thermoluminescence dosimetry has experienced a remarkable progress in the recent years. This concerns the detector materials, read-out techniques and accessories. In personnel and environmental monitoring improved automation provides high reliability, low malfunction rate, glow curve analysis and glow curve documentation; national and international standards as well as type tests guarantee more safe operation. There is one exception from the list of improvements limited to manual dosimetry: the operator. Its significance for the overall uncertainty has since long been accepted qualitatively but not yet been quantified. The objective of the present study was to evaluate figures on the magnitude of the operator's influence. The problem was approached by using the GSF precision TLD system and inviting a number of skilled and unskilled operators to perform a reproducibility matrix with 10 TLD samples always identical, and 10 reuses at a convenient dose level. Each operator was carefully instructed prior to the experiment. Professionally the operators involved had degrees ranging from technician to PhD. Surprising results were achieved for both TLD precision and the magnitude of mean TL response, also as a function of the operator's education. The results give rise to useful discussions and conclusions.

4. Self-Controlled TLD System for Environmental Monitoring

Since more than a decade GSF is performing regular large-scale dose measurements in the environment of nuclear facilities on behalf of the Bavarian authorities. This duty requires a high degree of reliability and, for incidental situations, readout frequency. To fulfill these needs an automated TLD system had been installed. In this system the four TL elements available in each badge were fitted out with particular functions:

a) two elements are for routine dose measurements of photons above 50 keV, according to standards requirements;
b) one element serves as a dose indicator in incidental situations where photons down to about 10 keV and neutrons may occur;
c) the fourth element has a special control function that is to check for influence parameters during the monitoring and transport period which may affect the reading values of the routine and incidental elements.

For this purpose the "control element" is irradiated with an appropriate predose prior to exposure in the environment. The control element is supposed to read the predose and in addition the environmental dose as obtained from the "routine elements"; any deviation from this additive dose is a hint for a possible irregularity during exposure or transport and needs a careful consideration of the badge concerned. It is this badge-specific self-control that offers a particular reliability for the achieved environmental monitoring data. Apart from the thorough description of the TLD system, measurement results and experiences are reported that originate from laboratory, intercomparison and field tests.

5. TLD Partial Body Dosimetry

Up to now partial body dosimetry suffers from the lack of an adequate dosimetry system allowing for:

a) assessment of also absorbed doses from beta radiation and soft energy X-rays, and
b) automatic readout on a large scale. Several institutions try hard, at present, to overcome these problems.

Certain progress is achieved in particularly thin-layer TL detectors, but there is not yet a satisfying system available. This holds essentially also for automation.

6. Phosphate Glass or Radiophotoluminescence Dosimetry

This dosimetry technique has great advantages but was not recognized sufficiently by the physicists, up to now. The reasons were quite different, e.g., lack of a reliable detector production, problems with an intrinsic radiophotoluminescence (pre-dose) and, as a consequence, almost no developments in readout devices. The situation has changed most recently: a new readout instrument is now under laboratory test, characterized by short-pulse laser excitation in the nanosecond range (Toshiba). The principle was intensively studied by the author and promises a complete discrimination of the disturbing pre-dose signal.

7. Dosimetry by Thermally Stimulated Exoelectron Emission

This technique has major perspectives in dosimetry of beta and low-energy photon radiation, since the radiation sensitive layer is of some thousand angström only. Efforts have been made at some major research centers worldwide (France, Japan, USSR, Germany) to develop a TSEE dosimetry system for routine. Particularly the CEA, Fontenay-aux-Roses and the University Gießen succeeded in developing new readout instruments. Simultaneously TSEE detectors are in preparation based on BeO and AlP_2O_3 layers deposited on graphite substrates. Nevertheless, reproducibility is still lacking precision and, most important, reliability. A routine application is hence not possible, at present.

8. Thermoluminescence of Dust

The non-acceptance of food irradiation due to national legislation in certain countries and the avoidance of excess irradiation demand for a reliable detection method to identify irradiated food. It was recently reported in *Nature* that irradiated spices emit thermoluminescence upon heating from ambient temperature to below combustion. The authors described this phenomenon as a promising candidate in the search for a simple test whether food has been irradiated or not. However, it was admitted that the results of these TL measurements taken directly from integer spice samples were not yet satisfactory, since the errors were quite large. Other authors even reported that by TL measurements unirradiated spice samples had been identified as irradiated, which makes the method doubtful particularly where legal consequences are concerned.

It is shown that most of the natural products contain minute amounts of wind-blown or intruding dust which can be separated and used to identify irradiated spices by measuring their thermoluminescence. No TL was found for dust samples of unirradiated spices (Göksu-Ögelman and Regulla (1989)). TL analysis of the dust samples was performed by using a TL readout equipment (Harshaw 2000) with a linear heating rate of 8°C s^{-1}, an upper readout temperature of 400°C, and dry nitrogen gas flushing of the measuring cell. Despite, there is no spurious signal due to oxidation of the organic components. The TL measurements up to 400°C and more did not cause detectable changes in the samples neither in appearance nor in radiation response; this observation argues that the separated dust is inorganic.

The thin-layer of the dust samples made it possible to resolve the structure of the TL glow curves. They revealed a single maximum between 225°C and 250°C typical for most of the investigated spices. Up to about 10 kGy a distinct increase in TL output was found, but not above.

The TL response of the dust samples proved a high reproducibility without any special annealing procedure. The achieved precision can be described by coefficients of variation equal to or smaller than 5% on a 90% confidence level, useful for a precise dose determination. On the basis of isothermal decay experiments at 60°C, 100°C, and 150°C, the half life of the TL signals was determined to about 5 years at room temperature.

For the TL effect it is essentially irrelevant whether irradiation has been applied to the spice or to the separated dust; the TL memory of the dust obviously survives the procedure of ultrasonic separation from the spice. Since the "thermoluminescent dust" does obviously not contain organic parts it gives access to available TL at higher temperatures. These TL components are long-lived and seem to allow an identification of irradiated spices even after years.

It is needless to say but nevertheless physically very interesting for the definite proof of the TL origin, that the integer spice grains once deliberated from the dust, do not any longer emit TL or at least to a largely reduced amount, depending on the type of spice. This phenomenon holds whether the grains have been irradiated before or after washing.

The method can probably be extended to other kinds of irradiated food or matter in general, where the presence of irradiated dust is most likely. A lower limit of detection around 10 Gy or less seems feasible.

9. Free Radical Alanine/ESR Dosimetry

Free radical dosimetry provides stable paramagnetic centers generated by ionizing radiation, the linear growth of the radical concentration with dose and high precision of measurement. Precision depends on ESR equipment, probe material and probe preparation, but not on the operator such as in thermoluminescence dosimetry. The reason is that high purity instead of doped materials are used with no need for thermal treatments.

Because of a low G-value, the application of free radical dosimetry has been restricted to higher-level dose assessment; but improvements of technology and know-how show perspectives that also cover therapy-level and accidental doses. Organic detector materials that provide tissue equivalent scatter and absorption are most interesting for phantom dosimetry in radiotherapy; since applicable without probe walls this technique is principally superior to methods at least with anorganic detectors. For this reason and because of the high precision that has meanwhile been proven, free radical dosimetry may probably be able to replace reference dosemeters, e.g., Fricke dosimetry, in determining absorbed dose. It must be recognized that free radical dosimetry is so far a relative method, but promises research and development possibilities towards absolute dosimetry. The free radical technique is able to sum up and conserve dose informations in the probe, e.g., for documentation of the overall therapy dose; based on these properties it lends itself also for quality control programmes, for X-ray and gamma radiation as well as for electrons and neutrons.

With alanine as a detector material an International Dose Assurance Service is since 1986 jointly operated by IAEA and GSF for the IAEA member states to meet the requirements of a world-wide harmonization in high-level dosimetry. Apart, the application to therapy-level dosimetry is in progress. The alanine/ESR system lends itself also to transfer dosimetry for which there is a need particularly at electron accelerators. Since a set of exposed alanine dosemeters can be mailed and read out repeatedly at different laboratories, preconditions are fulfilled for a global calibration network. Most primary or secondary standard dosimetry laboratories have established the alanine/ESR system already or they have firm plans to do so. A first standard document on alanine dosimetry is now being prepared in the United States by ASTM.

The state of the art has been summarized by the 2nd International Symposium on ESR Dosimetry and Applications, GSF Neuherberg, 10-13 Oct. 1988 (Regulla et al. (1989)). It showed that ESR dosimetry finds applications also in radiation accidental dosimetry, in neutron detection, identification and quality control of irradiated foods and in radiographic imaging, apart from archaeological and geological dating. A cross-section of subjects and results is presented on the physical, technical and applied aspects and an outlook is given on what seems to be promising in the future ESR dosimetry development.

10. Alanine/ESR Dosimetry in Photon and Electron Radiation Therapy

The finding of long-lived free radicals produced by ionizing radiation in organic crystals and the quantification of this effect by electron spin resonance (ESR) spectroscopy has proven to possess excellent dosimetric applicability at absorbed doses of 10 Gy and above. The method has metrologically been developed at GSF within a decade approaching traditional dosimetry techniques in accuracy and reliability. It is under study also by other secondary and primary standard dosimetry laboratories. Its use for therapy level dosimetry has been studied intensively within the present therapy-level dosimetry project, with the attention focussed to the dose range below 10 Gy and down to 0.1 Gy. It is shown that with the alanine/ESR dosimetry system improved, linearity and reproducibility fit therapeutic meeds. The response of the tissue-equivalent alanine probes shows no supra-linearity, which affects the application of thermoluminescence dosimetry in this field.

Since the alanine probes are tissue equivalent and need no walls for application they look most interesting also for use in therapy-level dosimetry. The investigations deal with the measurement of absorbed dose generated by high-energy electrons in different depth of a water phantom. The response of the alanine dosemeters has been evaluated for electrons at different therapeutic dose levels; linearity could be proven from 100 Gy down to less than 0.5 Gy. Finally, depth dose measurements have been carried out at a medical linear accelerator, Mevatron KD 7767 (Siemens) for electrons with initial energies of 6, 9, 12, 15, 18 and 21 MeV. It is shown that the absorbed dose values measured with the alanine/ESR system agree with the results of ionization chamber measurements within uncertainty limits of ±1% (confidence level: 90%).

11. ESR Dosimetry using Pharmaceuticals

A fatal radiation accident occurred in Norway, 1982, at a highly radioactive source of unshielded cobalt-60. Protection level and personal dosemeters failed in providing a realistic dose estimate as a necessary basis for medical cure. Chromosome aberration analysis and dose assessment by reconstructing the accidental situation resulted in doses inconsistent with the physiological symptoms. Reasonable dose values were evaluated at GSF upon request from nitroglycerol tablets which the victim ha carried with him; by means of ESR spectroscopy long-lived free radicals could be identified. The concentration of free radicals, by proper calibration converted into dose, proved to be appropriate for dosimetry. Procedures and evaluated dose results were reported; comparison was made with data from literature dealing with the same subject and using TL apart from ESR. For biological relevance, the doses as determined from the tablets have been converted into body and organ doses.

12. Quartz Dosimetry up to 1 Giga-Gray

The ESR spectroscopy of free radiation induced radicals is accepted for dosimetry in the high-level range. With alanine as the commonly used dosimetric material, the ESR spectroscopy has proven to be applicable for doses up to 0.5 MGy at irradiation and storage temperatures up to 70°C. The upper limits are given by saturation and thermal stability of the radiation induced paramagnetic centres and by thermal stability of the material itself. For most solid materials saturation of paramagnetic centers is reached at dose levels below 1 MGy.

Materials were investigated that allow for dosimetry at dose levels of 0.01 MGy - 0.01 GGy. A new dosimetric method uses quartz as a detector. The E'-centre was found to saturate not below 10 MGy and to be thermally stable up to at least 200°C for long period of time. Correlation was made between calibration dose and the concentration of paramagnetic centers achieved at different irradiation temperatures.

13. Accidental Dosimetry

Recent ESR technology together with high operational know-how gives access to biologically relevant accidental doses, i.e., some 0.1 Gy to several Gy provided that appropriate detector materials are used. The sensitivity of domestic articles to radiation and its possible applicability for dose evaluation under accidental circumstances has been studied. A selection of such articles has qualitatively and quantitatively been investigated by ESR spectroscopy with the scope whether free radicals are principally generated by irradiation.

14. LiF-Dosimetry in the Megagray Range and at Elevated Irradiation Temperatures

Spectrophotometry of radiation induced colour centers in optical grade LiF crystals was found to be appropriate for the assessment of gamma doses up to 50 MGy and at irradiation temperatures up to 250°C. The results were based on irradiation at 50°C, 100°C, 150°C, 200°C and 250°C under calibration conditions in a gamma cell irradiator using a specially designed oven. The effect of the irradiation temperature on the dose response and shape of optical spectra was shown. A method was described that allows for the estimation of the irradiation temperature based on changes in the optical spectra, and by applying various post annealing procedures.

2. References

Göksu-Ögelman, H.Y., Regulla, D.F., (1989) "Detection of irradiated food". Nature, Vol. 340, No. 6228, 23.

Regulla, D.F., Scharmann, A. and McLaughlin (Eds.) (1989) "ESR Dosimetry and Applications". Appl. Radiat. Isot., Vol. 40, No. 10-12.

Further references are available from the author upon request.

MICRODOSIMETRY HEALTH PHYSICS INSTRUMENTATION

Th. SCHMITZ
Institut für Medizin
KFA-Forschungszentrum Jülich
Postfach 1913,
D - 5170 Jülich, FRG

1. Introduction

Physicists very often are interested in the question "what does matter do to radiation". Health physicists as well as radiation biologists are, however, interested in the question "what does radiation do to biological matter" and what are the consequences. Ionizing radiation deposits energy in matter, as indicated by the name, through ionizing interactions with atoms and molecules. These interactions occur at random and it is of importance for the interpretation of biological effects of ionizing radiation to adequately describe the discontinuous deposition of energy. Absorbed dose and absorbed dose rates are defined as statistical averages and therefore disregard the resulting random fluctuations. The starting point of microdosimetry was the introduction of random variables by H.H. Rossi for the description of energy deposition. He and his colleagues proceeded then to develop techniques for measuring the random fluctuations of energy deposition in micrometer sites, to construct a novel conceptual and mathematical framework and to apply the new concepts and methods to radiobiology (for a description of the early concepts see e.g., Rossi, 1968).

The first half of this presentation on microdosimetry is divided into two parts. First, an introduction to the field of microdosimetry is given. This introduction is very brief and by no means complete. For a more detailed coverage, the Report No. 36 on Microdosimetry by the International Commission on Radiation Units and Measurements (ICRU (1983)) and the article by A.M. Kellerer (Kellerer (1985)) is suggested. For an introduction to the field of radiological physics, the book of F.H. Attix (Attix (1986)), and the article of G. Alm Carlsson (Alm Carlsson (1985)) is recommended. With regard to one of the main computational tools in microdosimetry, track structure calculations, the article by H.G. Paretzke (Paretzke, 1978) is of in-

M. Oberhofer (ed.), Advances in Radiation Protection, 171–197.

terest.

The second part introduces the experimental techniques to measure energy deposition by ionizing radiation in micrometer sites. This experimental discipline of microdosimetry is treated in greater depth, since it offers a valuable tool for health physicists. Broader coverage of this subject can be found in the ICRU Report 36 (ICRU (1983)) and the open literature, e.g. proceedings of a workshop on "Microdosimetric Counters in Radiation Protection" (Booz et al. (1984)), proceedings of a workshop on "Implementation of Dose Equivalent Meters Based on Microdosimetric Techniques in Radiation Protection" (Menzel et al. (1989a)).

The second half of the presentation deals with practical applications of microdosimetry. The major aspect will be the application of experimental microdosimetric techniques to radiation protection. An attempt is made to answer the question "what can microdosimetry do for the health physicist in his daily work" by presenting, necessarily, selected but representative examples. Interesting literature with respect to this subject are the proceedings of the workshop on the "Implementation of Dose Equivalent Meters...." (Menzel et al. (1989a)) and reports about the intercomparison of dose equivalent meters based on microdosimetric principles by Dietze et al. (1988), Alberts et al. (1988) and Menzel et al. (1989).

Other applications of microdosimetry will discussed afterwards, to give a more complete view of the field. Since this last topic again would fill several chapters literature is suggested for anyone who is further interested. General overviews are given by J. Booz (Booz (1987)) and D.T. Goodhead (Goodhead (1987)). In these publication, more references are stated. The latest picture on the state of the art will be given by the Proceedings of the last Symposium on "Microdosimetry" (Booz et al. (1990)).

At the end, a brief outlook on the future work will be given.

2. Microdosimetry

2.1. Patterns of Energy Deposition

The evaluation of the consequences of energy deposition to a biological system has to consider two aspects. First, where was the energy deposited and second, how much energy was deposited.

Figure 1 illustrates the first question. A piece biological matter is shown, build up by individual microscopic radiation sensitive volumes (e.g. cells), which is irradiated with different doses of photon radiation (^{60}Co). In case of a dose of 30 mGy, each of the sensitive volumes receives several energy deposition events (Figure 1, right) as

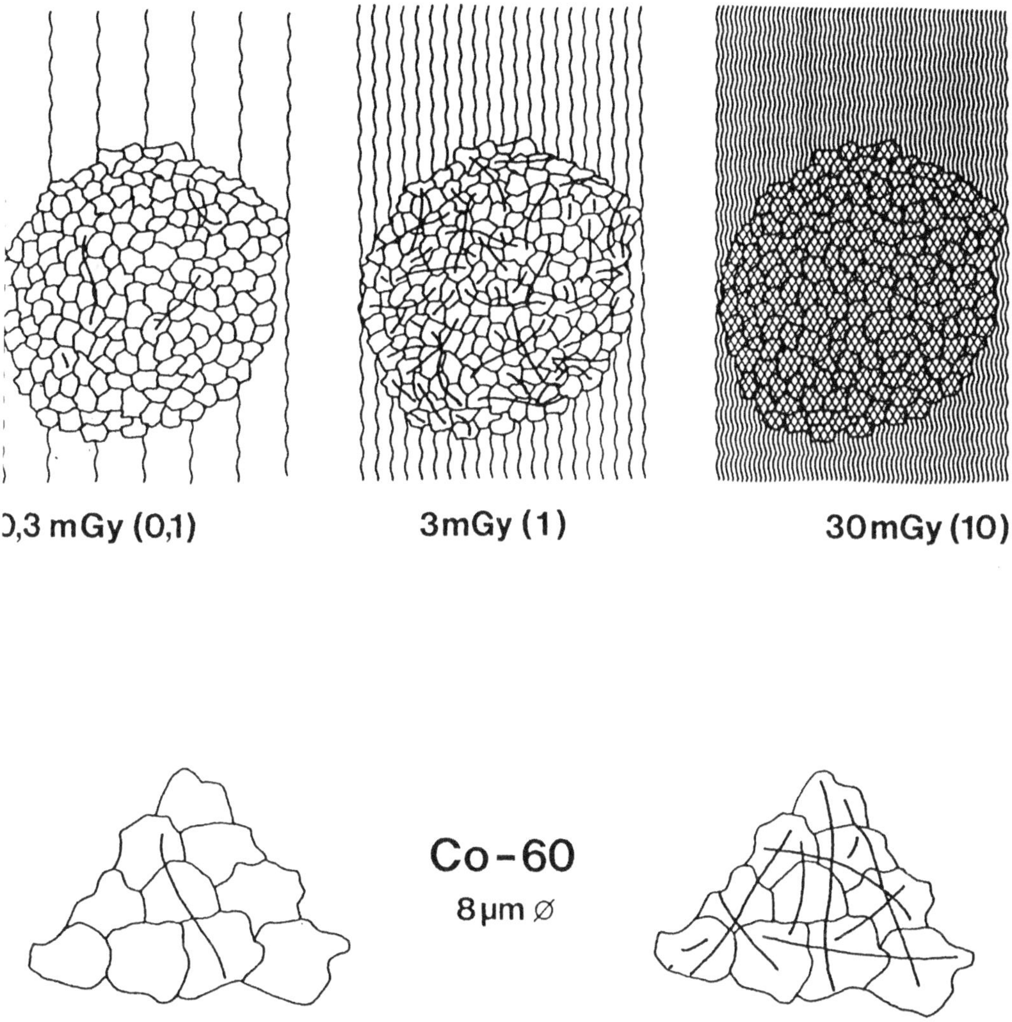

Figure 1. Schematic view of biological matter exposed to three different doses of ^{60}Co radiation (top). The matter is build up by individual radiation sensitive volumes (e.g. cell nuclei), which in this example have an average diameter of 8μm. The lines through sensitive sites denote tracks of electrons. For doses of 0.3 mGy and 30 mGy, some sensitive volumes are shown enlarged (bottom) (Booz et al., 1988).

indicated by the lines. At a dose of 0.3 mGy on the other hand only a few sensitive volumes experience energy depositions. It exists a special distribution of energy deposition events.

Figure 2 serves as illustration for the second question. The two dimensional representation of tracks which may be produced in water vapour by protons of different energy is shown. Each dot in these tracks represents an ionization. The special concentration of the ionizations changes with changing energy of the particle. If sensitive sites in an irradiated medium experience energy depositions, the amount of energy deposited varies (even if the incident irradiation is monoenergetic). Therefore, a distribution of the energy actually deposited has to be considered.

The consequence of the above considerations is demonstrated in Figure 3, for the case of neutron irradiation. The mean number of affected sensitive volumes (dashed line) and the energy deposited in an affected volume (dots, mean value indicated by the line) are given as a function of the macroscopic dose. It can be seen, that for low doses, the energy deposited (denoted by specific energy, see below) in an affected volume varies substantially. Furtheron, in the low dose area, the number of affected volumes increases almost linearly with increasing dose. At high doses, all volumes are affected and they experience similar amounts of energy.

The subject of microdosimetry is to describe the special and energy distribution of energy deposition of ionizing radiation and to recognize those features and parameters that are critical for a biological effect. The purpose of microdosimetry however, is not the unlimited generation of data, but their deliberate reduction to the most essential parameters (Kellerer (1985)).

2.1.1. Quantities Used in Microdosimetry. From the various quantities used in microdosimetry, only two are briefly shown. A more detailed description of quantities and concepts used in microdosimetry is given in the ICRU Report 36 (ICRU (1983)). Figure 4 illustrates the quantities specific energy, z, and lineal energy, y. Specific energy is the energy delivered to single sensitive volumes (e.g. cells in biological matter) divided by the mass of this volume. In contrast, the macroscopic dose, D, is the quotient of the energy delivered to a macroscopic piece of matter divided by its mass. In the limiting case of homogeneous distribution of energy deposition, absorbed dose, D, and specific energy, z, become equal (see Figure 3).

Lineal energy, y, is the quotient of the energy delivered to a sensitive site by a single charged particle divided by the mean chord length, l, of this volume. Whereas the name lineal energy seems to indicate that the special extension of charged particle tracks and the

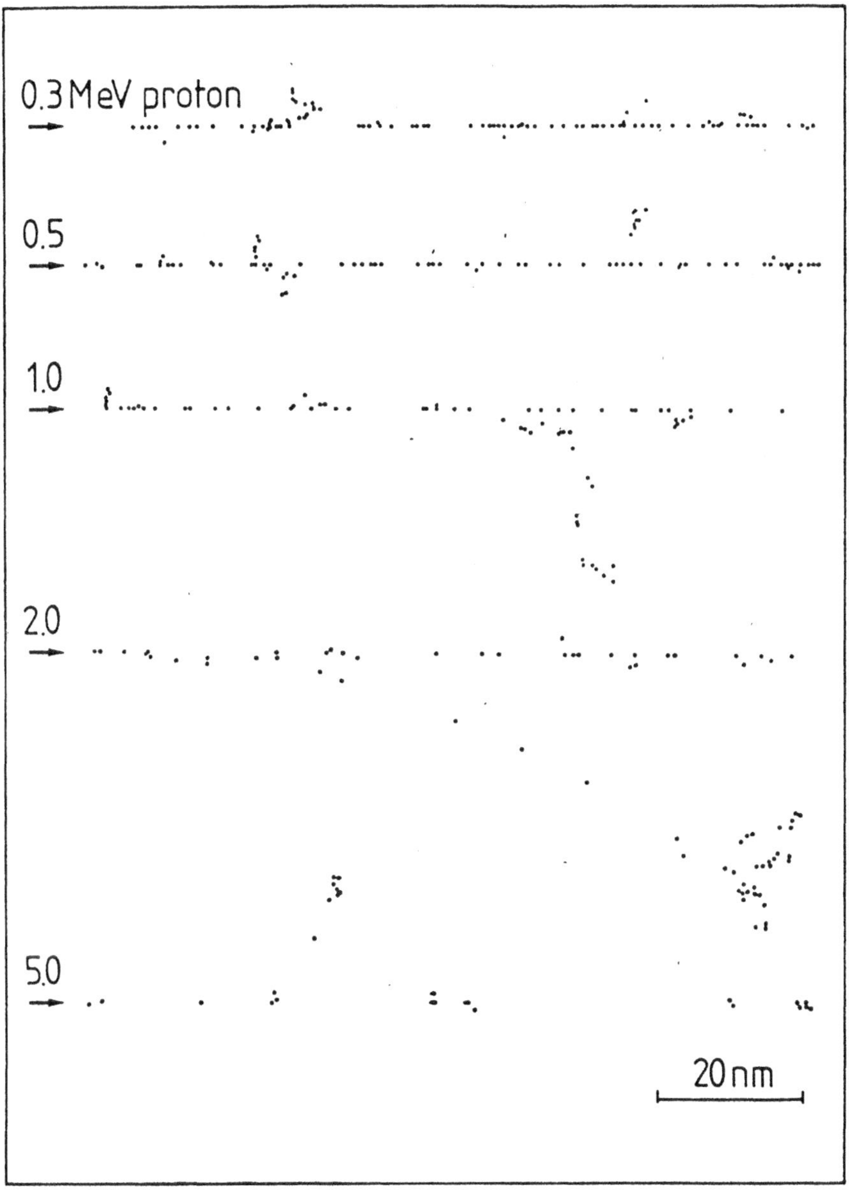

Figure 2. Two dimensional representations of tracks produced by protons of different energy in water vapour (Moca-14, 1984).

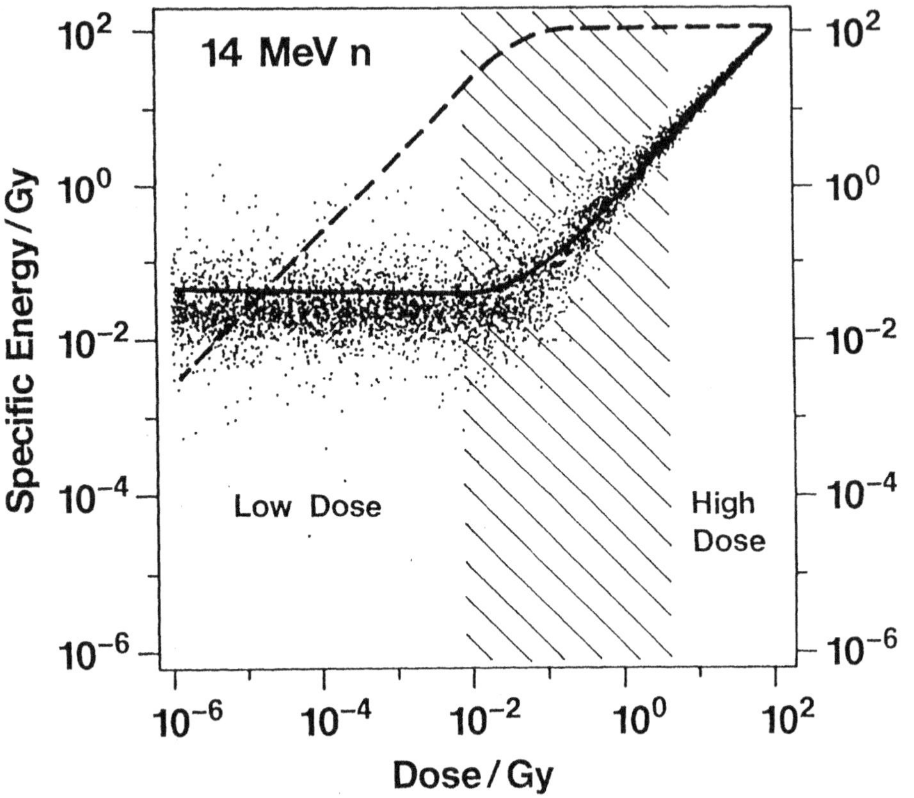

Figure 3. Mean energy deposited in affected targets (denoted as specific energy: solid line and dots, left ordinate) and fraction of affected targets (dashed line, right ordinate) as a function of absorbed dose. Each dot represents one casual value of the specific energy, z, which may be deposited in an affected target, the line represents the mean value of the specific energy in the affected targets (Booz et al., 1988).

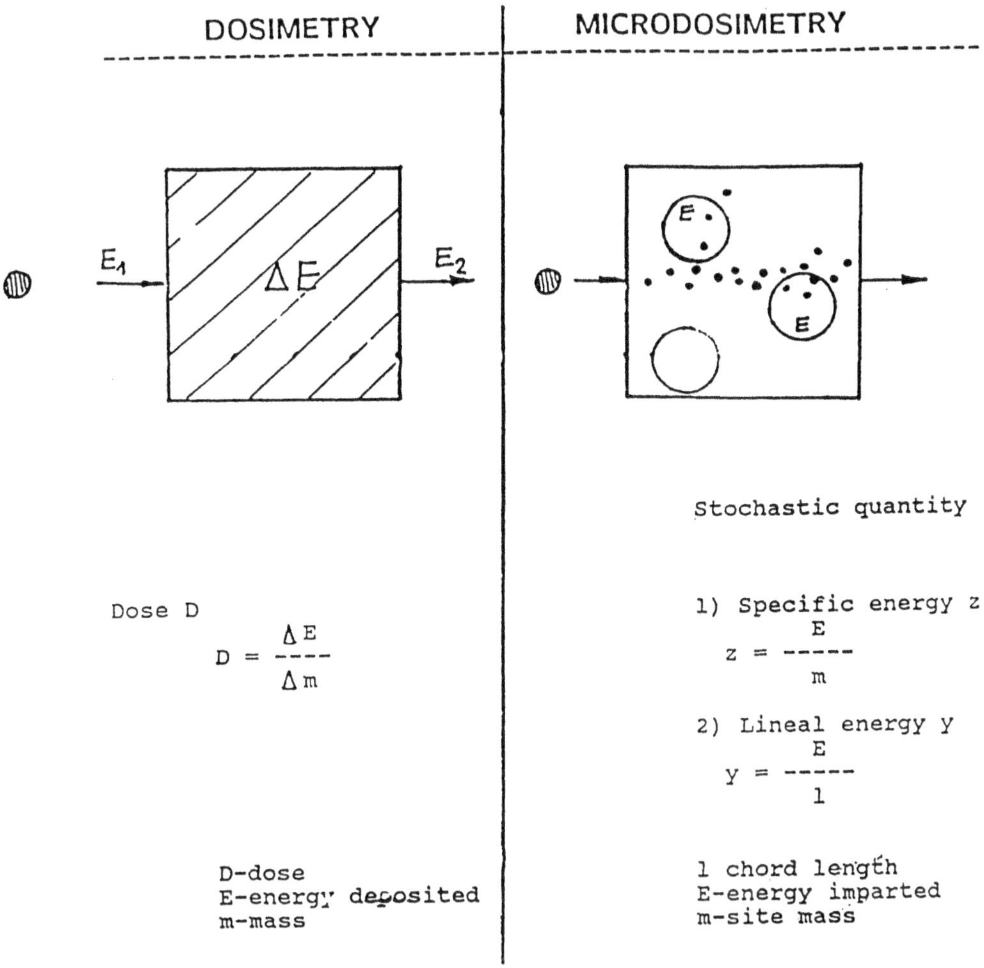

Figure 4. Dosimetry versus microdosimetry. Dosimetry disregrads the structure of biological matter and the stochastic nature of energy depositions by ionizing radiations. One frequently used quantity is the absorbed dose, D. Microdosimetry acknowledges the structure of biological matter and the stochastics of energy depositions. Two related quantities, which are often used, are specific energy, z, and lineal energy, y. They are stochastic quantities in contrast to absorbed dose.

corresponding pattern of energy depositions in the volume is disregarded, this is not the case. All the energy delivered in ionizations in the sensitive volume is taken into account.

Lineal energy, y, and specific energy, z, are related. If the distribution of lineal energies in a sensitive volume is known either from calculation or from measurement, the corresponding distribution of specific energy can be determined. One of the beautiful features of lineal energy is the mentioned measurability. The device used for this purpose is the low pressure tissue equivalent proportional counter (TEPC).

2.2. Proportional Couner Microdosimetry

Figure 5 shows a cross sectional view of a tissue equivalent proportional counter, which is the device to measure distributions of lineal energy in small volumes. This particular one has an outer shell of 20 mm of polyethylene. It serves as a vacuum container and neutron moderator. The inner cylinder, called the sensitive volume of the counter, consists of 1 mm of tissue equivalent plastic A-150 (ICRU (1983)) and has an effective height of 7 cm and a diameter of 7 cm. The counter is filled with tissue equivalent gas on methane basis (Rossi-Failla, ICRU (1983)).

Tissue volumes of several micrometer diameters can be simulated with a proportional counter by setting the pressure of the gas to such a value that the mean energy deposited by charged particles in the sensitive volume is the same as in the simulated tissue volume (1).

Figure 6 shows examples of measurements with this TEPC in narrow photon fields. The mean energy of photons varies between 26 keV and 1.2 MeV (^{60}Co) (2). Above 100 keV photon energy, the distributions get broader with increasing photon energy. This can be understood from the fact that the maximum energy of electrons increases, which are released in Compton interactions. The stopping power (energy lost per unit track length) of electrons decreases with increasing energy. Consequently, with increasing energy they deposit less energy when passing the sensitive volume and are detected at lower lineal energies. The distributions for 26 keV photons and 100 keV photons are very similar in shape. This indicates that the corresponding secondary electron spectra in the sensitive volume must be similar, even though they originate from photoelectric interactions in the case of 26 keV and from Compton interactions for 100 keV (Schmitz et al. (1989), Olko et al. (1989)).

---

(1) Furtheron, the equivalent diameter of the tissue volume simulated by the counter is called the simulated diameter.

(2) The experiments were performed in the standard photon beams at the Physikalisch Technische Bundesanstalt (PTB), Braunschweig, FRG.

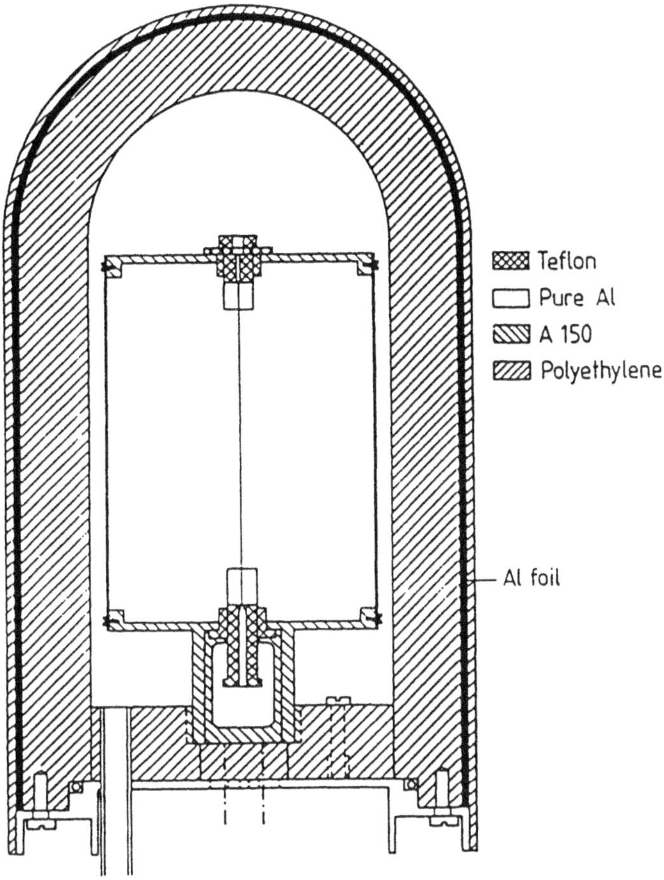

Figure 5. Cross sectional view of the KFA counter, a tissue equivalent proportional counter. The sensitive volume is made of tissue equivalent plastic, A-150, and was given a diameter and effective height of 7 cm. The containment consists of polyethylene.

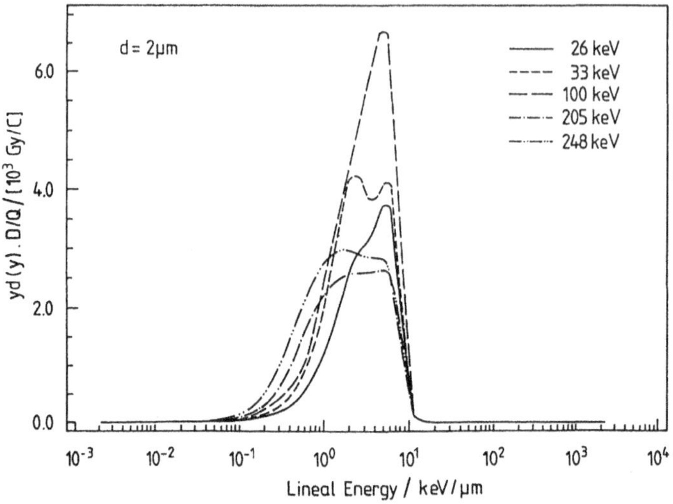

Figure 6. Examples of measurement results with the KFA counter, schematically shown in Figure 5, in photon fields of different energy. The simulated diameter was 2μm. Shown are microdosimetric dose distributions. The areas under the curves are normalized to one. In this semi-logarithmic representation, equal areas under a distribution represent equal contributions to absorbed dose (Schmitz et al., 1989).

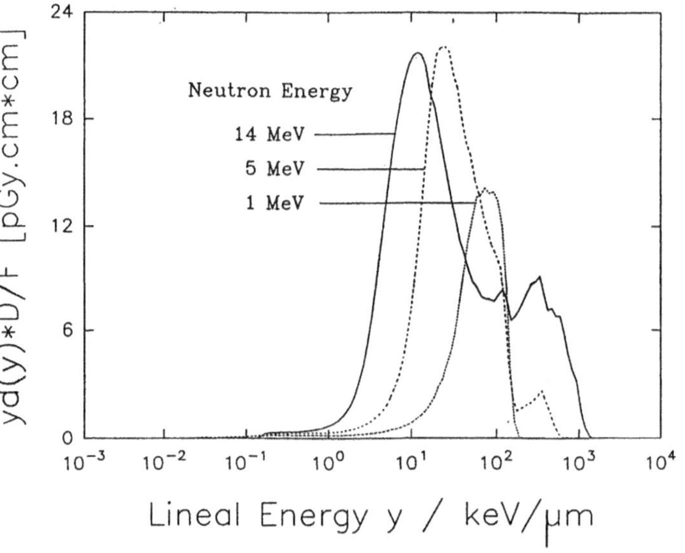

Figure 7. Same as Figure 6, but for quasi monoenergetic neutron fields of different energies (Dietze et al., 1988).

In Figure 7, measured distributions in narrow neutron beams of 14.8 MeV, 5 MeV and 1.2 MeV are shown (3). The most complex spectrum is the one for 14.8 MeV neutrons. Three different regions can be identified. The part around 10 keV/µm up to about 100 keV/µm is due to recoil protons. The peak around 300 keV/µm is mostly due to alpha particles and the shoulder around 1000 keV/µm is created by heavier ions like oxygen and carbon ions. The distribution of 5 MeV neutrons shows a maximum around 30 keV/µm which is created by recoil protons, whereas the part in the distribution above 200 keV/µm is due to heavy recoil particles. The distribution for 1 MeV neutrons shows mainly a recoil proton part around 80 keV/µm. The maxima of the recoil proton contribution in the spectra shift to higher lineal energies with decreasing neutron energy. This is due to the increasing stopping power of the protons with decreasing energy (see for example Figure 2).

In Figure 8, an example of a measurement in the mixed neutron and gamma radiation field of a heavy water moderated ^{252}Cf is given. The part of the distribution between 0.1 keV/µm and about 10 keV/µm is mostly due to electrons released in photon interaction, whereas the part around 100 KeV/µm is due to protons released in neutron interactions.

The examples given show two features of the proportional counter measurement. First, the device is sensitive to both, neutrons and photons. Second, the measured spectra reflect characteristics of the incident radiation field. The spectra also reflect the influence of the simulated diameter and the construction of the counter (Dietze et al. (1988)) as will be shown below. Nevertheless, the sensitivity to photons and neutrons together with the diagnostic properties, available due to the measured distributions, lend the device to the application as area monitor in the field of radiation protection.

3. Application of Microdosimetry

Microdosimetric principles are applied in various fields of radiation physics and radiation biology (Booz (1987)). Two aspects are selected within the scope of this presentation. Radiation protection and the method to relate microdosimetric distributions to biological effects by the use of so-called response functions. The latter is demonstrated for the case of neutron therapy and radiation biology.

(3) The data presented were collected during an international intercomparison of various TEPC systems. The intercomparison was held at the PTB Braunschweig in 1987 and 1988. Results of this intercomparison are given by Dietze et al. (1988) and Alberts et al. (1988) and Menzel et al. (1989).

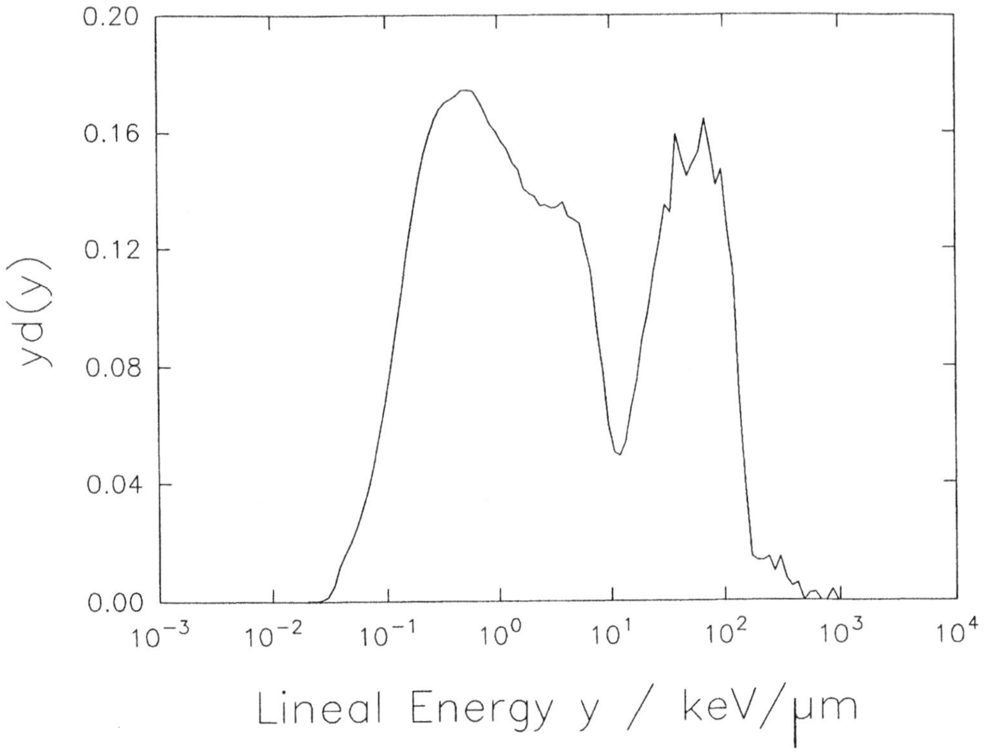

Figure 8. Dose distributions measured in the field of a heavy water moderated ^{252}Cf Source (Dietze et al., 1988).

3.1. Radiation Protection

Absorbed dose can be calculated from proportional counter measurements, by applying cavity chamber theory. According to this theory, the absorbed dose, D, in the wall of the chamber can be calculated from the total amount of charge collected in the gas

$$D = \frac{q \cdot W}{e \cdot m} \qquad (1)$$

where q is the total charge in the cavity, W is the amount of energy expended by charged particles to create an ion pair, e is the electronic charge and m is the mass of the gas.

The proportional counter measures pulse height distributions, often simply called microdosimetric spectra. From these distributions, absorbed dose can be calculated according to

$$D = C_D \cdot N \cdot y_F \tag{2}$$

where y_F denotes the frequency mean lineal energy of the y distribution and N the total number of events. C_D is a factor which is determined either from calculation or by a calibration measurement (ICRU (1983), Dietze et al. (1988)).

The mean quality factor, Q, is determined according to

$$Q = \int q(y) \cdot d(y) dy. \tag{3}$$

d(y) is the probability density of the tissue absorbed dose in y and the function q(y) approximates $q(L_\infty)$ (ICRP (1973)) by setting $y = L_\infty$. The total dose equivalent is then determined according to

$$H = Q \cdot D. \tag{4}$$

The quantity to be measured by an area dose meter for penetrating irradiations is the ambient dose equivalent, H*(10), which is defined in 1 cm depth in the ICRU sphere in an expanded and aligned field (ICRU (1985)). The ICRU sphere is a tissue sphere of 30 cm in diameter. H*(10) is the operational quantity in radiation protection. The limiting quantity is effective dose equivalent, H_e, which is a weighted sum over organ doses (ICRP (1973)).

3.1.1. Dose Equivalent Response. The dose equivalent response is the quotient of the dose equivalent reading of the proportional counter divided by ambient dose equivalent, H*(10). Figure 9 compares the responses of two different proportional counter systems (BIO (4), KFA (5)) with the response of a conventional rem meter of the Leake type in monoenergetic neutron fields (3). The main difference between the KFA system, which uses the counter shown in Figure 5, and BIO, is the thickness of the wall. KFA uses a total wall thickness of 20 mm, whereas BIO applies 2.5 mm. The response of the two systems is very similar above 1 MeV. Below 1 MeV, the response of KFA decreases faster

---

(4) BIO is the laboratory TEPC system used by the colleagues of the group of Strahlenbiophysik, Zentrum für Umweltforschung, Universität des Saarlandes, FRG.

(5) KFA counter is a transportable dose equivalent meter for use in radiation protection, developed at the Institut für Medizin, Kernforschungsanlage Jülich, FRG. This system is available through the company Reaktorwartungsdienst & Anlagenbau, Eleonorenstrasse 1, D-5170 Jülich, FRG.

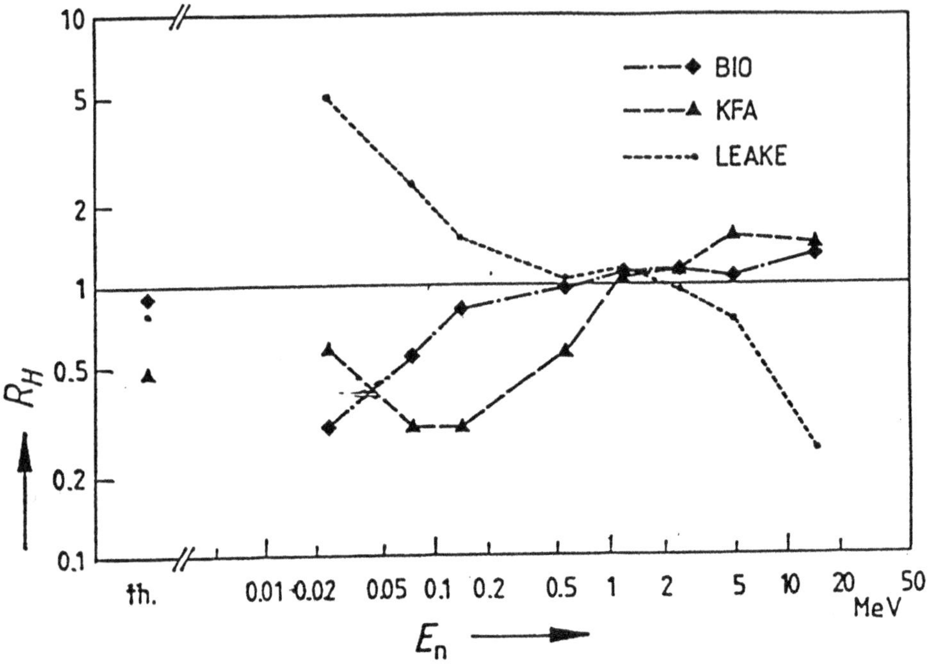

Figure 9. Ambient dose equivalent response of two different tissue equivalent proportional counters and a Leake counter as a function of incident neutron energy (please see text) (Taken from: Alberts et al., 1988).

than the one of BIO until around 100 keV incident neutron energy. At 24.5 keV neutron energy, the response of KFA is higher than the one of BIO and at thermal neutron energy this is the other way round. In comparison to the LEAKE counter, the variation of the response of both TEPC systems is considerably smaller.

The difference between KFA and BIO is mainly due to the difference in wall thickness (Menzel et al. (1989)). The question, which arises is why KFA applies such a thick wall. The KFA counter was designed for use as area monitor in mixed neutron and gamma radiation fields, for neutron energies between thermal and 20 MeV and for photons between 30 keV and several MeV. The ambient dose equivalent, H*(10), is defined in 1 cm depth in the ICRU sphere as mentioned above, in other words, the reference point is below a layer of 1 cm of tissue. Neutrons interact with tissue in a complex way and specifically at energies below about 500 keV neutron energy the transport of neutrons in tissue has to be considered. Figure 10 demonstrates this point. The figure shows as a

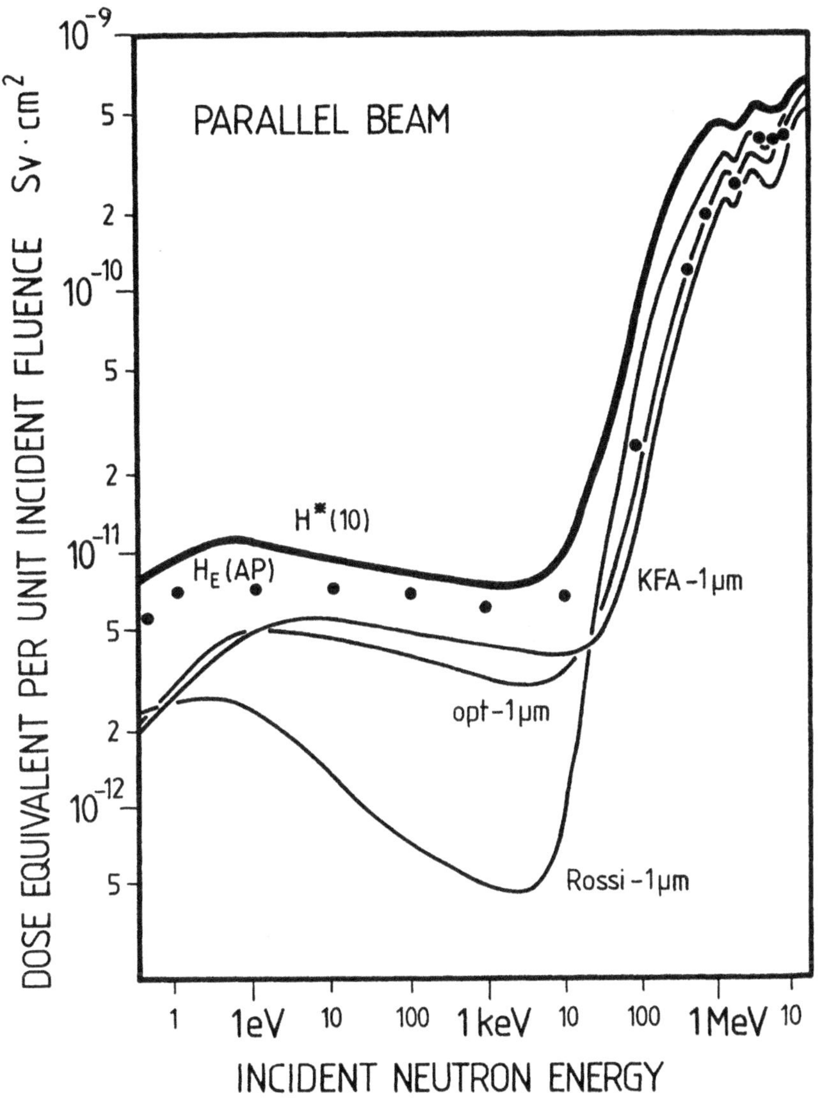

Figure 10. Calculated dose equivalent per incident neutron fluence as a function of incident neutron energy. H*(10) denotes the ambient dose equivalent, H_E(AP) denotes the effective dose equivalent for anterior posterior irradiation. KFA-1 μm, reading calculated for the KFA counter with a wall thickness of 20 mm, opt-1 μm, reading calculated for the KFA counter with an optimal 15 mm wall thickness. Also given is the reading for a conventional Rossi counter (Rossi-1 μm) (Booz, 1984).

function of incident neutron energy and normalized to unit incident neutron fluence the ambient dose equivalent, effective dose equivalent and the dose equivalent readings of the KFA counter (KFA-1 μm), the KFA counter with the optimal wall thickness of 15 mm (opt-1 μm) and a Rossi counter (Rossi-1μm) (Booz 1984, Morstin 1985, Kawecka (1985)). The BIO counter is similar to the Rossi counter. It can be seen from this figure that below 100 keV neutron energy down to about thermal energy the use of a thicker wall improves the dose equivalent reading of a TEPC with respect to ambient dose equivalent.

In the experimental response the increase of the dose equivalent reading of KFA in comparison to BIO due to the influence of thicker wall of KFA can be seen for the neutron energy of 24.5 keV (Figure 9). Figure 11 presents the dose and dose equivalent distributions for 24.5 keV neutrons for the KFA counter using 5 mm, 15 mm and 20 mm wall thickness. In the case of the *dose* distributions (Figure 11a), for 5 mm wall thickness, a recoil proton part can be seen at around 10 keV/μm. This part is less pronounced for 15 mm wall thickness and almost vanishes at 20 mm wall thickness. In the distributions for 15 mm and 20 mm wall thickness, a part appears at lineal energies between 0.1 keV/μm and 10 keV/μm. This part is due to photons. The neutron field is moderated in the wall of the counters to thermal energies. Thermal neutrons are captured by hydrogen or nitrogen. Photons of 2.2 MeV are created by the capture in hydrogen. In the interaction with nitrogen, 580 keV protons are created, which contribute to the part in the distributions at around 100 keV/μm. The increase due to these protons in the distributions for 15 mm and 20 mm wall thickness can be more clearly seen in the corresponding *dose equivalent* distributions (Figure 11b). It has to be mentioned that this part of the distributions is also influenced by a high neutron energy contribution in the incident beam, and only the increase as a function of the wall thickness is due to moderated neutrons.

The responses of BIO and KFA show that tissue equivalent proportional counters are able to measure dose equivalent and that their overall ambient dose equivalent response is better than that of a conventional rem counter of the LEAKE type. The response of proportional counters can be influenced by changing the wall thickness. There are also other methods to influence the response (Menzel (1989)). A combination of such methods may be used to further improve the response of TEPC area dose meter.

Even though the dose equivalent reading of a TEPC offers already a good estimation of ambient dose equivalent, H*(10), it nevertheless underestimates H*(10) in the neutron energy range below 500 keV. For the case of the KFA counter, it can was shown that by using a correction factor, the dose equivalent reading never underestimates the

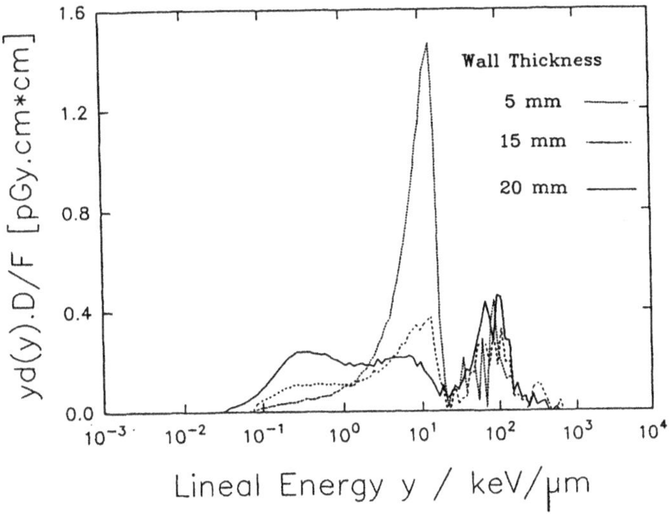

a) *Dose* distribution

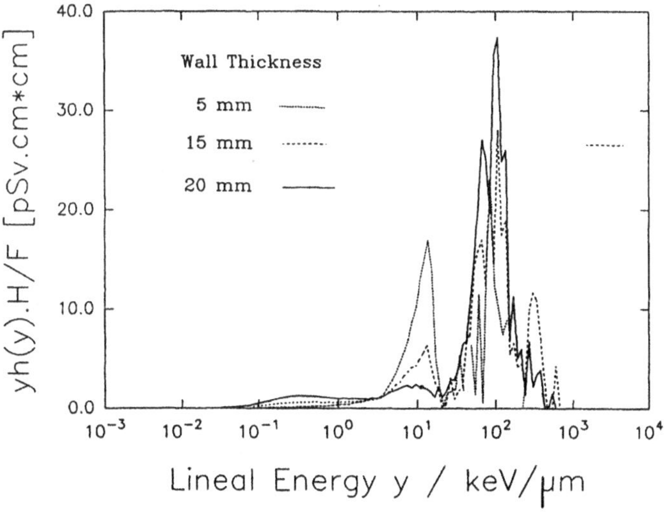

b) *Dose equivalent* distribution

Figure 11. *Dose* distributions (a) and *dose equivalent* distributions (b) measured in a 24.5 keV neutron field. The KFA counterr was used with a wall thickness of 5 mm, 15 mm and 10 mm (for explanation please see text).

effective dose equivalent, $H_e(AP)$, for anterior-posterior irradiation (Booz (1989)). Effective dose equivalent is the limiting quantity in radiation protection.

3.1.2. Protection Measurements. Figure 12 shows an example of two measurements with the KFA counter, which were performed in front of the door of an experimental bunker at a heavy ion accelerator. In one case, the door was a left open a little bit, to allow cables to be put through (thick line in Figure 12). During the second measurement (thin line in Figure 12) the door was closed completely. The dose distributions are normalized to unit dose rate. The reduction of dose rate, if the door is closed completely, can easily be read from this figure.

The dose distributions also give an indication on the energy range of the initial neutrons. The bulk must have been of energies below about 1 MeV, which contributed to the parts at lineal energies above 10 keV/μm up to 100 keV/μm. There is however a significant contribution of high energy neutrons (with energies above ca. 10 MeV), which contributed to the part in the distribution between 6 keV/μm and 10 keV/μm. In addition, photons had been in the initial field.

The dose equivalent rates during the measurements were 14.7 μSv/h and 5.2 μSv/h for the open and closed door respectively. The photon contributions to dose equivalent were about the same in both cases and had a value of about 0.6 μSv/h. The quality of the radiation decreased from 6 to 4.7 when the door was closed.

This example demonstrates that tissue equivalent proportional counters are able to give informations on the incident radiation field as well as on dose rates and quality factors.

3.2. Radiation Biology and Therapy

Within this chapter, only the method of employing biological response functions is briefly described. For the full view of applications the literature suggested in the introduction should be consulted.

Figure 13 presents the principal idea of response function. It is assumed that at low radiation doses, a biological effect, E, is directly correlated to energy deposition events in critical targets. The variation of energy depositions in critical targets is described by dose distributions in terms of lineal energy, d(y). The correlation to the biological effect is described by the response function r(y). This response function can be unfolded, if for a given biological endpoint sufficient biological data in different radiation fields together with corresponding microdosimetric informations are available.

Figure 14 shows an example of such a response function (here termed probability of quantal response), for the case of chromosomal

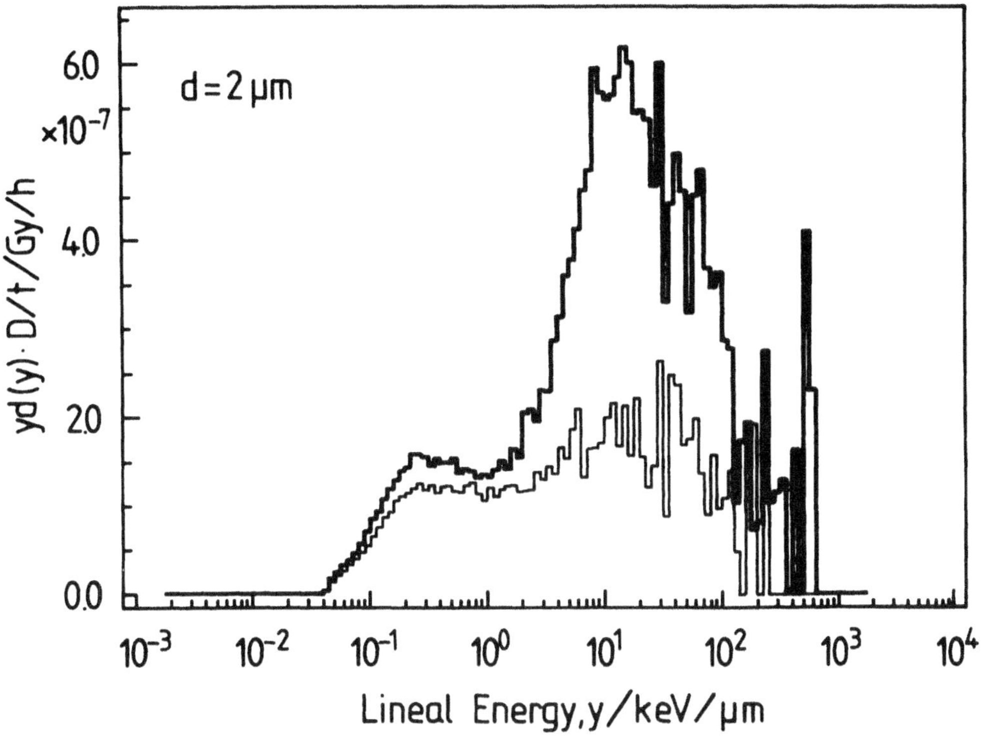

Figure 12. Dose distributions measured in front of the door of an experimental bunker at a heavy ion accelerator. Thick line: the door was open a little, to allow cables to be put through; thin line: the door was completely closed. The distributions are normalized to time so that the area under the curves give the dose rates. The reduction in dose rate, when the door is completely closed, can easily be read from the figure. The incident field contained neutrons in a broad energy range. A significant contribution to dose comes from neutrons of energies above 10 MeV, as can be deduced from the measurement when the door was slightly open. The part in the spectrum between 6 keV/μm and 10 keV/μm is dominated by these neutrons. When the door is closed, the neutron dose is significantly reduced, as can be seen from the comparison of both measurements.

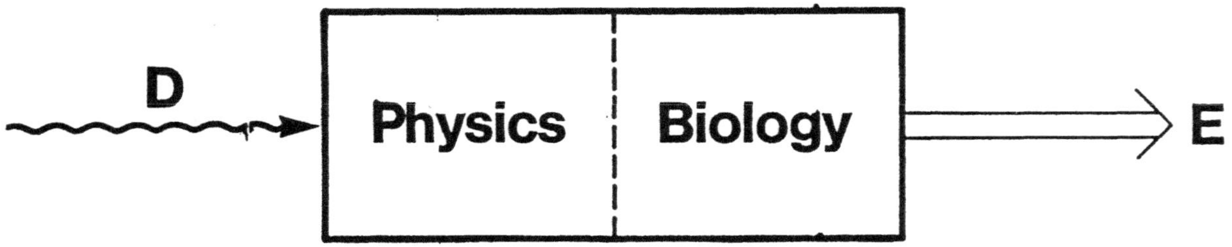

$$: E = D \cdot R = D \cdot \int d_1(y_1)\, r(y_1)\, dy_1$$

Figure 13. Concept of response function. First row: the biological system is considered as a black box receiving the absorbed dose D and showing the effect E. The concept is to separate physics and microdosimetry on the one hand and biology on the other hand in order to learn more about the biological response. Second row: Evaluation of the response function, $r(y_1)$, as proposed by Bond. R is the integral response, y_1, is the lineal energy referred to a sphere of 1 µm and $d_1(y_1)$, the microdosimetric distributions for single events (after Booz et al., 1988).

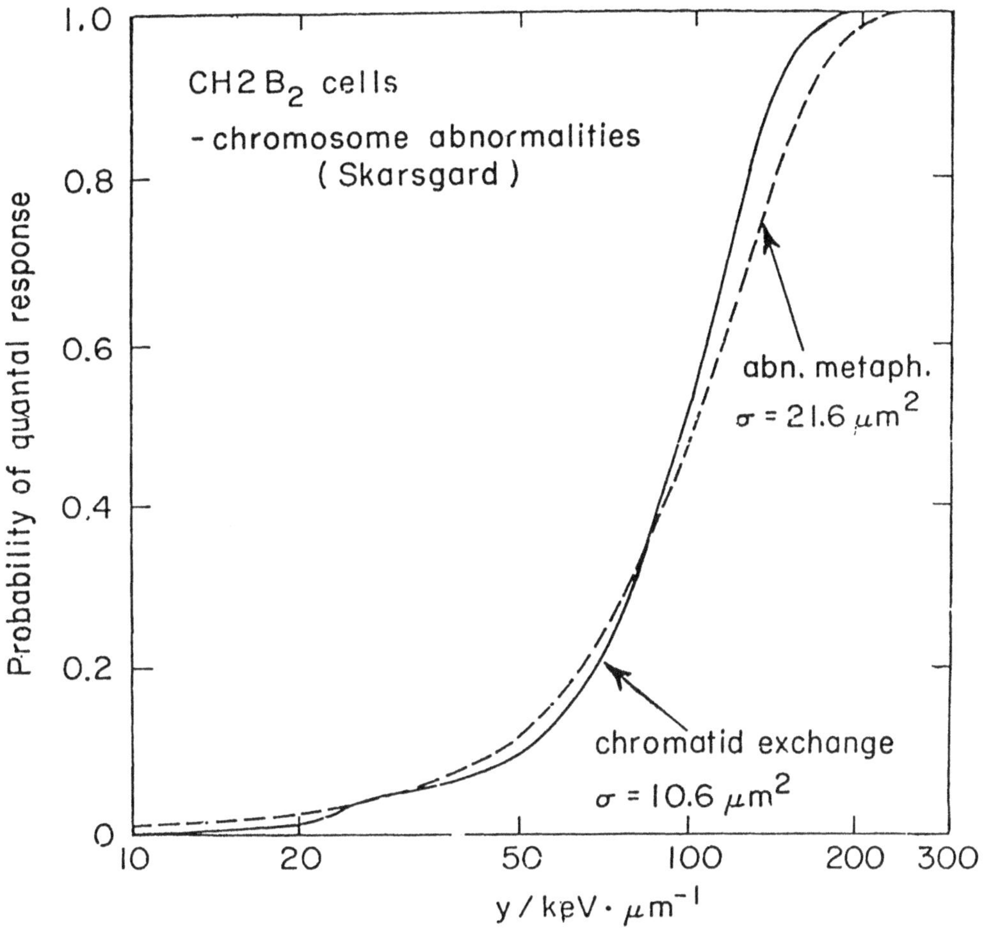

Figure 14. Example of response functions for chromosome abnormalities in hamster cells as a function of lineal energy (Morstin et al., 1989).

abnormalities in $CH2B_2$ cells (Morstin et al. (1989)). For this unfolding, the data of Skarsgard (1967) were taken. If these results are used as weighting functions for dose distributions measured in unknown fields with a TEPC, then the reading would give an estimation of the risk of chromosomal abnormalities in $CH2B_2$ cells.

The concept of response functions was also utilized to characterise radiation fields used in radiatiotherapy (6). Dose distributions were measured in various neutron therapy fields (Figure 15). Using biological experiments which were available for these fields, a response function was unfolded. This response function was then applied to weigh the dose distributions. The weighted dose distributions are given in Figure 16 and can be understood as a characterisation of the quality of the therapeutic beams (Pihet et al. (1990)). The authors point out however, that these results should be applied with great care, if neutron beams outside the range in their work are considered an if other doses than 8 Gy for which the biological results were achieved are used (Pihet (1989)).

4. Outlook

Two topics are considered in this outlook, the application of tissue equivalent proportional counters and the development of response functions.

Due to its sensitivity to neutrons and photons and due to their diagnostic abilities, proportional counters are applicable to a variety of problems. One problem of interest to our group in Jülich (we are not the only group) is space dosimetry. Therefore, we will optimize the counter for the radiation fields in space. Such an optimization is performed with the help of computer simulations of counter measurements. These simulations consider the transport of radiation in the wall of such a counter and in the end the energy deposition in the sensitive volume. In these simulations, also the geometry of the device is taken into account. In fact, computer simulations of this kind are another discipline under the roof of microdosimetry and they are not only applicable to simple geometries like the one of proportional counters, but also to geometries as complex as a human body.

The unfolding of biological response functions is another field of interest. More biological experiments in different fields together with microdosimetric measurements are needed. If biological experiments can

---

(6) The work referred to was performed by P. Pihet and H. Menzel within a microdosimetric intercomparison programme of European Neutron Therapy Centers (Pihet et al. (1990)).

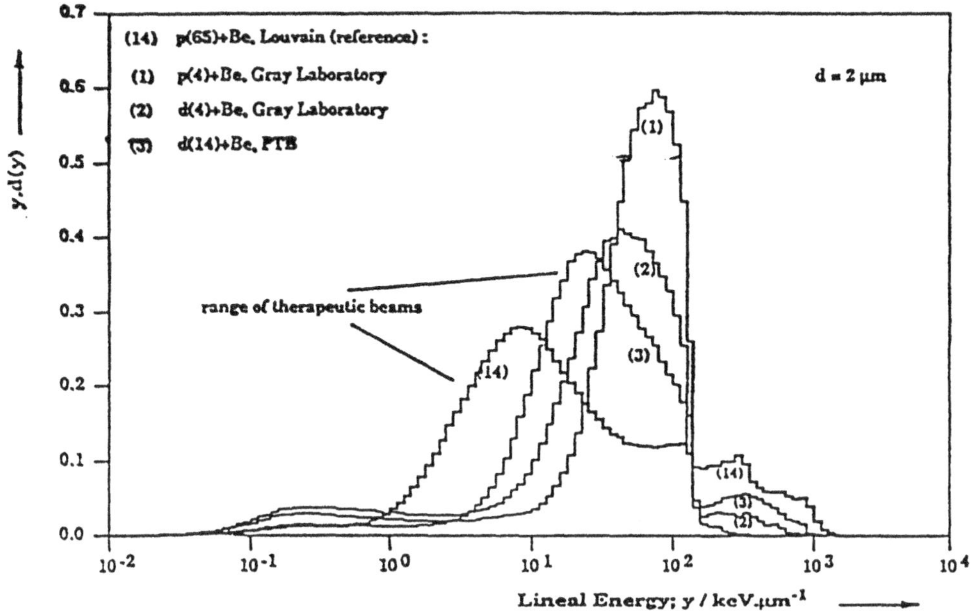

Figure 15. Dose distributions of neutron beams used in therapy (taken from: Pihet et al., 1990).

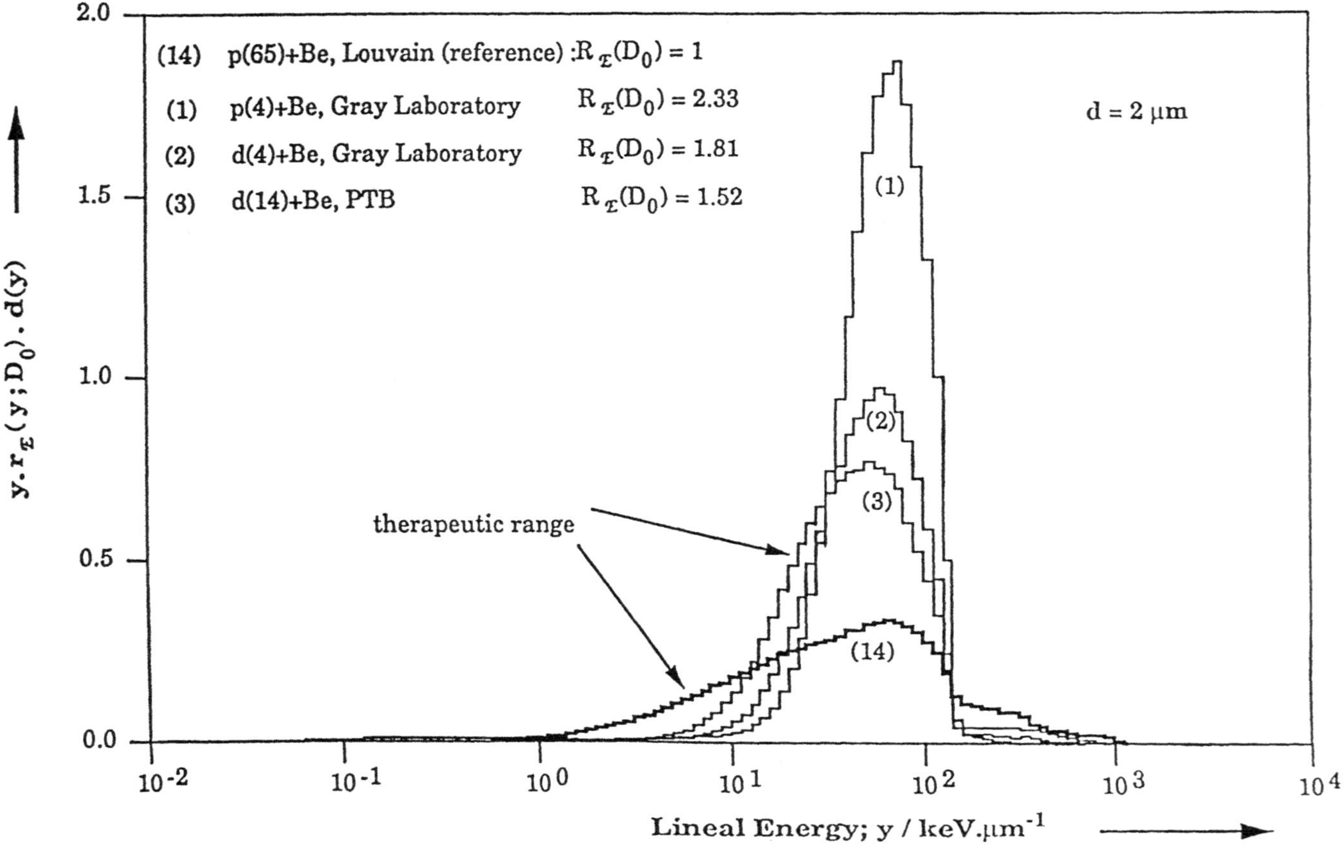

Figure 16. Same as Figure 15, but the dose distributions were weighted with a response function, determined by using results of biological experiments available for the neutron beams (taken from: Pihet et at., 1990).

be performed for endpoints, which are relevant to radiation protection, this method could give radiation protection a stronger biological back up.

5. Suggested Reading, References

Alberts, W.G., Dietz, E., Guldbakke, S., Kluge, H., and Schuhmacher, H. (1988) "Radiation Protection Instruments Based on Tissue Equivalent Proportional Counters: Part II of an International Intercomparison", PTB-Report PTB-FMRB-117, Physikalisch Technische Bundesanstalt, Braunschweig, FRG.

Alm Carlsson, G. (1985) "Theoretical Basics for Dosimetry", in "The Dosimetry of Ionizing Radiation" Vol. I, pp. 1-75, Kase, R.K., Bjärngard, B.E., Attix, F.H. (Eds.), Academic Press Inc., NY.

Attix, F.H. (1986) "Introduction to Radiological Physics and Radiation Dosimetry",John Wiley & Sons, NY.

Booz, J. (1984) "Advantages of Introducing Microdosimetric Instruments and Methods into Radiation Protection", Radiat. Prot. Dosim. 9, pp. 175-183.

Booz, J., Edwards, A.A., and Harrison, A. (Eds.) "Microdosimetric Counters in Radiation Protection", Proceedings of a Workshop held in Homburg, FRG, Radiat. Prot. Dosim. 9, No. 3.

Booz, J. (1987) "Applications of Microdosimetry", Radiat. Res. 2, pp. 175-183.

Booz, J., and Feinendegen, L.E. (1988), "A Microdosimetric Understanding of Low Dose Radiation Effects", Int. J. Radiat. Biol. 53, pp. 13-21.

Booz, J., Olko, P., Schmitz, Th., Morstin, K., and Feinendegen, L.E. (1989), "The KFA Counter, its Photon and Neutron Responses and its Potential for Future Developments", Radiat. Prot. Dosim. 29, No. 1-2, pp. 87-92.

Booz, J., Dennis, J.A., and Menzel, H.G. (Eds.) (1990) "Proceedings of the 10th Symposium on Microdosimetry", held in Rome, May 21st-26th 1989, Radiat. Prot. Dosim. 31, No. 14.

Dietze, G., Edwards, A.A., Guldbakke, S., Kluge, H., Leroux, J.B., Lindborg, L., Menzel H.G., Ngyen, V.D., Schmitz, Th., and Schuhmacher, H. (1988) "Investigation of Radiation Protection Instruments Based on Tissue equivalent Proportional Counters. Results of an EURADOS Intercomparison",Commission of the European Communities, Luxemburg, EUR 11867EN.

Goodhead, D.T. (1987) "Relationship of Microdosimetric Techniques to applications in Biological Systems" in "The Dosimetry of Ionizing Radiation" Vol. II, pp. 1-89, Kase, R.K., Bjärngard, B.E., Attix, F.H. (Eds.), Academic Press Inc., NY.

ICRP (1973) "Data for Protection against Ionizing Radiation from External Sources, Supplement to ICRP Publication 15, Recommendations of the International Commission on Radiological Protection, Annals of the ICRP 1, Vol. 3.

ICRU (1983) "Microdosimetry", Report No. 36 International Commission on Radiation Units and Measurements, Bethesda, Maryland.

ICRU (1985) "Determination of Dose Equivalent Resulting from External Radiation Sources", Report No. 39, International Commission on Radiation Units and Measurements, Bethesda, Maryland.

Kawecka, B., Morstin, K., and Booz, J. (1984) "Optimization of the Design of Microdosimetric Dose Equivalent Meters", Radiat. Prot. Dosim. 9, pp. 203-206.

Kellerer, A.M. (1985) "Fundamentals of Microdosimetry", in "The Dosimetry of Ionizing Radiation", Vol. I, pp. 78-162, Kase, R.K., Bjärngard, B.E., and Attix, F.H. (Eds.), Academic Press Inc., NY.

Menzel, H.G., Lindborg, L., Schmitz, Th., Schuhmacher, H., and Waker, A.J. (1989) "Intercomparison of Dose Equivalent Meters Based on Microdosimetric Techniques: Detailed Analysis and Conclusions", Radiat. Prot. Dosim. 29, Nos. 1-4, pp. 55-68.

Menzel, H., Paretzke, A.G, and Booz, J. (Eds.) (1989a) "Implementation of Dose Equivalent Meters Based on Microdosimetric Techniques in Radiation Protection", Proceedings of a Workshop held in Elmau, FRG, Radiat. Prot. Dosim. 29, Nos. 1-2.

Moca-14 (1984) "Monte Carlo Track Structur Code", Paretzke, H.G., Wilson, W.E. (Authors).

Morstin, K., Kawecka, B., and Booz, J. (1984) "Numerical Estimates of the Response Functions of Microdosimetric Proportional Counters Exposed to Neutrons", Radiat. Prot. Dosim. 9, pp. 199-202.

Morstin, K., Bond, V.P., and Baum, J.W. (1989) "Probabilistic Approach to Obtain Hit-size Effectiveness Functions which Relate Microdosimetry and Radiobiology", Radiat. Res. 120, pp. 383-402.

Olko, P., Schmitz, Th., Morstin, K., Dydejczyk, A., and Booz, J. (1989) "Microdosimetric Distributions for Photons", Radiat. Prot. Dosim. 29, Nos. 1-2, pp. 105-108.

Paretzke, H.G. (1987) "Radiation Track Structure Theory", in: Kinetics of Nonhomogeneous Processes, Freeman, G.R. (Edt.), pp. 89-170.

Pihet, P., Menzel, H.G., Schmidt, R., Beauduin, M., and Wambersie, A. (1990) "Biological Weighting Functions for RBE Specification of Neutron Therapy Beams. Intercomparison of 9 European Centers", presented at the 10th Symposium on Microdosimetry, Rome, 21-26 May 1989, Radiat. Prot. Dosim. 31, Dos. 1-4.

Rossi, H.H. (1968) "Microscopic Energy Distribution in Irradiated Matter" in "Radiation Dosimetry" 2nd Ed., Vol. I, pp. 43-92, Attix, F.H., Roesch, W.C. (Eds.), Academic Press, NY.

Schmitz, Th., Kramer, H.M., and Booz, J. (1989) "Assessment of the Photon Response of a TEPC; Implementation of Operational Quantities for Dose Equivalent", Radiat. Prot. Dosim. 29, Nos. 1-2, pp. 69-73.

Skarsgard, L.D., Kihlam, B.A., Parker, L., Pujara, C.M., and Richardson, S. (1967) "Survival, Chromosome Abnormalities and Recovery in Heavy Ion- and X-Irradiated Mammaliam Cells", Radiat. Res. Suppl. 7, pp. 208-221.

REDUCTION OF PATIENT EXPOSURE IN DIAGNOSTIC RADIOLOGY A TASK OF THE RADIATION PROTECTION RESEARCH PROGRAMME OF THE COMMISSION OF THE EUROPEAN COMMUNITIES

H. SCHIBILLA
Commission of the European Communities
DG XII, Directorate General: Science, Research and Development
Rue de La Loi, 200
B-1049-Bruxelles, Belgium

1. Introduction

Radiation protection of the patient is a special task of the CEC Radiation Protection Research Programme: about 80% of the population exposure are due to the use of ionizing radiation in diagnosis, nuclear medicine and therapy; 90% of the workers, who are in contact with ionizing radiation at their workplace, are in the medical field. There is sufficient reason to make every effort to avoid unnecessary exposure of the patient, the medical staff and the involved public.

These general remarks are underlined by the results of surveys of radiological practices in the various areas of diagnostic radiology: entrance surface doses have been compared in paediatric radiology (Fendel (1989)); for the same examination differences of 1 to 30 or 40 have been found. For the diagnosis of congenital hip displasia the difference even reached 1:186 [0.007 to 1.292 mGy; (Fendel (1989))].

Quality assurance measures, applied to a number of parameters, particularly important for image quality and patient exposure, help to reduce patient exposure in optimizing, at the same time, the quality of diagnostic information. In mammography, for example, technical improvement of equipment and radiographic techniques during the last 10 years - and when used under optimal conditions - reduced entrance surface doses by about 60% [from 18 mGy to 6 mGy, (Bouhnik (1987))]. However, during a trial in 1987 with a series of quality criteria for diagnostic radiographic images, entrance surface doses between 2 and 60 mGy per examination were found; Maccia (1989). Since mammography for breast cancer screening will be introduced more generally the causes of these differences must be studied with a view to establishing the most reasonable dose ranges.

Dose reduction can be achieved even at rather low cost by modifying some specific technical factors or equipment or by reducing the

M. Oberhofer (ed.), Advances in Radiation Protection, 199–208.

number of examinations (Rosenstein (1989)):

e.g.: - maintaining optimum temperature of the photographic developer can reduce dose by upto 20% (Russel et al. (1987));
- using screen-film combinations with high efficiency intensifying screens, such as rare earths, requires upto 50% less dose (Croft et al. (1988));
- using carbon fiber or other low attenuation materials, placed between the patient and image receptor, reduces doses by 10 to 30% (Hufton et al. (1986)).

2. European Community Actions in Radiation Protection of the Patient

Against this background the European Community Radiation Protection Programme concentrates an important part of its efforts on the use of ionizing radiation in medicine. Indeed it was the application in diagnostic radiology, which historically started to stress the necessity for radiation protection. I only recall the foundation of the International Commission on Radiological Protection, which took place at the 2nd International Congress of Radiology in Stockholm, 1928.

Diagnostic radiology requires special attention because in several industrial countries nearly every inhabitant is, on average, examined once a year by X-rays or radionuclides. Diagnostic procedures and new equipment are continuously being developed or improved and their use is indispensable for the benefit of the patient. To give some numbers: 320 million people in the European Community undergo about 200 million radiological examinations every year, involving more than 500 million radiographic films. These diagnostic procedures contribute about 300 to 1500 μSv per year to each member of the general public (Table 1).

TABLE 1. Diagnostic Radiology in the European Community

~ 320 million inhabitants	
~ 200 million radiological examinations per year	
~ 500 million radiographic films per year	
300 to 1500 μSv:	contribution to the exposure of each person of the general public per year in the various Member States
500 μSv :	average value for the European Community

There is a considerable task for radiation protection research to look for ways and means to optimize the protection of the patient. Optimization means in this context: to avoid unnecessary exposure of the patient by optimum use of equipment and techniques, and by proper justification of the radiological examination and procedure.

We have to bear in mind that the EURATOM - Treaty stipulates since 1957: the study of harmful effects of radiation on living organisms by the study of adequate prevention and protection measures and by appropriate safety standards.

In 1961, the first multiannual research and training programme on radiation protection was started, including research on the medical applications of nuclear energy, with emphasis on the development of methods for nuclear medicine. The various research sectors of the CEC Radiation Protection Programme covered subject areas which contributed to the establishment of those data, that were necessary to define the problems due to the medical use of ionizing radiation; these subject areas dealt with dosimetry, studies on dose-effect relationship, radiation carcinogenesis, genetic effects of ionizing radiation, epidemiology studies and evaluation of radiation risks. In addition to these more basic research areas specific actions had to be initiated in order to recognize the full extent of the problems in medical radiology: this took place only 15 years after the start of the 1st programme, when research needs for the radiation protection of the patient were clearly defined, during the *1976-1980* research programme, and gave rise to the 1st Study Group of contractors on *medical exposure*.

At the end of the seventies and the beginning of the eighties (1980-1984 Programme) priority was given:

- to the description of various radiological practices in the Member States;
- to the comparison of the various irradiation conditions and to the study of their improvement and possible standardization.

In this period research on the criteria and parameters for quality assurance of radiological equipment, of the imaging process and of image interpretation got more and more importance.

In order to compare the numerous individual and national initiatives for quality assurance, the CEC organized the 1st International Scientific Seminar on "Criteria and methods for quality assurance in medical X-ray diagnosis" 1984, in Udine (1985). At that moment, the approaches to quality assurance measures were of very different character and of heterogeneous scientific and technical level.

The goals of quality assurance were defined as an optimization of every aspect of radiological practice, or as the establishment and maintenance of good radiological practice. The propagation of good radiological practices was understood in accordance with the ICRP concept, presented in ICRP publication No. 34: adequate image quality, low cost and low exposure of patient and personnel.

The conclusions of the Udine-Seminar emphasized the following:

- harmonization and normalization of radiological procedures;
- promotion of quality assurance measures by:
  - more research on the most important parameters, associated to the various radiological techniques;
  - research on image quality specification taking more into account the clinical requirements;
  - elaboration of simple test schemes;
  - establishment of training programmes;
  - improvement of coordination among the various bodies involved, for example, health care authorities, scientists working on quality assurance measures and radiation protection, manufacturers and the user in the radiological departments; and this, on the regional, national and international level.

The changing priorities during the various programmes indicate the evolution in this area: the growing awareness of what has to be studied, what has to be improved and by which means. What is less evident, is the experience that, especially in this field, research on its own does not achieve the final aim of radiation protection of the patient and specifically the reduction of patient exposure in practical, daily routine work at the X-ray department. Here, the role of the European Community becomes particularly evident: beside the stimulation and support of research work on high priority subjects there is the task to coordinate the individual efforts, to initiate cooperation and to create a common scientific language so that research results can be better compared and evaluated. This will help to agree upon the most pertinent results which could then be used to optimize radiological practices in a similar way everywhere. This important aspect is also emphasized by the regulatory activities of the CEC on radiation protection of the patient (Council Directive (1984)).

The implementation of the requirements of the Directive in the day-to-day radiation protection practice requires various accompanying measures. In particular, it will be useful, if not necessary, to produce some guidelines setting particular provisions of the Directive in a more explicit and practical context.

These guidelines will be based on recent experiences and research results and will certainly need further research efforts. Therefore some of the achievements of the Radiation Protection Research Programme should be highlighted, which contributed to make radiation protection measures in diagnostic radiology more comprehensive, better comparable and hopefully more efficient.

2.1 Highlights of Achievements in Research on Reduction of Patient Exposure

a) Dose assessment

The CEC Radiation Protection Research Programme contributed to the establishment of more coherent methods for the assessment of organ and tissue doses. The comparison of various methods of direct or indirect dose assessment gave preference to the less time consuming determination of the *mean energy imparted*, based on the exposure-area product. It was further agreed to take the *entrance surface dose* as the reference quantity for comparing patient exposure. Methods of dose calculation have been improved by the development of complex inhomogeneous mathematical phantoms and sex and age specific phantoms, including a baby, for a more accurate approach to real patients. Tables of conversion factors have been set up, which allow the calculation of organ dose, using technical parameters of typical radiographic techniques (Drexler et al. (1984)).

b) Assessment of radiological practices

A highlight of achievement of special importance concerns the assessment of radiological practices in the Member States. Hopefully data will be available for all the Member States in the near future. Surveys on radiological practices carried out 15 years ago were done in different ways and it was difficult, if not impossible, to compare their results. Progressively, surveys are done now in a nearly standardized way by selecting similar criteria such as representative numbers and types of examinations, test-instrumentation, methods for dose assessment and so on. The results then indicate more clearly the differences in habits and practices and explain, at least partly, the differences among the obtained values.

Another example is the comparison of the results of surveys carried by contractors in France, North-East Italy and Britain (Contento et al. (1988)). This comparison gave some indications on the possible reasons for the differences in the quantities on patient exposure per examination and for population exposure: GSD and EDE, as we can see in Table 2. Some of the differences listed in (Contento et al. (1988)) are:

Differences in radiological practices:

- different percentages of private practices in general radiology;
- different age distribution of patients examined by x-rays;
- different number of examinations per 1000 inhabitants per year;
- different number of films per examination;
- different technical parameters used for producing the radiographic images (voltage in kV, beam filtration, different use of rare-earth screens, anti-scatter grids, etc.).

TABLE 2. Practices in diagnostic radiology and quantities for population exposure.

CEC Member States	Number of X-ray-examin. per 1000 inhabitants per year (1)	Mean number of film per examination	GSD μSv	EDE μSv (2)	References	Contract Nr. BI6-
Feder. Rep. Germany	1200	2.2	400	1000	F.E. Stieve 1989	-
France	825	3.6	295	1580	C. Maccia et al., 1988	(132)
Great Britain	444	2.3	120	282	B. Wall et al., 1981	(135)
Ireland	460	2.2	164	-	J. Cunningham et al., 1988	
North-East Italy	665	3.1	248	763	G. Contento et al., 1988	(136)
Spain (Madrid region)	680	2.2	252	1210	E. Vano et al., 1988 L. Arranz, 1989	(214)
Portugal	480	2.3	-	-	Galvao et at., 1989	(299)

1) Without dental examinations GSD=Genetically Significant Dose
2) Remainder organs included EDE=Effective Dose Equivalent per caput

These differences do not vary each with the same trend in the three countries, but might - at least partly - be responsible for the differences stated for GSD and EDE. An example to illustrate this is the high number of new-borns and infants examined for hip dislocation in France. This young age group influenced the calculation of GSD since this quantity is related to the potential child expectancy of the population under consideration.

c) Optimization of image quality and patient exposure

The research and regulatory activities of the CEC concerning radiation protection of the patient have stimulated a greater awareness of the possibilities for optimizing radiological procedures and thus to avoid unnecessary patient exposure while improving image quality at the same time. Consequently, it was indicated to look more closely into the practical implementation of this awareness in daily routine work of radiological departments. This means also to evaluate the data obtained so far. Special attention was focussed on two subjects areas:

c.1) Quality Control

The first subject area concerned the appropriate quality control of technical and physical parameters for constantly high performance of the radiological equipment and of image processing. A Workshop was organized in Brussels in 1988 (Brussels (1989)) which was attended by scientists working on radiation protection and medical physics, radiologists and radiographers, manufacturers of radiographic equipment and representatives of public health authorities, regulating bodies and competent international organizations.
These participants at the Workshop discussed the requirements to measurement methods, instrumentation and techniques for the quality control of those technical and physical parameters which have the greatest impact on image quality and patient exposure. For consistent clinical information and reasonably low exposure to the patient it would be necessary to agree upon the limiting values of the measured parameters as well as upon the tolerances of the measurement methods. Similar limiting values of a set of parameters and similar tolerances of measurement methods would mean a certain standardization of quality control techniques. It became clear that before arriving at standardized quality control conditions a number of intercomparison actions should be carried out, for example on dosemeters or on phantoms used for quality control.

c.2) Specification of image quality

The second subject area concerned quality criteria for diagnostic

radiographic images. To this topic the CEC would like to give a real European dimension. If we succeed to define commonly acceptable ranges for image quality and patient exposure then we will be able to produce comparable normal clinical radiographs. Barry Wall called it the Euro X-ray.
Indeed, in the light of the preparation of the large European Internal Market 1993, where every citizen will be able to move freely, the need to continue medical diagnostic examinations should be met everywhere under similar conditions. The Euro X-ray would be the contribution from the side of Diagnostic Radiology to this project.
A CEC - Study Group for "Quality Criteria of Diagnostic Radiographic Images" has drawn up a list of quality criteria for radiographs of six of the most frequent conventional examinations: chest, skull, lumbar spine, pelvis, urinary tract and breast. These quality criteria include:

- diagnostic requirements expressed by image criteria, which have been defined by radiologists;
- example of good radiographic technique, defined mainly by those who are in charge of quality assurance and quality control;
- guidelines for good imaging performance, including important image details and entrance surface doses for a standard-sized patient, defined together by radiologists, medical physicists and those in charge of radiation protection.

These quality criteria are a contribution to an objective and quantifiable description of image quality. The study of the relationship between the quality criteria and patient exposure will help to define reasonable dose ranges for the respective radiological examination and techniques.
The first list of quality criteria was checked in a trial carried out on about 1000 radiographs in 10 European Countries. The results of this trial and further approaches to the optimization and quantification of image quality were discussed at a Workshop in Oxford 1988 (Oxford (1989)).
As a consequence, the list of quality criteria was revised. The new version has been published as an Appendix to the Proceedings of the Workshop. It has been translated into the 9 official languages of the European Community and is being sent for promulgation to the various professional societies, organisations and institutions.

2.2. Future Aims And Perspectives

Radiologists, radiographers, medical physicists and other professionals engaged in quality assurance and radiation protection of the patient will be invited to contribute their own experiences to those quality criteria, so that finally quality criteria might be defined that can be generally accepted.

Research priorities have been established for the 1990-1991 period and recently selected research projects will deal with:

1. Quality criteria for special examinations and techniques: special attention will be given to those diagnostic procedures where high sensitivities of individual tissues are involved as in paediatric radiology, where large numbers of persons are examined, as by mammography, those applications of new techniques which are rapidly increasing, such as computed tomography, or those which are newly introduced such as digital radiography.
2. Impact of quality criteria on patient exposure: possible standardization of quality criteria for radiographic images and consequences for exposure of patients and medical staff.
3. Dose assessment for special irradiation conditions taking into account the variations of parameters of equipment performance, imaging processes, image interpretation and patient characteristics.
4. Evaluation of dose reduction measures and establishment of knowledge data bases and expert systems.
5. Evaluation of data for the quantification of individual risks with a view to selecting the most efficacious procedures.

3. Conclusions

In conclusion, I want to emphasize the particular role of the Commission of the European Communities for strengthening the cooperation among research workers:

Radiation protection of the patient requires more than any other sector of the Radiation Protection Programme, cooperation among various scientific disciplines such as medicine, radiology, physics, mathematics, engineering, radiation protection and health care as well as regulations. This multidisciplinary approach fits precisely the CEC goal of creating a *Europe of research workers* and could achieve a better integration of diagnostic requirements, radiation protection principles and technological feasibility.

4. References

Arranz, L., (1989) personal communication.

Bouhnik, H., et al. (1987) Report EUR 11059, "Radiation Protection-37, p. 311-341.

Brussels (1989) "Technical and Physical Parameters for Quality Assurance in Medical Diagnostic Radiology: Tolerances, Limiting Values and Appropriate Measuring Methods", Brussels, 23-25 February 1988, Proceedings edited by B.M. Moores, F.E. Stieve, H. Eriskat, H. Schibilla, British Inst. of Radiol. - Report No. 18.

Contento, G., Malisan, M.R., Padovani, R., Maccia, C., Wall, B.F., Shrimpton, P. (1988) "A comparison of Diagnostic Radiology Practice and Patient Exposure in Britain, France and Italy", Br. J. Radiol. 61: 143-152.

Council Directive laying down basic measures for the radiation protection of persons undergoing medical examination or treatment, 3 September 1984, 84 1466 EURATOM, O.J. L 265, 5.10.1984, p. 1.

Croft, J.R. and Hudson A.P. (1988) "Optimisation in medicine with illustrative exercise", National Radiological Protection Board of the United Kingdom, Didcot.

Cunningham, J.D., Howett, D., Hone, C. (1988) "Survey of Diagnostic Radiology in the Republic of Ireland", the Nuclear Energy Board, Dublin.

Drexler, G., Panzer, W., Widenmann, L., Williams, M. Zankl (1984) "The calculation of dose from external photon exposures using reference human phantoms and Monte Carlo methods. Part III: Organ doses in X-ray diagnosis", GSF-Report S-1026, ISSN 0721-1694.

Fendel, H., et al. (1989) "Specific principles for optimisation of image quality and patient exposure in paediatric diagnostic imaging", BIR-Report 20 (British Institute of Radiology).

Maccia, C., Benedettini, M., Lefaure, C. (1988) "Doses to the Patient from Diagnostic Radiology in France", Health Physics, 54(4): 397-408.

Maccia, C., Moores, B.M., Nahrstedt, U., Padovani, R., Wall, B.F. (1990) "CEC Quality criteria for diagnostic radiographic images and patient exposure trial", Report EUR 12952.

Oxford (1989) "Optimization of Image Quality and Patient Exposure in Diagnostic Radiology", Oxford, 27-29 September 1988, Proceedings edited by B.M. Moores, B.F. Wall, H. Eriskat, H. Schibilla, Br. Inst. Radiol. - Report No. 20.

Rosenstein, M. et al. (1989) "Reducing unnecessary radiation dose to patients in diagnostic radiology - inexpensively and easily applicable technical means", revised draft ICRP/88/C3-15.

Russel, J.G.B. and Hudson, A.P. (1987) "Developer temperature in diagnostic departments. British Journal of Radiology. 60: 782.

Stieve, F.E. (1989) Personal Communication.

Vano, E., Gonzales, L. (1989) "Progress Report 1989 of the Radiation Protection Programme of the CEC", in press.

Wall, B.F., Rae, S., Darby, S.C., Kendall, G.M. (1981) "A Reappraisal of the Genetic Consequences of Diagnostic Radiology in Britain", Br. J. Radiol. 54: 719-730.

QUALITY CONTROL IN DIAGNOSTIC RADIOLOGY

B. M. MOORES
Integrated Radiological Services Ltd. (IRS)
Radiation Protection Service
42, Rodney Str.
Liverpool - United Kingdom

1. Introduction

This article provides practical guidance on the implementation and operation of a routine quality assurance programme for those personnel concerned with the daily provision of diagnostic radiology services.

The tests presented here are in the authors'opinion those which could most usefully be performed on a routine basis by staff actually working within a department of diagnostic radiology.

Tests such as focal spot size measurements, exposure time, etc., which have been omitted are deemed to be status tests. These tests are undertaken upon installation of a new item of equipment or when undertaking detailed investigation of possible faults. They may or may not be performed by staff working in an X-ray department, depending upon local circumstances.

1.1. What is meant by quality assurance?

The World Health Organization (WHO) has defined quality assurance in medical X-ray diagnosis as:

> "... an organized effort by the staff operating a facility, to ensure that the diagnostic images produced by the facility are of sufficiently high quality so that they consistently provide adequate diagnostic information at the lowest possible cost and with the least possible exposure of the patient to radiation".

Assuming that this definition is comprehensive, then quality assurance is concerned with the practical interplay of three related aspects of radiological practice, namely, costs, risks and benefits.

Quality assurance affects every aspect of radiological practice and may be viewed as the most effective operational framework for the

M. Oberhofer (ed.), Advances in Radiation Protection, 209–236.

application of cost-risk-benefit principles. These principles are founded upon the application of scientific initiatives, since, without detailed knowledge of performance of equipment (and staff), it is impossible to meet the aims implicit in the WHO definition of quality assurance.

1.2. Is quality assurance necessary?

The need for QA in radiological practice is clearly indicated by considering:

a) The nature and extent of radiological practice.
b) The magnitude of radiation exposure involved.
c) The cost of providing radiological services.

1.2.1. The nature and extent of radiological practice. At present there are approximately 50 different types of X-ray examinations undertaken with varying frequency, and each type of examination has a particular set of radiographic requirements in terms of kV, mAs, field size, etc. There are approximately 22 million examinations undertaken annually within the UK, 180 million in the USA and possibly 150 million in Europe. A variety of geographical, sociological and economic factors influence the patterns of radiology not only within individual countries but also between countries.

Many different types of radiological equipment are available from numerous manufacturers. Their methods of specifying the performance of any item of equipment vary enormously; however, these are not the only variables. There may be more than 300.000 radiographers and 30.000 radiologists world wide, and methods of training vary within and between countries. Consequently professional judgement and the standard of image quality requested, as well as the visual acuity of individual radiologists, can all vary enormously.

1.2.2. The magnitude of radiation dose. Measurements of radiation doses resulting from radiological examinations indicate that the mean entrance dose may vary by more than a factor of 1000 over the whole range of examinations. Variations of 20-100 have also been recorded for particular examinations. Based on published data, wide variations in the genetically significant dose have been observed at both national and international levels. Medical exposure is now the largest single man-made source of radiation and accounts for approximately 20% of the total radiation dose received by the general public.

1.2.3. The cost of providing radiological services. The running costs for provision of radiological services within the UK is probably in the region of £ 270 million per annum, including salaries, service/maintenance and films/chemicals. The capital costs associated with provision of these services is in the region of £ 300 million for the UK. In both North America and Europe these costs are in the region of ten times the UK costs.

Quality assurance has a direct bearing on all of the above three factors and, in view of their scope and magnitude, indicates the necessity of applying QA routinely to radiological practice.

1.3. The framework of a QA programme

The components which need to be considered in any QA programme and their interrelationship are indicated in Figure 1. Aspects of these are now discussed individually.

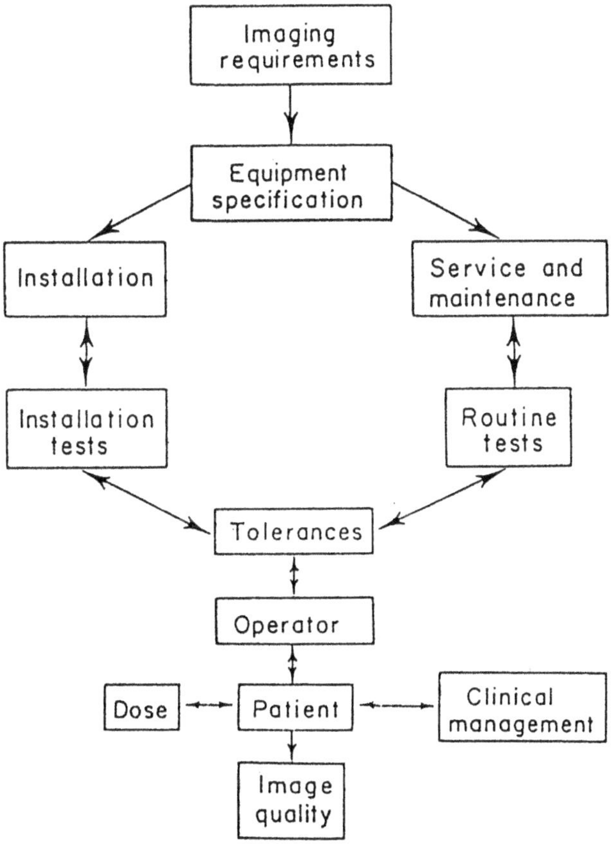

Figure 1. Basic components and operational framework for a QA programme.

1.3.1. Specification of performance. Parameters need to be identified which describe the performance of each item of X-ray equipment. These parameters should relate either directly or indirectly to the quality of the diagnostic image and should be associated with the appropriate level of patient dose. Obviously, measurement of these parameters must be possible on equipment which is already installed in the X-ray department.

In order to define standards of acceptable performance, the value of each parameter needs to be specified in terms of both the acceptable level and allowable tolerances. These may need to differ with the age and type of equipment under consideration.

1.3.2. Choice of equipment. To ensure that the desired clinical objectives can be met, it is essential that the appropriate equipment is acquired in the first instance. This will be achieved if the required standard of performance, in terms of relevant parameters, is specified in detail beforehand by the user. These specifications should all be made before the purchase and must form part of the contract which the equipment supplier is required to meet: e.g. for tubes and generators; the accuracy of kV, linearity of X-ray output with changing mA and exposure time. This should avoid dissatisfaction with equipment that is purchased and subsequently fails to meet expected performance standards, simply because those standards had not been clearly identified in the first place.

1.3.3. Acceptance/status tests. As soon as new equipment is installed, and before clinical use, it is essential that detailed status checks of all relevant parameters are carried out in order to:

a) Ensure that the standard of performance meets the conditions specified in the contract.
b) Establish the baseline level of performance against which the extent of any future deterioration can be assessed.

1.3.4. Routine/constancy tests. Once equipment has been accepted, it is necessary to be assured that its performance continues at an acceptable level consistent with the aims of QA. To ensure that this occurs it is obviously necessary to carry out regular routine checks which are relatively quick and easy to perform. These will indicate either that the performance continues to be satisfactory or that a degree of deterioration has occurred which requires further attention. It should be noted that, in some circumstances, status tests and constancy tests may be identical.

1.3.5. Equipment maintenance. Analysis of the results of the routine checks carried out by an X-ray department over a long period will indicate the degree of consistency of equipment performance. This information will be of value when decisions' are made about equipment replacement. Analysis may also indicate the effectiveness of routine maintenance and the extent to which it is required. Experience with weekly tests of image intensifier performance in a number of departments has revealed instances where a significant deterioration in image quality occurred as a consequence of maintenance action. This would seem to suggest that there could be some advantage in reducing the extent of maintenance on equipment which is performing satisfactorily. Decisions of this nature can only be taken sensibly in the light of routine checks.

Obviously some degree of equipment maintenance is essential, particularly in the interests of safety, but, in view of the expense that this entails, it is important that the health care industry gets value for money. It could prove cost-effective to explore other ways by which appropriate maintenance can be provided.

1.3.6. Equipment replacement. There comes a time when, despite all the efforts of the service engineer, the performance of any item of equipment can no longer be maintained at a standard necessary to achieve the desired clinical objective at an acceptable level of patient dose. It is important, therefore, that minimum acceptable levels are defined for those parameters which are used to describe equipment performance. When measurements lie below these values, equipment may need to be replaced in the interests of image quality, patient dose or the effective use of resources.

1.3.7. Optimum use of equipment and resources. In addition to checking that the equipment is functioning satisfactorily, it is also necessary to check that equipment and resources are used with optimum efficiency. This can be achieved through:

a) *Film reject analysis* - which indicates wastage of manpower and resources, as well as identifying the possible causes.
b) *Patient dose survey* - which gives an indication of the extent to which efficient use is being made of radiation in diagnostic imaging. (In this respect, patient dose is analogous to petrol consumption as a measure of efficient car performance).

Both these surveys are required in order to assess the degree of effectiveness of action taken as a result of the QA programme.

2. The quality assurance programme

2.1. *Implementing a QA programme*

No two X-ray departments are ever the same and the way in which a QA programme is organized will be governed by the conditions which prevail in the department concerned. Irrespective of local conditions, any programme must incorporate the following aims:

a) To ensure that diagnostic images of optimum quality are always produced to a standard compatible with the desired clinical objective.
b) To ensure that the doses to both patients and staff are always kept to a minimum compatible with the type of clinical examination undertaken.
c) To maximize the departmental cost-effectiveness by minimizing wasted effort and resources.
d) To meet those requirements of any legislation for radiation protection which specifically apply to the medical applications of radiation.

In order to maximize the effectiveness of the QA programme and ensure that these aims can most easily be achieved, priority should be given to the formation of a departmental QA committee.

2.2. *Who is involved in a QA programme?*

2.2.1. The departmental QA Committee. Initially it is suggested that the committee should consist of:

- Radiologist(s).
- Superintendent radiographer(s).
- QA radiographer(s).
- Physicist(s).

The QA radiographer is intended to be the person(s) to whom the responsibility has been delegated for coordinating the departmental QA team who actually carry out the QA tests and procedures. The purpose of the committee is:

a) To manage and direct the QA programme.
b) To decide what needs to be done, when, how and by whom.
c) To coordinate the QA procedures and data collection.
d) To review and evaluate the data produced as a result of QA procedures.
e) To decide what action is required in the light of these results with regard to:
   (i) management and use of equipment;
   (ii) organization and management of staff and clinical procedures;
   (iii) future QA activities.

f) To consider the relevance of the results of any local and/or national analysis of QA data provided by appropriate organizations.

2.2.2. The staff of the X-ray department. The cooperation of all the staff in the X-ray department is essential for a QA programme to have any chance of success. Prior to initiating such a programme, it is necessary to discuss and explain in detail to everyone the aims of the programme and the part which they will be expected to play.

2.2.3. X-ray equipment manufacturers/service engineers. In circumstances where tests have shown that the performance of equipment may be less than satisfactory, it is inevitable that the manufacturers will be involved. It is in their interests that the equipment performs satisfactorily. Their support and cooperation in the QA programme should be both expected and welcomed.

2.2.4. Others involved. If hospital works or engineering departments are involved in equipment management and/or maintenance within the X-ray department, they may also need to be included in the QA scheme.

If, as is hoped, QA action results in financial savings, the redistribution of any such savings back into the X-ray department may need to be discussed with the hospital manager at an early stage. There is a distinct possibility that savings derived as a direct result of implementing QA may otherwise be transferred solely for the benefit of other departments. In this case a major incentive for continuation of the programme has been jeopardized.

2.3. Setting up the QA programme

Having established the QA committee and discussed the introduction of a QA programme with the staff of the X-ray department, the next priorities are:

a) Establish the existing conditions within the department by conducting film reject analysis and patient dose surveys.
b) After the surveys have been completed, undertake status tests in order to establish the existing standards of performance of the equipment within the department and record the findings. Equipment which may need to be considered includes:
   - Automatic processors.
   - Films, screens and cassettes.
   - Darkroom and film storage conditions.
   - Film viewing conditions.
   - Tubes and generators.
   - Image intensifier systems.

- Tomography systems.
- Automatic exposure control systems.

On the basis of the results of a) and b) above, identify deficiencies in equipment performance and its use, which significantly affect image quality, patient dose and wastage of resources. Decide what course of action is required to correct the deficiency and, after corrective action has been taken, repeat the status tests to ensure that a satisfactory standard has now been achieved. It is envisaged that throughout the initial stages of setting up the QA programme (which may take up to six months in each department), guidance from staff trained in QA procedures will be required.

2.4. Continuing the QA programme

Establish a programme of routine/constancy checks to be undertaken by the radiographic team on those items of equipment that require frequent monitoring - e.g. automatic processors (daily); film stock/storage (weekly); tubes and generators (weekly); image intensifiers (fortnightly).

Establish a programme for those items of equipment that require less frequent monitoring - e.g. alignment and collimation of X-ray tubes (quarterly); tomography systems (six-monthly); screens/cassettes, viewing conditions, automatic exposure control systems (all annually).

If, as a result of QA tests, abnormal performance is suspected and if appropriate:

a) Arrange for further checks to be carried out, using more sophisticated test equipment.
b) Increase the frequency of the checks.
c) Once the abnormality has been identified and confirmed, decide what corrective action is necessary.
d) After corrective action has been taken, repeat the checks.

Abnormal performance may also be suspected during normal clinical use. In either case, the routine tests should be repeated in order to confirm or allay suspicions. In addition, tests should always be performed immediately after equipment maintenance has been carried out and preferably before the departure of the service engineer.

Records of the results of routine tests must always be kept:

a) For immediate information and future reference.
b) For inclusion in any local/national analysis.
c) As a possible influence on long-term management and resource utilization.

2.5. Effectiveness of the programme

Assess the effectiveness of the QA programme at regular intervals by repeating film reject analysis and patient dose surveys. Institute regular meetings of the QA committee to review and discuss the results of the programme and, if appropriate, modify the type of tests being carried out and/or their frequency.

Once the programme is properly established and records have been accumulated over a sufficient period of time (e.g. a minimum of twelve months), it would be beneficial for the QA committee to evaluate those results in detail, with a view to discussions with manufacturers' representatives concerning the cost-effectiveness of service and maintenance. As stated in the introduction, it may be that the results have a direct bearing on the effectiveness of routine maintenance as well as on the frequency with which it is carried out.

Test equipment employed in the QA programme will itself need to be tested and, if necessary, calibrated in order to ensure consistent and accurate performance. This applies especially to instrumentation such as X-ray monitors, sensitometers and densitometers. Provision should be made at the start of the QA programme to facilitate this requirement. In some instances baseline calibration can be provided by the suppliers of test equipment, and suspected changes in performance can then easily be verified by subsequent tests. In other instances, some form of baseline performance test may need to be developed within the department when equipment is received.

In all cases careful monitoring of the results produced during a QA programme can play an important role in monitoring the performance of test equipment. Similarly, establishing simple cross-calibration facilities with other groups employing similar equipment can play a part in ensuring consistent performance of test equipment. All test equipment should be inspected for damage or obvious faults as well as tested for consistency, at least annually.

3. Reject Analysis

3.1. Aims

- To quantify the amount of waste film as a percentage of the total film used in the department - reject rate(s) (see Appendix, section 1).
- To establish the baseline for a QA programme and thereafter monitor the effectiveness of that programme.

- To provide a means of early detection of changes in reject rates and possible identification of causes.

3.2. Apparatus

- Bins/boxes in which to collect waste films.
- Reject and data charts (see Appendix, section 7).
- Departmental records to ascertain appropriate workload information (see Appendix, section 2).
- Access to exposure counters for each X-ray tube (if possible).
- Facilities for computer analysis and appropriate software (essential for Method 2).

3.3. Method

The information required from reject analysis, as well as the duration and frequency of the analysis, will vary according to the needs of the department concerned, and indeed those needs will vary from time to time. For this reason, a selection of relatively simple techniques will be described under Method 1, with stages ranging from a very simple numerical count of waste film (which might usefully be implemented as a permanent exercise); progressing to analysis by room, analysis by reason and analysis by room and reason. These techniques, because of their simplicity, provide only limited information and it is essential that this limitation is appreciated.

A fully comprehensive analysis is described in Method 2 but, because of the complexity of the calculations, it is suggested that computer facilities will be required; a microcomputer would be sufficient. The comprehensive analysis may be considered as analogous to the acceptance testing of X-ray equipment and is recommended as a baseline for QA programmes. Such a baseline provides a means of monitoring changes in reject rates and ultimately will be extremely useful in determining the cost-effectiveness of the QA programme.

Normally it should not be necessary to apply this analysis to repeat films (see Appendix, sections 1 and 6), unless it is suspected that they are in sufficient numbers to warrant investigation. Should this be the case, then allowances must be made when designing the reject/repeat charts and collecting data (see Appendix, section 6).

3.3.1. Method 1 - Stage 1 (Simple numerical count).

- Provide bins-boxes at appropriate sites throughout the department: e.g. adjacent to the processor and in the darkroom.
- Discuss the project with all staff, including the radiologists and darkroom technicians, before implementation, and stress the impor-

tance of using only the designated bins. Decide upon the duration of the project.

- Determine whether the radiologists are in the habit of discarding any radiographs during reporting sessions and, if so, provide a suitably labelled bin for this purpose in the reporting area.
- Collect, count and record the contents of the bins at regular intervals (e.g. daily or weekly).
- Ascertain from departmental records the total number of radiographs put through for reporting each day. If such records are not already available, provision must be made to rectify the situation before implementing the project. If step 3 applies, remember to adjust the figures obtained from the departmental records before final calculations are made.
- This simple analysis can usefully be expanded to provide further information by sorting the waste film into broad-based, self-evident categories (e.g. anatomical areas, blank films, test film). It is unwise to attempt to classify rejected radiographs into actual X-ray examinations merely by viewing because, in the absence of adequate information, in some cases this might lead to erroneous conclusions.
- Collect all data together, calculated reject rates from the formulae shown (see Appendix, section 5) and record the results.

3.3.2. Method 1 - Stage 2 (Analysis by room or location).

- Provide enough bins/boxes so that there is one for each X-ray room, darkroom, theatre, etc. Each bin should be clearly labelled and placed in a suitable location.
- It is extremely important to discuss the project with all staff, including radiologists and darkroom technicians. Stress that discarded films must be placed in the correct bin and not just the nearest one.
- Determine whether the radiologists are in the habit of discarding any radiographs during reporting sessions and, if so, designate a suitably labelled bin for the reporting area.
- If possible, note the exposure counter readings for each X-ray tube before implementing the project, then again each time the collecting bins are emptied for counting. Remember to allow for, and record, any exposures used for tube warm-up, test procedures, breakdowns and service visits. If this is not feasible, an alternative means of determining the number of films used in each room will need to be found (see Appendix, sections 2 and 3).
- Collect, count and record the contents of each reject bin at regular intervals (e.g. daily or weekly).
- Ascertain from the departmental records the total number of radiographs put through for reporting each day. If such records are not already available, provision must be made to rectify this situation

before starting the analysis programme. If step 3 applies, remember to adjust the figures before final calculations are made.

- This analysis can usefully be expanded to provide further information by sorting the rejects from each room or location into broad-based, self-evident categories (e.g. anatomical areas, blank film, test film). It is unwise to attempt to classify reject radiographs into actual X-ray examinations merely by viewing because, in the absence of adequate information, in some cases this might lead to erroneous conclusions.
- Collect all the data together and calculate reject rates from the formulae shown in the Appendix, section 5).

3.3.3. Method 1 - Stage 3 (Analysis by reason).

- Determine the categories of 'reason' to be used in the assessment of rejected film (see Appendix, sections 7 and 7.1, Tables 2 and 4).
- Provide sufficient bins/boxes so that there is one for each of the chosen categories of 'reason'; label each clearly. The bins should be sited in an appropriate place (e.g. the viewing area). If more than one darkroom/viewing area exists in the department, a complete set of bins will be required for each.
- It is very important to discuss the project fully with all staff, including radiologists and darkroom technicians, before implementation. Stress the importance of discarding rejects into the correct bin and not just the nearest one!
- Determine whether the radiologists are in the habit of discarding any radiographs during reporting sessions and, if so, include a suitably labelled bin for this purpose in the reporting area(s).
- Collect, count and record the contents of the bins at regular intervals (e.g. daily or weekly).
- Ascertain from departmental records the total number of films put through for reporting each day. If such records are not already available, provision must be made to rectify this situation before beginning the programme. If step 4 applies, remember to adjust the figures before final calculations are made.
- This analysis can be usefully expanded to provide more information by sorting the contents of each reject bin into broad-based, self-evident categories (e.g anatomical areas). It is unwise to try to classify radiographs into X-ray examinations merely by viewing because, in the absence of adequate information, in some cases this might lead to erroneous conclusions.
- Collect all data together and calculate the reject rates from the formulae given (see Appendix, section 5).

3.3.4. Method 1 - Stage 4 (Analysis by room/location and reason

- This is a combination of stage 2 and 3 and, as such, requires considerably more time and effort than any of the previous stage.
- Determine what categories of 'reason' are to be used for assessment of rejected films (see Appendix, section 7 and 7.1, Table 2 and Figure 2).
- Provide sufficient bins/boxes for the chosen categories of 'reason' and label each clearly. Position the bins in an appropriate place (e.g. in the viewing area). If the department is large enough to have more than one darkroom/viewing area, ensure that each has a complete set of bins.
- Devise a colour or number coding system using small adhesive labels, so that each X-ray room or location has a colour/number (do not forget mobiles and theatres). Put the labels and associated codes in a convenient place near to each set of reject bins.
- It may be felt that the use of adhesive labels is an unnecessary expense; the alternative is to use a felt-tip or Chinagraph pencil. In this case it is essential that staff remember to annotate every film they discard with the room or location it came from and that they do so legibly.
- Discuss the project with all staff, including radiologists and darkroom technicians, before implementation. Explain what is expected them and stress the importance of ensuring not only that they use the correct 'reason' bin when discarding film but also that they use the correct adhesive label (or write clearly) to indicate its origin.
- Determine whether the radiologists are in the habit of discarding radiographs during reporting sessions and, if so, provide a suitably labelled bin for this purpose in the reporting area.
- If possible, note the exposure counter readings for each X-ray tube before implementing the project, then again each time the reject bins are emptied for counting. Remember to allow for and record any exposures for test procedures, breakdowns and service visits. if this is not feasible, an alternative means of determining the number of films used in each room will need to be found (see Appendix 3, sections 2 and 3).
- Collect, count and record the contents of each reject bin at regular intervals (e.g. daily or weekly). Sort the contents of each 'reason' bin into room/location and record the findings before moving on to the next.
- The analysis can be expanded to provide further information by sorting the rejects for each reason and room/location into broad-based, self-evident categories of anatomical area. It is unwise to try to classify radiographs into actual X-ray examinations merely by viewing

because, in the absence of adequate information, in some cases this might lead to erroneous conclusions.

- Ascertain from departmental records the total number of radiographs put through for reporting each day. If such records are not already available, provision must be made to rectify the situation before implementing the programme. If step 7 applies, remember to adjust the figures before final calculations are made.
- Collect all data together and calculate the reject rates from the formulae given (see Appendix, section 5).

3.3.5. Method 2 - Stage 1 (Pilot study for the full analysis)

- Unless the department is a small one (i.e. less than four rooms), it is advisable to conduct the pilot study in just one area of the department. Discuss the project in detail with the staff involved in that area, explaining the objectives and what is required of them. Study the sample reject sheets and code charts (see Appendix, section 7). Determine the duration of the pilot study (e.g. four weeks).
- Following the discussion, design reject sheets appropriate to the department or area concerned and the type of work done in that area (see for example, Figure 2).
- Provide an appropriately labelled bin/box in which to collect the rejected films and place it in a suitable location (e.g. adjacent to the processor).
- Each of the radiographers involved is given a reject sheet to be completed throughout the day. Encourage them to discuss and make notes on how the sheets might be improved for the main study that is to be conducted later.
- It will be necessary for the radiographers to put their names on their reject sheets for the very practical reason that there will probably be queries concerning the information presented and/or problems with legibility. Once the sheets have been collected and checked for queries, it is strongly recommended that all means of identification are removed immediately. Reject analysis should be conducted in order to establish causes of waste film and not as a means of assessing the performance of an individual.
- Collect and count the rejected films at regular intervals (e.g. daily or weekly) and record the results. Make note of any films discarded by radiologists and service engineers and also the 'cleaning films' used for the processor.
- Collect and examine the radiographers' reject sheets at regular intervals as in step 6.
- Deal with any queries that might arise from the reject sheets as soon as possible after collection, then remove any means of identification before coding the information (see Appendix, section 7). The radio-

graphers involved should be included in this part of the procedure, open discussion with exchange of ideas and information concerning the project will frequently result in improved motivation of staff. This motivation and the good-will created in the pilot study will go a long way towards the success of the full scale survey to be conducted later.

- Obtain from the departmental records all relevant data for the area involved in the project. If such records are not already available, provision must be made to rectify the situation before the project is implemented.
- Prepare a suitable computer program for analysis of all data. A micro-computer will be quite adequate for this purpose.
- Analyse the data and use these results as a guide to modify or improve the technique to be used in the full scale survey.

3.3.6. Method 2 - Stage 2 (Full survey)

- Study the inevitable errors/omissions incurred during the pilot study and utilize this information when designing new reject sheets appropriate to the whole department.
- Determine the duration of the survey, making sure that it is conducted for long enough to provide sufficient information for constructive data analysis (three to six months). Future full surveys may not need to be quite so lengthy as they are likely to be comparative rather than baseline surveys.
- Organize a departmental meeting to discuss the findings of the pilot study and the objectives of the forthcoming survey. include the radiologists and darkroom technicians in the meeting and possibly the office staff. Encourage the radiographers involved in the pilot study to discuss their experiences; the trust and good-will established with them at that time should help to promote the full survey to the rest of the staff.
- Provide enough bins/boxes in which to collect the rejects and place these in suitable locations throughout the department (e.g. viewing areas).
- Determine whether the radiologists are in the habit of discarding any radiographs during reporting sessions and, if so, provide a suitably labelled bin in the reporting area(s).
- Provide each radiographer with reject sheets to be completed throughout the day.
- If possible, note the exposure counter reading for each X-ray tube before implementing the survey, then again each time the reject bins are emptied for counting. Remember to allow for, and record, any exposure used for test procedures, breakdowns and service visits. If this is not feasible, it is essential that an alternative means of

determining the number of films used in each room/location is found (see Appendix, sections 2 and 3).

- Collect, count and record the contents of the reject bins at regular intervals (e.g. daily or weekly).
- Ensure that departmental records provide all the necessary information that will be required for the survey and, if not, be sure to rectify this situation before implementing the survey. If step 5 applies, remember to adjust the figures before final calculations are made.
- Collect and examine the reject sheets at regular intervals as in step 8. Deal with any queries that might arise as soon as possible after collection, then remove any means of identification before coding.
- Prepare a computer program for analysis of all data.
- Analyse the results and use them as the baseline for future surveys of reject rates (see Appendix, section 5).

3.4. Implementation

3.4.1. Frequency

- A full reject analysis should be implemented to provide a baseline for the QA programme before any other aspect of the programme is introduced.
- A full survey of six to eight weeks' duration should be undertaken at least annually.
- A simple numerical count might usefully be established as standard procedure at all times.
- More detailed analysis (if not the full survey) should be undertaken whenever there is a change in radiographic procedure/equipment.

3.4.2. Data record keeping

- Information required from department records:
  - . Total number of radiographs put through for reporting each day.
  - . Number of radiographs per patient examination.
  - . Number and types of examinations performed each day.
  - . Which room the examination was performed in.
  - . Number of films used each day for 'other' reason (e.g. duplication, QA tests).
- Information required from reject sheets to include:
  - . Radiographer's name (for initial reference only).
  - . Number of rejects.
  - . Examination type.
  - . View.
  - . Room number/location.

. Reason for rejection.
. Any other useful information.
- Exposure counter readings for each X-ray tube (if possible/practical). If this is not feasible then it is essential that departmental records provide the information as to where each radiograph was carried out and exactly how many films (including rejects and repeats) were used per room/location.
- Weekly and overall reject rates.
- Reject rates for:
  a) Each room/location.
  b) Each reason.
  c) Each X-ray examination type.
  d) All permutations of a), b) and c).

3.4.3. Tolerances. These should be determined by the individual department.

3.4.4. Action
- Seek to keep reject rates as low as can reasonably be expected.
- Any sudden changes (increase) in reject rates would indicate a problem that needs to be located and corrected.
- Be alert for trends that may be developing slowly through careless techniques, such as poor radiographic positioning.

4. Patient Dose Measurements

4.1. Aims

- To measure the entrance and exit doses on patients during the course of the X-ray examination.
- To investigate the causes of relatively high doses or wide ranges of doses for particular examinations in:
  a) Different X-ray rooms.
  b) Different X-ray departments (where possible).
  c) Between different types of X-ray equipment and/or techniques.
- To monitor any measures which have been taken in order to consistently maintain at a minimum, compatible with the clinical objectives of the X-ray examinations being performed.

4.2. Apparatus

A thermoluminescent dosimetry (TLD) system which uses tissue equivalent dosemeters in small sachets (e.g. LiF, $Li_2B_4O_7$), similar to those used for extremity monitoring of staff. The quantity and nature of the TLD material should be such that it does not produce a discernible image on the imaging device (e.g. radiograph) and is capable of measuring doses of at least 50 μSv.

4.3. Method

- It is probably not feasible to include all examinations; therefore, select the type of radiographic examination(s) to be investigated. For example:
  a) *Chest* - because of the large contribution to somatic dose to the population arising from the high frequency rate of this examination.
  b) *AP lumbar spine* - because of the large contribution to somatic and genetic doses to the population arising from the high exposure involved.
- It is probably appropriate to investigate doses received during screening examinations (e.g. barium meals/enemas) but there is no single position suitable for entrance dose measurements, owing to the positioning techniques involved. It will be necessary, therefore, to use three TLD sachets, one located on the skin surface of the patient's left side and one located on the skin surface of the right side at the same level as the first; the third sachet must be positioned in the midline (either anteriorly or posteriorly, whichever is appropriate), at the same level as the other two.
- For each type of examination to be investigated, specify the exact site(s) that the TLD(s) is (are) to be positioned on the patient's skin surface.
- Record the following information on each occasion that a dose measurement is made:
  . Patient size: small/average/large.
  . Male/female.
  . Weight.
  . kV, mAs, focus-film distance (FFD).
  . X-ray room/machine.
  . Automatic exposure control: yes/no.

  In the case of screening examinations, it is also necessary to record screening kV, mA and exposure time, name of the clinician performing the examination and whether the equipment is an over/undercouch intensifier system.
- Measurement of patient exit doses may also be undertaken in order to provide useful information concerning the doses delivered to the imaging device. This will not be possible, however, unless the sensitivity of the TLD is such that doses of at least 1 μSv can be mea-

sured. Furthermore, particular care is necessary in choosing the anatomical site for such measurements.

- Having measured the doses recorded by TLDs, analyse the results in order to establish the mean entrance doses and the overall spread in these doses, for each type of examination investigated.
- Compare these values with:
  a) Those normally to be expected.
  b) Those obtained in other similar X-ray departments.
  c) Those for different clinicians (screening examinations).
- Examine the values in relation to the details recorded concerning the patient and exposure conditions. Investigate further any wide discrepancies between the values recorded for different X-ray machines.

4.4. Implementation

4.4.1. Frequency

- Carry out measurements annually on the most frequently performed X-ray examinations.
- If any major changes in processing, image receptor or radiographic technique are implemented, undertake additional patient dose measurements.

4.4.2. Data record keeping

- Record the mean entrance dose resulting from any given examination.
- Record the variation in entrance dose resulting from any given examination.
- Record all relevant information (see 4.3, item 4).
- For fluoroscopic examinations, record the appropriate screening factors and the name of the radiologist undertaking the examination.

4.4.3. Tolerance. The need for (and possible levels of) dose limits for individual examinations has yet to be officially recognized by appropriate international bodies.

4.4.4. Action.

- Determine the variations in patient dose arising from particular investigations and compare them with any available data.
- Monitor variations by repeating the survey at regular intervals - at least annually.
- Extend the survey to include other types of X-ray examinations.

APPENDIX

1. Definitions

1.1. Reject film

A reject film is one that has been deemed useless and to be thrown away. It may be a rejected radiograph, a clear film or even an unprocessed film.

1.2. Repeat film

A repeat film is a radiograph which has been repeated purely to provide improved information, e.g. by altering exposure factors or improving patient positioning. It is *not* an additional view. This second radiograph is put through for reporting together with the original.

1.3. Retake

This is the term used to accommodate both reject and repeat films.

2. Departmental records

2.1. In order that the maximum amount of information can be collected together in one place for ease of access, it is suggested that the departmental patient ledger provides the ideal centralized information point. In some departments the required information may already be recorded as standard procedure; other departments may need to make some improvements in the amount of information collected and recorded.

2.2. The radiographers will be required to provide the following information on *every* request card once they have completed the X-ray examination(s):

a) Number of films used for each examination (excluding rejects).
b) Radiographer's name or initials.
c) Room number/department.
d) Exactly what examinations have been carried out if different from, or in addition to, that which was originally requested.

Example 1

'Pelvis' requested. The request card should be annotated as follows: 1 film). Rm 2, Main. J. Smith.

Example 2

'Lumbo-sacral spine and pelvis's requested. In this case, after viewing the processed radiographs it is found that hip and femur views will also be required in order to provide adequate information for clinical diagnosis. The request card requires to be amended and 'hip and femur' added as well as the following:
L. spine - 2 (films), sacral spine - 2, (films), pelvis - 1, hip - 2, femur - 1, Rm 2, Main. J. Smith.

2.3. The information provided by the radiographer should be entered in the departmental patient ledger (along with all the other information normally recorded) so that it can be collated each week to provide the following data:

- Total number of examinations performed.
- Total number of films used.
- Number of examinations of each type (e.g. chests).
- Number of examinations of each type in each room (e.g. chests in room 2, Main).
- Number of films for each examination type (e.g. total number of films used for chest radiography).
- Number of films per examination type per room (e.g. number of films used for chest examinations in room 2, Main).

2.4. Remember to make provision for recording those films used in the department which are not going through for reporting (QA test films, any duplicating films, etc.). These can be added at the end of each day as 'miscellaneous'.

3. Exposure counters for X-ray tubes

3.1. Some X-ray units provide easy access to the exposure counters and some do not. Consideration must be given to whether or not counter reading is practical and **safe.**

3.2. If it is decided that this is a practicable exercise, these figures will provide a valuable back-up to the data obtained from the departmental ledger.

3.3. When comparing counter readings with ledger figures for the number of films taken in a given room, remember that the counter readings will include exposures for tube warm-up and rejected films as well as any QA tests performed; they will also include any exposure made by service engineers. Whenever possible, remember to take readings before and after a service/breakdown visit.

3.4. Check that exposure counters record *only* when an exposure is made and not on an aborted 'prep'.

4. Reject sheets

4.1. Some departments find it advantageous to have a senior radiographer on duty to check each film as it comes out of the processor(s). If this is the case, he/she could also take responsibility for completing the reject sheets at the time of checking the films; however, it is essential that all the information required is ascertained from the person who conducted the X-ray examination.

4.2. Sample reject sheets and code sheets are given at the end of this appendix (Tables 1 - 4 and Figure 2).

5. Reject rates

5.1. Glossary

Let: A = total number of rejected films.

B = total number of 'good' films, i.e. radiographs put through for reporting, plus any others under 'miscellaneous'.

A_l = number of rejected films for any given room/location.

A_e = number of rejected films for any given examination/category.

A_r = number of rejected films for any given reason.

A_{le} = number of rejected films for any given room/location, for any given examination.

A_{lr} = number of rejected films for any given room/location, for any given reason.

A_{re} = number of rejected films for any given reason, for any given examination/category.

B_l = number of 'good' films used in any given room/location.

B_e = number of 'good' films used for any given examination.

B_{el} = number of 'good' films used for any given examination, in any given room/location.

etc.

5.1.1. Overall reject rate as a percentage of the total number of films used:

$$\frac{A}{A + B} \times 100\%$$

5.1.2. Reject rate for any given examination/category as a percentage of the total films used:

$$\frac{A_e}{A + B} \times 100\%$$

5.1.3. Reject rate for any given examination/category as a percentage of the total number of rejects:

$$\frac{A_e}{A} \times 100\%$$

5.1.4. Reject rate for a given room/location as a percentage of the total films used:

$$\frac{A_1}{A + B} \times 100\%$$

5.1.5. Reject rate for a given room/location as a percentage of the number of films used in that room:

$$\frac{A_1}{B_1 + A_1} \times 100\%$$

5.1.6. Reject rate for a given examination in a given room:

$$\frac{A_{e1}}{B_{e1} + A_{e1}} \times 100\%$$

5.1.7. Reject rate for a given examination, for a given reason, in a given room:

$$\frac{A_{er1}}{B_{er1} + A_{er1}} \times 100\%$$

etc.

Note: Once information is available for each room/location it becomes a simple matter to compile the reject rates for the whole department by adding together the relevant totals for each room. For example:
Total number of rejected films (A) = A_1 (1) + A_1 (2) + A_1 (3), total number of 'good' films for a given examination $B_e = B_{e1} + B_{e1}$ (2) + B_{e1} (3).

$$B_e = B_{e1}\ (1) + B_{e1}\ (2) + B_{e1}\ (3)$$

6. Repeat films

6.1. If it is decided that the numbers of repeat films in the department are sufficiently high to warrant investigation, the method of doing this is exactly the same as for rejects and should be carried out simultaneously as a 'retake analysis'.

6.2. Include two extra columns in the reject (retake) sheet, one labelled 'reject' and the other 'repeat"; the sheets can then be completed in exactly the same way as for the reject analysis.

6.3. Remember that 'repeats' are put through for reporting and therefore the total number of repeats (as calculated from the retake sheets) will need to be deducted from the total number of films put through for reporting (this figure should be available from the patient ledger).

6.4. Formulae for repeats are worked out in the same way as for rejects.

7. Sample reject sheets and sample code charts

7.1. Terminology

7.1.1. The terminology used in the sample charts has been chosen deliberately! It is more common to use the terms 'over-exposed' and 'under-exposed' when referring to the blackening of a radiograph; however, since this is frequently interpreted by the radiographer to mean that he/she has selected the wrong exposure factors, the terms 'too dark' and 'too light' have been used. In practice, the dialled exposure factors may be correct but the output/processing conditions/screen speeds are perhaps unsatisfactory.

7.1.2. The term 'good' film has been used to distinguish between those radiographs that were taken correctly first time and 'repeats'.

Table 1. Reject Analysis: Sample X-ray Examination Code Chart

Code	Examination
01	Chest.
02	Ribs.
03	Sternum.
04	Shoulder (including scapula, clavicle).
05	Humerus (including elbow).
06	Lower arm (including radius, ulna, wrist, hands).
07	Skull (including base).
08	Sinuses.
09	Jaw/Facial bones (*NB: OPG = 'tomo' view code).
10	STVs/PNS (including brass filter views for larynx).
11	Other views (including mastoids, petrous bones, pituitary fossa, etc.
12	Cervical spine.
13	Thoracic spine.
14	Lumbar spine.
15	Sacral spine.
16	Pelvis/Hip.
17	Femur.
18	Knee.
19	Lower leg (including tibia, fibula, ankle, foot.
20	Abdomen.
21	IVU.
22	Gall bladder (all types).
23	Cystograms.
24	Larynx (remember 'tomo' view code).
25	Special procedures (e.g. angiograms, embolizations, etc.).
26	Screening procedures (as done by radiologist - enter overcouch films under appropriate anatomical site).
27	Special procedures.
28	Ultrasound (all).
29	Isotopes (all).
30	CT scanning (all).
31	Miscellaneous (e.g. QA, duplicating films, subtraction films).
32	Blank film (including 'cleaners' for processor).
33	Other (not listed above).

List and allocate a code number for each X-ray examination performed in the department. Remember that *every* sheet of film will need to be accounted for, including 'cleaners' for the processor, duplicating film and QA test film as well as test film used by the service engineers.
It is likely that most examinations will fall into anatomical areas but some will need to be classified under the technique used. Depending on clinical demand, in some cases it may be more convenient to group some of the less frequently performed examinations which are anatomically related (e.g. code 11 above). Any omissions or modifications that need to be made should quickly reveal themselves during the pilot study.

Table 2. Reject Analysis - Sample Code Chart

Examination	View	Code	Room	Code	Reason	Code	Auto. exp. device used	Week no.
Code 01 to 33	PA	1	Main/Rm 1	1:1	Too dark	1	No = 1	Codes
(see exam	AP	2	Main/Rm 2	1:2	Too light	2	Yes = 2	01 to 52
code chart,	Lat.	3	Main/Rm 3	1:3	Positioning	3		
Table 1)	Obl.	4	Main/Rm 4	1:4	Movement	4		
			Main/Dk Rm	1:5	Other	5		
	Submento-vert.	5			Equipment (mechanical)	6		
	Tomos	6	OP/Rm 1	2:1	Processor (mechanical)	7		
	Scans	7	OP/Rm 2	2:2				
	Other	0	OP/Rm 3	2:3	Blank film	8		
			OP/Dk Rm	2:4	Test film	9		
			Theatre	3:1				
			Mobile 1	3:2				
			Mobile 2	3:3				

For computer purposes, it will be necessary to allocate code numbers not only for the X-ray examinations but also for 'view', 'department/room', 'reason' and 'date' (i.e. week number). Depending upon the number of categories under each column (see above), remember to allow for one or two digit spaces per column.

Table 3. Reject Analysis - Coded Version of Reject Sheet shown in Figure 2

Exam	View	Room	Reason	AED	Week no.
01	2	1:3	4	2	04
01	6	1:4	3	1	04
16	2	1:1	3	2	04
26	4	1:2	5	2	04
26	2	1:2	8	2	04
20	2	1:3	7	1	04

Utilizing the examination codes from Tables 1 and 2, translate each reject sheet into the appropriate coding sequence The illustration above demonstrates the coded version of the reject sheet example shown in Figure 2.

Table 4. Correlation of Information from Patient Ledger

Code	Examination	No. exams done	No. good films taken
01	Chest	50	120
02	Ribs	3	4
03	Sternum	0	0
04	Shoulder	0	0
05	Humerus	0	0
06	Lower arm	75	160
etc.			

Produce data sheets for each room; these can then be completed each week from the patient ledger. A standard form should be drawn up which can be used for all rooms.

Name: ______________________ Date: ______________

Dept : ______________________

Examination	View	Room	Date	Too dark	Too light	Positioning	Movement	Other	Auto. Exp. device used	Mechanical error only: Equipment	Mechanical error only: Processor	Comments	Leave blank — Wk. code: 04
CHEST	AP	3	27TH				✓		✓			patient breathed.	✓
CHEST	Tomo	4	"			✓			X				✓
PELVIS	AP	1	28th			✓ (arrow from Other)		✓	✓			patient's fingers over hip joint	X positioning error
SCREENING	OBL	2	29TH					✓	✓			double exposure	✓
	AP	2	"					✓	✓			blank film	✓
ABDO.	AP	3	"						X		✓	film stuck in rollers	✓

Figure 2. Sample reject sheet to be completed by radiographer(s). Design a suitable reject sheet so that it provides as much useful information as possible, while remaining simple and yet quick to complete. It is suggested that a column for comments is included as well as one for the assessor's notes. In this example, the pelvis was wrongly allocated to 'Other', the radiographer having decided that fingers overlying the joint constituted an artefact. Provision of the 'Comments' column showed up this error, allowing the assessor to discuss the case with the radiographer and point out that, in fact, lack of attention in positioning technique was the reason. The coding was then adjusted accordingly.

DOSE ASSESSMENT IN DIAGNOSTIC RADIOLOGY

B.F. WALL
National Radiological Protection Board
Chilton, Didcot, Oxon OX11 ORQ
United Kingdom

1. Introduction

Radiation dosimetry (Greening, (1985))in diagnostic radiology differs from occupational and environmental dosimetry in that the exposures are deliberate and are intended to be in some way beneficial to the recipient and hence:

(i) The radiation field conditions are usually well defined.

(ii) The objective of the dose measurements is to assess the level of harm associated with the exposure, not to check compliance with statutory dose limits.

The objective requires knowledge of the doses received by radiosensitive organs within the patient, and the well defined nature of the exposure conditions makes this knowledge more easily attainable than in most situation encountered in occupational and environmental dosimetry.

There is no need to resort to the crude assumptions made in operational dosimetry where people are reduced to homogeneous spheres and effective dose equivalent is approximated by "ambient", "directional" or "individual" dose equivalents.

Exposures in diagnostic radiology are typified by "Source Quantities" that are usually indicated on the equipment, by "Field Quantities", some of which can be measured with suitable instruments and by "Patient Dose Quantities" which are mostly incapable of direct measurement but which are the quantities that we really want to know (Figure 1).

The quality and quantity of the radiation emerging from an X-ray tube is a function of the tube current, exposure time, applied potential and filtration which give rise to a radiation field propagating through air that can be specified in terms of the photon or energy fluence. The photons interact with air to produce measurable values of ex

M. Oberhofer (ed.), Advances in Radiation Protection, 237–263.

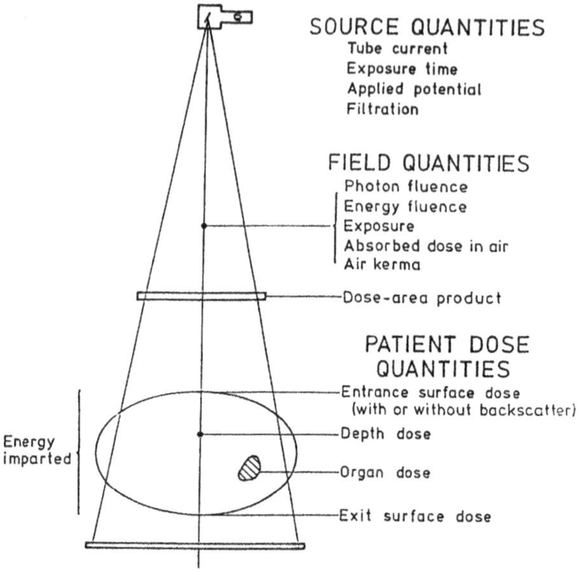

Figure 1. Source, field and patient dose quantities.

posure, absorbed dose in air or air kerma at any point in the X-ray beam. The interaction of the X-ray beam with the patient produces a distribution of absorbed dose within the body but the only directly measurable patient doses are generally those on the surface where the beam enters and leaves the patient. The doses of ultimate interest in radiological protection are those to the radiosensitive organs within the body.

Organ doses cannot usually be measured directly because of the obvious difficulties of introducing dosemeters into appropriate parts of the patient. They can be calculated however by taking appropriate models of the patient's anatomy and the incident radiation field and allowing for the amount of attenuation and scattering between the volume of interest and a point, frequently on the patient's skin, where a direct dose measurement can be made. Models of the patient can vary in complexity from a tank of water to one derived from computed topography (CT) scans of an individual patient.

2. Dose Quantities in Diagnostic Radiology and Methods of Measurement

2.1. Field Quantities

2.1.1. Exposure, Absorbed Dose in Air and Air Kerma. The radiation emitted by an X-ray tube can be quantified in terms of the radiated energy fluence, Ψ, but is most conveniently measured by observing the amount of charge produced by ionisation in a volume of air situated in the X-ray beam. The quantity exposure X is the charge produced per unit mass of air, dQ/dm, and is related to the energy fluence by the mass energy absorption coefficient for air $(\mu_{en}/\rho)_{air}$, the mean energy expended per ion pair formed W_{air} and the electronic charge e.

$$X = \frac{dQ}{dm} = \Psi(\mu_{en}/\rho)_{air}\,\frac{e}{W_{air}}$$

Cavity ionization chambers used to measure exposure rely on the establishment of electronic equilibrium in the walls surrounding the cavity, whose thickness must consequently equal or exceed the maximum electron range and introduce insignificant attenuation of the primary photons. This condition is easily met in diagnostic radiology since the secondary electron range is always less than 0.2 mm in unit density, low atomic number materials, and is about 1000 times lower than the mean free path of photons of the same energy.

Absorbed dose is defined by ICRU (ICRU (1980)) as the mean energy imparted ($d\bar{\varepsilon}$) by ionising radiation to material of mass dm ie. $D = d\bar{\varepsilon}/dm$.

[The mean (or expectation) value of the energy imparted ($\bar{\varepsilon}$) is specified to emphasise that absorbed dose is a non-stochastic quantity, whereas energy imparted ($\bar{\varepsilon}$) is a stochastic quantity whose values vary discontinuously in space and time.]

The mean energy imparted to air by electrons from a mass dm is:

$$\frac{dQ}{e} \cdot W_{air} = X\,dm\,\frac{W_{air}}{e}$$

but the energy is not deposited in the identical mass of air from which the secondary electrons originated.

However if electronic equilibrium is established the energy deposited in each mass dm of air is the same, so that the energy imparted per unit mass will be $X \cdot W_{air}/e$.

Hence, absorbed dose to air, $D_{air} = X\,W_{air}/e$ (under conditions of electronic equilibrium).

W_{air}/e is a constant for electrons in air, with a value of 33.85 J/C, except at very low energies which are of little consequence in diagnostic radiology (ICRU (1979)). Consequently the numerical relationship between exposure measured in the SI units of coulombs per kilogram and absorbed dose to air measured in gray (1 Gy = 1 J/kg) is 33.85 Gy/C kg^{-1}. The approximate 1:1 relationship between exposure expressed in the old unit, the roentgen (R), and absorbed dose expressed in rad, no longer applies. Since the relationship between ionisation chamber measurements made in free air and the absorbed dose to patient is frequently of interest it would be preferable for the ionisation chambers to be calibrated in terms of absorbed dose to air rather than exposure so that the large factor of 33.85 is avoided.

Historically national standards for photon dosimetry below energies of 2 MeV have been established in terms of the quantity exposure expressed in roentgens. However, in 1982 the British Committee on Radiation Units and Measurements (BCRU (1982)) proposed that the quantity "air kerma" rather then "absorbed dose to air" should replace exposure for measurements in air. Kerma, which is an acronym for Kinetic Energy Released per unit Mass, relates to the first stage of the process that occurs when indirectly ionising particles like photons impart energy to matter, ie. the transfer of energy from the primary photons to the kinetic energy of secondary electrons. In common with exposure, kerma relates conceptually to initial interactions with infinitesimal isolated masses of material, unlike absorbed dose which is influenced by secondary electrons coming from a finite volume of material surrounding the point of interest. This is why air kerma is currently preferred by national standards laboratories, since very accurate absorbed dose measurements would require tightly specified experimental conditions and the application of a number of small correction factors, if charged particle equilibrium could not be assured. However, charged particle equilibrium is easily achieved with the relatively low energy photons used in diagnostic radiology and in this situation absorbed dose and kerma are equal when both are expressed in grays. (Strictly it should be added that this is so only if bremsstrahlung production is negligible, which it is for diagnostic quality X-rays in air).

Dosemeters used in diagnostic radiology, whose calibration is traceable to the national primary standard of air kerma can consequently be regarded as measuring absorbed dose to air. To express doses to patients in terms of kerma rather than absorbed dose would only serve to increase the confusion that already exists with the multiplicity of old and new quantities in current use.

Suitable ionization chambers and electrometers for measuring X-ray tube output are available from a number of manufacturers. Some examples are shown in Figure 2.

2.1.2. Dose-area product. The dose rate across the X-ray beam in a direction perpendicular to the central axis is frequently non-uniform due primarily to the heel-effect in the anode of the X-ray tube. Large area ionisation chambers are available which can be mounted on the X-ray tube diaphragm housing to intercept the entire X-ray beam and integrate

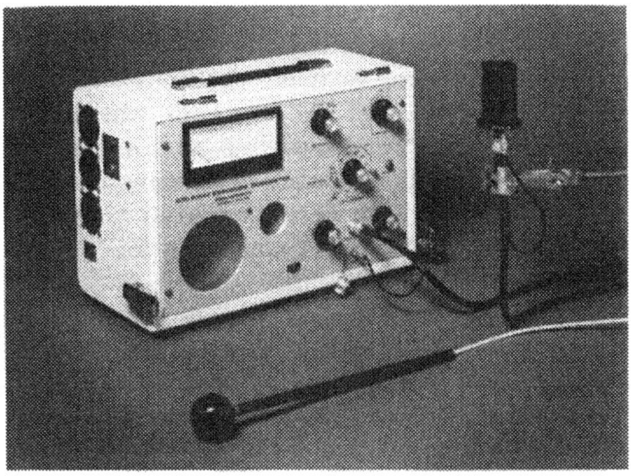

The Vinten (formerly Pitman Instruments) 37D electrometer with their cylindrical 35 cm^3 ionization chamber (held in clamp) and the Extradin (Exacting Radiation Instrumentation Warrenville, Illinois) spherical 30 cm^3 ionization chamber (lying on table) (photo. courtesy of NRPB).

The Radcal Corporation (Monrovia, California) model 2025 radiation monitor, with digital signalling between electrometer and readout. Two ionization chambers connected by short stems to the electrometers are packed in the case together with the display unit.

Figure 2. Examples of dosemeters used in diagnostic radiology.

the absorbed dose over the whole beam area. Their response ($\int D_{air}$ da) is independent of the distance from the tube focus. They can consequently be positioned anywhere between the patient and the X-ray tube and provide useful information on the size of the beam as well as the intensity of the radiation. By providing an instantaneous readout of the total dose-area product associated with any examination procedure, in a manner which involves no interference with the patient or the procedure, these devices are particularly useful for training radiologists to adopt dose saving techniques.

They provide a single measurement of the total amount of radiation used in even the most complex X-ray procedures involving radiography and fluoroscopy and although it is difficult to relate $\int D_{air}$ da to individual organ doses, it can be used to obtain an estimate of the total energy imparted to the patient (see 2.2.6.).

Large area transmission ionization chambers that are used to measure the exposure-area product also require careful calibration. The Diamento-M manufactured by PTW-Freiburg, West Germany is an example of this type of chamber and is shown in Figure 3 overleaf. Sensitivity factors provided by the manufacturer frequently represent a compromise between the various ways that the instrument can be mounted on an X-ray imaging system. For example attenuation of the X-ray beam in the table top for undercouch X-ray tubes alters the required calibration compared with overcouch mounted tubes, and an average value may be quoted by the manufacturer which does not apply accurately to either situation. The possible effects of scattered radiation and non uniformity in the dose rate across the X-ray beam should be appreciated and an independent calibration of the ionisation chamber and its electrometer and display unit after installation on the X-ray imaging system being monitored is essential. A useful evaluation of the Diamentor exposure-area product meter including suitable calibration procedures is reported by Shrimpton and Wall (Shrimpton and Wall (1982)).

2.2. Patient Dose Quantities

2.2.1. Entrance Absorbed Dose in Free Air (frequently abbreviated t *Entrance Dose*). This is absorbed dose to air at the point where the ra diation beam axis would first intercept the surface of the patient. I is measured in the absence of the patient i.e. does **not** include scat tered radiation from the patient.

Unit: gray   1 Gy = 1 J/kg

2.2.2. Entrance Surface Absorbed Dose with Backscatter (frequently ab breviated to *Entrance Surface Dose* or *Entrance Skin Dose*). This is ab

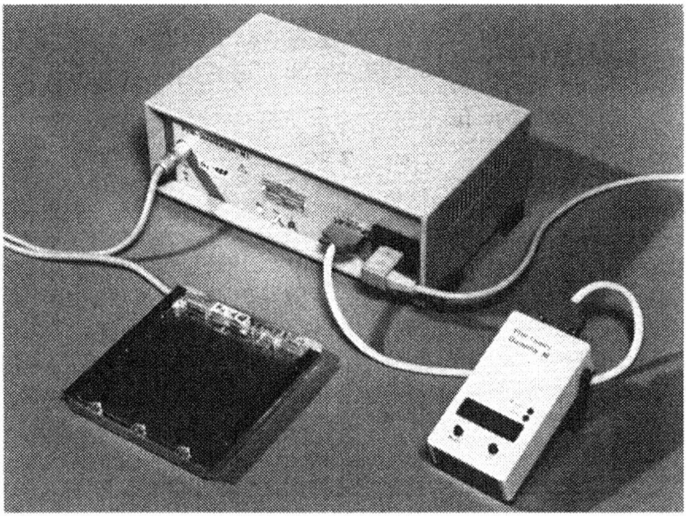

(a) The central unit (top) is connected by a long cable to the transparent large area ionization chamber (left) and by a shorter cable to the display unit (right).

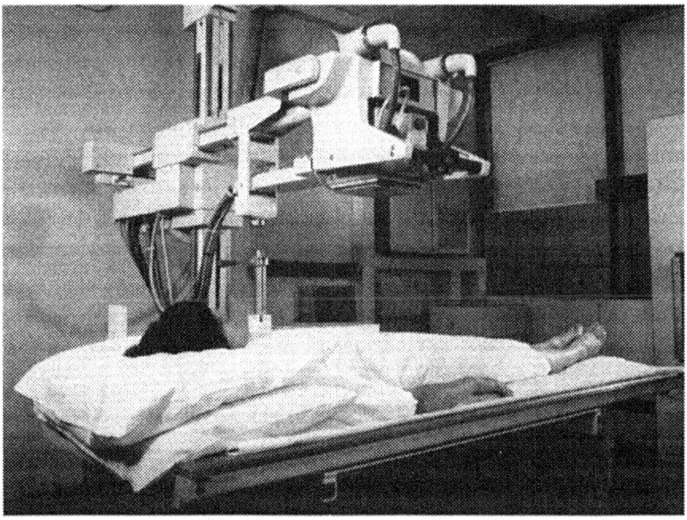

(b) The ionization chamber is unobtrusively attached to the light beam diaphragm housing (photo. courtesy of NRPB).

Figure 3. The Diamentor-M exposure-area product meter (manufactured by PTW, Freiburg).

sorbed dose at the point where the radiation beam axis first intercepts the surface of the patient. It is measured on the surface of the patient i.e. including scattered radiation from the patient.

Unit: gray   1 Gy = 1 J/kg

Absorbed dose is a property of the absorbing medium as well as the radiation field and the exact composition of the medium should be clearly stated when absorbed dose measurements are quoted. For measurements on the surface of a patient the absorbing medium is commonly taken to be soft tissue or water. Many different compositions for soft tissue have been used in the past but that of striated muscle given by ICRU (ICRU (1970)) has the advantage of being precisely defined and is closely matched by water in its X-ray absorption and scattering properties.

Absorbed dose to muscle is related to absorbed dose to air by the ratio of the mass energy absorption coefficients of muscle and air, i.e.:

$$D_{muscle} = D_{air} \frac{(\mu_{en}/\rho)_{muscle}}{(\mu_{en}/\rho)_{air}}$$

While mass energy absorption coefficients depend on photon energy, average values weighted for typical X-ray spectra do not vary appreciably from one spectrum to another. As a consequence the ratio of the mass energy absorption coefficients of muscle to air is 1.05 for a spectrum generated at 50 kVp with 2.0 mm Al total filtration and 1.07 for a spectrum generated at 140 kVp with 4.0 mm Al filtration. Absorbed dose to muscle can consequently be assumed to differ from absorbed dose to air by a factor of 1.06 for all typical diagnostic X-ray spectra with an error of no more than ± 1%.

The entrance surface absorbed dose to muscle ($D_{surface}$) will be related to be absorbed dose to air measured in free air by:

$$D_{surface} = D_{air} \frac{(\mu_{en}/\rho)_{muscle}}{(\mu_{en}/\rho)_{air}} (1/FSD)^2 \cdot BSF$$

where D_{air} is the absorbed dose to air at distance 1 from the tube focus, FSD is the focus to surface distance and BSF is the backscatter factor.

The word "surface" is preferred to "skin" since the medium in which the dose is measured is unlikely to have the exact composition of skin (which is slightly different from other soft tissues) and it avoids confusion with the mean dose to the entire skin which is sometimes given as an "organ" dose.

The backscatter factor (BSF) depends primarily on the radiation quality and the beam area and ranges from about 1.2 to 1.45 for the conditions normally used in diagnostic radiology. It is therefore important to state clearly whether backscatter has been included or not, in any statement of patient entrance surface doses.

Thermoluminescence dosemeters (TLDs) can be conveniently used to measure the entrance skin dose directly in patients under routine clinical conditions as shown in Figure 4. Such in vivo measurements provide a realistic assessment of the dose received under everyday practical conditions taking into account the skill and experience of the radiographer, the quality and suitability of the imaging equipment and differences in shape, size and pathology of the patients.

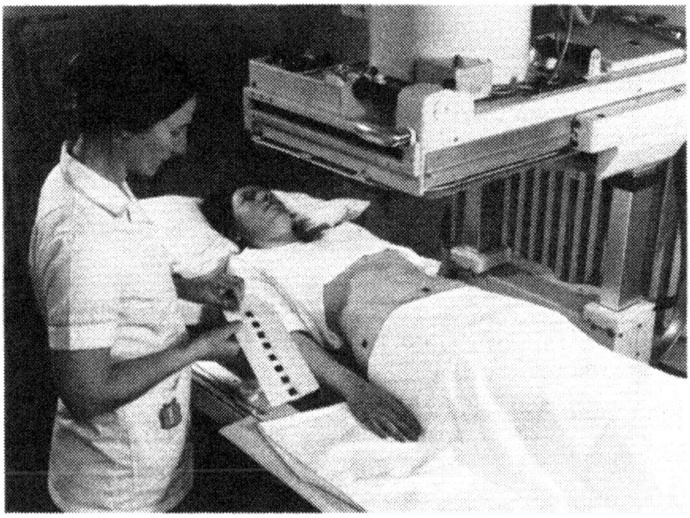

Figure 4. Application of TLDs to measure entrance skin dose.

The ideal TLD phosphor for use in diagnostic radiology would possess the following characteristics:

(i) Be available in the appropriate physical form;

(ii) have a relative photon energy response that closely follows that of soft tissue from 10 to 150 keV;

(iii) exhibit negligible fading of the trapped dose information over periods of several weeks;

(iv) accurate to within ± 10% from 10 μGy (1 mrad) to 1 Gy (100 rad).

None of the commonly available phosphors meet all of these requirements but there are four which go a long way towards satisfying most of them. Each one could be the material of choice, depending on the relative importance of the difference characteristics for a particular application.

Energy response curves for the four phosphors based on the ratio of the mass energy absorption coefficients of the phosphor to that of ICRU muscle (normalized to unity at 1 MeV) are shown in Figure 5 below and their main characteristics are summarized in Table 1.

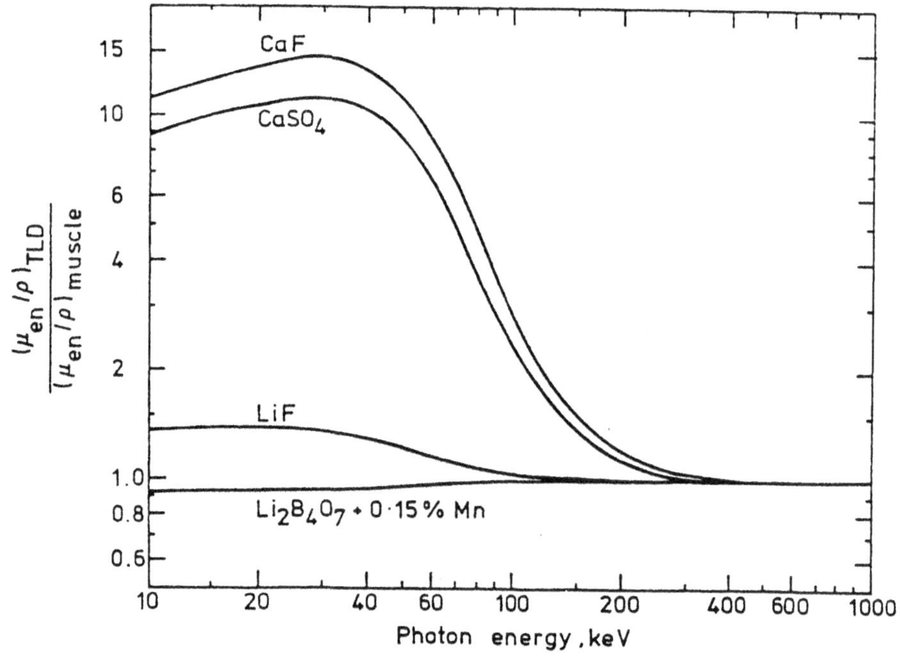

Figure 5. Energy response curves for CaF_2, $CaSO_4$, LiF and $Li_2B_4O_7$.

Lithium borate

Lithium borate ($Li_2B_4O_7$) is the most tissue-equivalent phosphor available and when doped with 0.15% by weight manganese has a relative photon energy response that differs by less than ± 5% from that of ICRU muscle, for photon energies between 20 keV and 1 MeV. There will be some fall in response below 20 keV in addition to that shown in the previous figure due to self-attenuation of low energy photons in the dosemeter, the extent of this fall-off being dependent on the physical form of the dosemeter. However this will be a significant problem only in mammography since X-ray spectra emitted by tubes with filtration

TABLE 1. Phosphor main characteristics.

Phosphor Dopant	$Li_2B_4O_7$ Mn	$Li_2B_4O_7$ Cu	LiF Mg, Ti	$CaSO_4$ Dy	CaF_2 Dy
Rel. energy response (30 keV/1 MeV)	0.95	0.85	1.35	11.0	14.5
Fading(*)	5% m^{-1}	5% m^{-1}	5% y^{-1}	1% m^{-1}	10% d^{-1}
Rel. sensitivity to 1 MeV photons	0.5	2	1	15	13
Detection (μGy) Threshold (mrad)	100 10	10 1	50 5	1 0.1	1 0.1

(*) Fading: % loss at 20°C per year (y), month (m) or day (d).

equivalent to 2.5 mm Al or more that are used in all other areas of radiography, have only a very small component below 20 keV.

Despite a high intrinsic thermoluminescence efficiency $Li_2B_4O_7$: Mn has the drawback of an emission spectrum that peaks at 600 nm (orange) which is well away from 400 nm (blue), the wavelength of maximum quantum efficiency for the photomultipliers commonly used in TLD readers. Consequently the measured light output per unit dose for $Li_2B_4O_7$: Mn pellets (available from Alnor Instrument AB, S61182 Nykoping, Sweden) is about 50% of that from comparable LiF dosemeters. However background noise from unexposed dosemeters can be made sufficiently low and consistent for reliable dose measurements down to 10^{-4} Gy (10 mrad).

$Li_2B_4O_7$: Mn possesses a relatively simple glow curve with a rapidly fading low temperature peak at around 80°C and a stable high temperature dosimetry peak at about 220°C. With suitable discrimination or a pre-heat to exclude the low temperature peak from the readout, fading can be reduced to less than 5% in one month even under ultraviolet irradiation. The simplicity of the glow curve obviates the need for a complicated annealing regime since there appear to be no deep traps from which electrons are transferred during the readout cycle to effect the subsequent response, as is the case with lithium fluoride. After low doses the readout cycle itself is sufficient to empty all the dosimetry traps and return lithium borate to its original background and sensitivity values. A short (15-30 min) anneal at 300°C is prudent following doses greater than a few tens of mGy.

More attempts have been made to shift the thermoluminescence of lithium borate towards the blue end of the spectrum by changing the dopant. Copper has proved to be the most successful alternative to manganese and lithium borate powder doped with 0.02% by weight of copper has been developed by Takenaga et al. (Takenaga et al. (1983)) with a peak spectral emission at 368 nm. This material is up to eight times more sensitive than $Li_2B_4O_7$: Mn depending on the spectral response of the reader (Wall et al. (1982)) and is consequently capable of measuring doses down to 2.19^{-5} Gy (2 mrad). Unfortunately it needs to be kept in the dark to avoid rapid fading and is not yet commercially available in a solid form. If they do become available, pellets or chips of this material would be very attractive for medical dosimetry.

Lithium fluoride

The previous table and figure show that lithium fluoride is less tissue equivalent than lithium borate with a relative energy response that exceeds that of muscle by about 35% at 30 keV. However, for the broad X-ray spectra experienced in diagnostic radiology, the response of this phosphor will probably only change by a few percent as the tube potential, filtration or the scattering conditions vary between the limits normally met in practice.

The most common lithium fluoride phosphor is doped with 300 ppm magnesium and 10-20 ppm titanium. It has an emission spectrum that peaks at 400 nm which conveniently coincides with the maximum quantum efficiency of most photomultipliers. However the glow curve structure is rather complicated with at least six peaks between room temperature and 300°C. Consequently although the measured light output per unit dose is higher that for lithium borate, a complicated and time consuming annealing regime is required to reduce unwanted luminescence caused by electron transfer between the peaks during readout. A 400°C anneal for 1 hour followed by an 80°C anneal for 16-24 hour is a standard procedure leading to reproducible sensitivity and suppression of the low temperature peaks. Reliable dosimetry down to doses of 5.10^{-5} Gy (5 mrad) should be possible if the correct procedures are adopted. Since the low temperature peaks fade rapidly, the pre-irradiation anneal helps to reduce fading and if dosemeters are given a post-irradiation pre-heat at around 140°C (just prior to readout at 260°C) the effect of the loss of stored dose information can be eliminated.

LiF: Mg,Ti is the most thoroughly investigated and the most frequently used thermoluminescence phosphor and is available in many physical forms. Square hot pressed chips are probably the most popular since they are convenient to handle, sufficiently robust and with appropriate selection and treatment possess adequately reproducible characteristics.

Calcium sulphate and Calcium fluoride

Both these materials offer the advantage of much higher sensitivities than those of lithium fluoride and lithium borate but at the expense of considerably inferior soft tissue equivalence. Their light output per unit dose can exceed that of lithium fluoride by factors of between 15 and 70 depending on the dopant, the spectral response of the reader and the photon energy. On the other hand, as the previous figure demonstrates, the over-response at 30 keV compared with 200 keV amounts to a factor of about 11 for calcium sulphate and 15 for calcium fluoride. The variation in response for the broad X-ray spectra typical of diagnostic radiology will not be so pronounced but as shown by Hasman and Groothedde (Hasman and Groothedde (1985)) can still result in differences of over 20% for diagnostic X-rays generated at 60 and 100 kVp.

2.2.3. Depth Dose. Following the well established procedures used in radiotherapy, doses at internal points along the radiation beam axis can be related to the dose measured at the entrance surface using empirical percentage depth dose (PDD) data, obtained in a water phantom.

$$D_{depth} = D_{surface} \frac{PDD}{100}$$

Alternatively, an absorbed dose in air measurement at a known distance from the focus (l) may be used together with tissue-air ratio (TAR) data to give the dose at depth d.

$$D_d = D_{air} \frac{(\mu_{en}/\rho)_{muscle}}{(\mu_{en}/\rho)_{air}} [l/(FSD + d)]^2 \cdot TAR$$

The PDD is a function of the beam quality, the field size at the surface, the FSD and depth, while the TAR is independent of FSD provided that the field size is specified at the depth of interest.

Depth doses estimated from PDD and TAR data take no account of the inhomogeneities that exist within the body in the form of local shielding by bone or reduced attenuation in air cavities or lung. More realistic heterogeneous phantons can be used to simulate the patient using the wide range of tissue equivalent materials that are now available.

Anthropomorphic phantoms allow for the measurement of the mean absorbed dose to specific organs, as well as the doses at specified depths below the surface.

2.2.4. Organ Doses. According to ICRP 26 (ICRP (1977)) the absorbed dose, or more strictly the dose equivalent, averaged over an entire organ or tissue is the most appropriate quantity for predicting the probability of radiation effects occurring.

The dose equivalent, H, is related to the absorbed dose, D, by:

$$H = DQN$$

where Q is the quality factor and N is the product of all other modifying factors specified by ICRP.

Unit: sievert 1 Sv = 1 J/kg

Since Q is unity for diagnostic quality X-rays and ICRP has at present assigned the value one to N, values of absorbed dose expressed in grays are identical with values of dose equivalent expressed in sieverts. Both units have been used in organ dosimetry but the quantity has usually been expressed simply as "organ dose".

The word "dose" is generally used as an abbreviation for "absorbed dose" rather than an abbreviation for "dose equivalent" so it can be argued that "organ doses" should be expressed in grays. If sieverts are used the quantity should be stated as "organ dose equivalent".

For compact organs close to the surface like the breasts, testes or thyroid, reliable estimates of the mean organ dose can often be obtained from direct measurements on the surface with due allowance being made for the relatively small degree of attenuation as the X-ray beam traverses the organ. (Attenuation in the breast will be appreciable for the low energy X-rays used in mammography and factors for converting surface dose measurements to the mean dose to the sensitive glandular tissues of the breast have been published for the special conditions prevailing during mammography (Rosenstein et al. (1985))).

Larger organs at depth are generally not amenable to direct dose measurements on patients and measurements or calculations in phantoms have to be invoked. The Rando phantom shown in Figure 6 is one of the most commonly used physical phantoms simulating the head and trunk of an average male adult. In proportion, it is equivalent to an "average man" 1.75 m tall and weighing 73.5 kg. It is made from tissue-equivalent synthetic rubber and contains a complete human skeleton, lung-equivalent material and airways corresponding to the maxillary sinuses, nasopharynx and trachea. It is composed of a number of 25 mm thick transverse sections each containing a matrix of 5 mm diameter holes spaced 3 cm apart. Each hole can accomodate a dosemeter or a solid plug of tissue-quivalent material. The complete phantom contains over 3000 holes and additional ones can be drilled if required.

TLDs are ideally suited for dose measurements in these phantoms since their small size and relative cheapness allows a large number to be distributed inside the phantom. Powdered phosphor contained in small capsules is shown in the figure but solid TLD pellets or chips can equally well be placed in the holes with a tissue equivalent plug on either side.

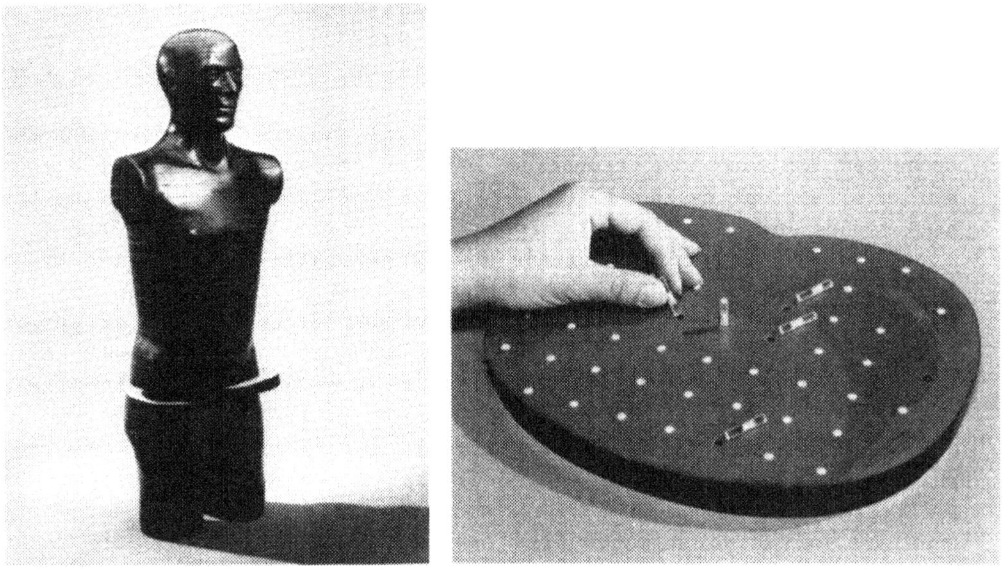

Figure 6. The Rando phantom (left picture) consists of 25 mm thick transverse slices (right picture) each containing a rectangular grid of plugs which can be replaced by capsules containing TLDs (photographs by courtesy of NRPB).

However the number of measurements required for a complete assessment of all relevant organ doses can easily become prohibitively large. For example, if a variety of clinical situations or exposure conditions need to be simulated and the dose to widely distributed organs like the bone marrow is required, many hundreds of dose measurements would be necessary.

Mathematical modelling of the transport of X-ray photons through the patient avoids labour intensive dose measurement schedules and the expense of anatomically accurate physical phantoms. Once the necessary computer codes have been developed, exposure conditions can be readily altered to provide organ dose information for any desired clinical situation.

Considerable computational expertise is needed to develop and operate the codes and a large capacity, high speed computer is required, but the results of such calculations are available in the literature (Rosenstein (1976), Rosenstein et al. (1979), Keriakes and Rosenstein (1980), ICRP (1982), Dresler et al. (1984), Jones and Wall (1985)) and can be used to convert measured values of entrance surface dose into organ doses for a wide range of routine X-ray exposures.

All the computational procedures quoted above use the Monte-Carlo method for simulating the transport of photons through the body and a mathematical phantom to describe the patient like that shown in Figure 7.

The phantom consists of an elliptical cylinder representing the trunk and the arms, two truncated circular cones representing the legs and an elliptical cylinder capped by half an ellipsoid representing the head (Jones and Wall (1985), Cristy (1980), Kramer et al. (1982)).

The skeleton and organs are mostly generated from further truncated elliptical cylinders or ellipsoids specified by quadratic inequalities. Such crude formulations for the organ boundaries permit reasonable computational efficiency in carrying out the Monte-Carlo calculations but produce the limitations in anatomical accuracy that are evident in the figure. Exact delineation of the X-ray beams used in clinical diagnosis becomes more difficult than with realistic physical phantoms, leading to greater uncertainty in the doses calculated for small organs close to beam edges. However, the mean doses to large organs are probably estimated more accurately by the Monte-Carlo technique where the total energy deposited in the organ is calculated, rather than averaging dose measurements made at a limited number of points within the organ.

2.2.5. Effective Dose Equivalent (H_E). ICRP in its system of dose limitation (ICRP (1977)) has recommended that the total risk for non-uniform irradiation of the body can be equated to the risk from a uniform irradiation by use of quantity "effective dose equivalent" (H_E). It combines organ doses in a manner that weights the dose to each relevant

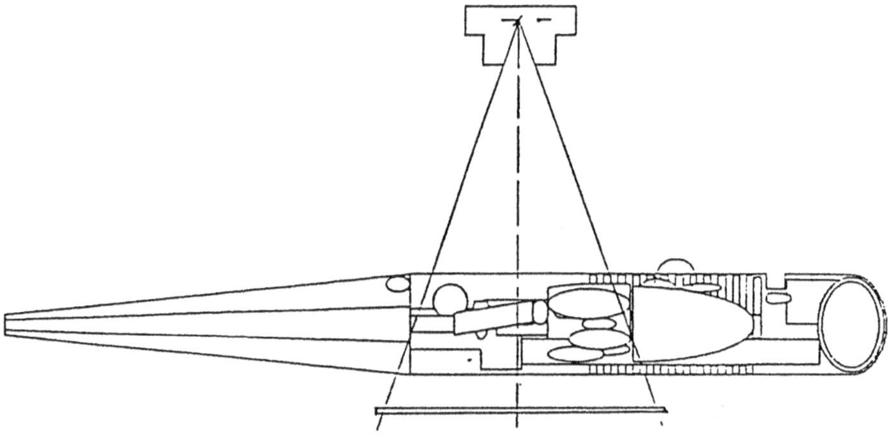

Figure 7. Mathematical phantom.

organ (i) by a factor w(i) representing the proportion of the risk resulting from organ (i) to the total risk when the body is irradiated uniformly.

Thus:

$$H_E = \sum_i H(i)\ w(i)$$

Only the risk of induction of fatal cancers or serious hereditary defects in the first two generations are considered appropriate in this system which was designed specifically for limiting doses to workers. The recommended values of w(i) are tabulated in Table 2. They are based on estimated risk factors for various organs or tissues that have been averaged over both sexes and all ages in a working population.

TABLE 2. Weighting factors.

Organ of tissue	w(i)
Gonads	0.25
Breast	0.15
Red bone marrow	0.12
Lung	0.12
Thyroid	0.03
Bone surface	0.03
Remainder	0.30

It is recognised that the risk associated with a given exposure will vary with the age and sex of the individual exposed and consequently that these factors which are recommended as appropriate for the **protection** of any worker, regardless of these sources of variability, will not necessary be appropriate for **estimating risks** to any individual patient or group of patients whose age distribution could be quite different from a population of workers.

The total stochastic risk factors assumed by the ICRP of $1.25\ 10^{-2}\ Sv^{-1}$ for radiation-induced fatal cancers and $4\ 10^{-3}\ Sv^{-1}$ for serious hereditary effects in the first two generations of a working population, that are inherent in the concept of H_E will frequently **overestimate** the risks for typical elderly patient populations.

2.2.6. Total Energy Imparted to Patient. For many types of X-ray examination, especially complicated ones involving many films or fluoroscopy, the measurement of the mean dose to specific organs can be very difficult and subject to large uncertainties. These dosimetric problems combined with the lack of precise knowledge concerning the radiosensitivities of the different organs have encouraged the adoption of the alternative approach of measuring the total energy imparted to the patient or the integral dose as it used to be known (Pauly (1978), Gustafsson (1979), Bengtsson and Jensen (1979), Wall and Shrimpton (1981)). In essence this approach assumes that radiosensitive and radioresistant tissues are homogeneously mixed throughout any volume of the patient that receives a significant dose, so that the overall risk can be related simply to the total energy imparted to that volume. This is obviously an oversimplification of the true situation since the required homogeneous mixture of tissues is unlikely to exist for all X-ray fields. However the distribution of sensitive organs within the trunk may in many circumstances be sufficiently uniform for the errors involved in this approach to be not greater than those associated with using a suitable combination of specific organ doses and weighting factors to express the overall risk.

There is evidence, from surveys where organ doses and total energy imparted to the patient have been measured simultaneously, that an approximately linear relationship holds between estimates of detriment and values measured for the energy imparted for a number of types of X-ray examination.

Figure 8 shows estimates of the probability of incidence of all cancers plotted against energy imparted for nine types of X-ray examination investigated during a recent national patient dose survey in England (Shrimpton et al. (1986)). The cancer incidence figures are

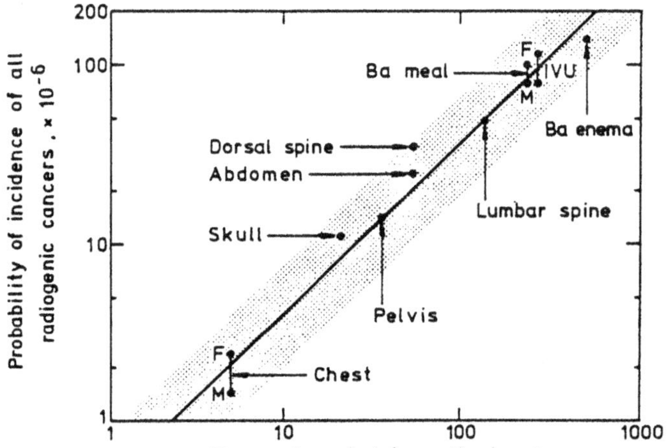

Figure 8. Probability of cancer incidence as a function of imparted energy.

based on risk factors for fatal and non-fatal cancers and average values for the organ doses measured at 20 English hospitals for each examination.

For those types of examination involving a significant breast dose the estimated cancer risk is higher for females than males. This is evident for chest X-rays, barium meals and intravenous urograms (IVU) in the figure where the estimated cancer risk is shown by a vertical bar, the top of which corresponds to the risk for females and the bottom to that for males.

The straight line showing the best fit to all the points on the graph, has a slope of 45° which is indicative of a linear relationship between two variables plotted on log-log paper. All points on the graph lie inside the shaded region which is within a factor of two of this line, which corresponds to a constant risk factor of 3.8 10^{-4} J^{-1}. A probability of 3.8 10^{-4} cases of cancer (some of which will not be fatal) for every joule of energy imparted to the patient could consequently be taken as a rough guide to the somatic health detriment for these types of X-ray examination. The risk will be lower if only fatal cancers are considered and/or full expression is not assumed and it will be higher if genetic or fetal effects are included. For example if the risk of fatal cancers and serious hereditary defects in the first two generations are considered, as in the effective dose equivalent concept of ICRP Publication 26, the risk factor drops to 2.3 10^{-4} J^{-1} and a similarly linear relationship between this measure of detriment and the total energy imparted to the patient can be demonstrated (Shrimpton (1985)).

A major advantage of the simple concept of the total energy imparted to the patient is the ease with which it may be estimated during even complex X-ray examinations:

$$\text{Energy fluence } \Psi = \frac{dR}{da}$$

where dR is the radiant energy incident on a sphere of cross-sectional area da.

Absorbed dose to air:

$$D_{air} = \Psi(\mu_{en}/\varrho)_{air} = \frac{dR}{da}(\mu_{en}/\varrho)_{air}$$

Radiant energy:

$$R = \frac{\int D_{air}\, da}{(\mu_{en}/\varrho)_{air}}$$

The numerator is the quantity measured by a large area ionisation chamber of the type discussed in 2.1.2.

An appreciable fraction of the energy incident on a patient will not be absorbed, the majority of the escaping energy being lost through backscattering from the incident surface. Monte-Carlo calculations of the total energy imparted to an anthropomorphic phantom from diagnostic X-ray fields indicate that the fraction of the incident energy imparted to the phantom varies from 0.60 to 0.87 depending primarily on beam quality and to a lesser extent on the irradiation geometry and patient physique providing that an insignificant proportion of the beam misses the patient altogether.

The total energy imparted to a patient during an X-ray examination can consequently be obtained from a measurement of the total dose-area product and knowledge of the fractional energy absorption (FEA) and an appropriately weighted mean value of $(\mu_{en}/\varrho)_{air}$ for the X-ray spectrum used.

$$\text{Energy imparted to patient} = \frac{\int D_{air}\, da}{(\mu_{en}/\varrho)_{air}} \cdot \text{FEA}$$

The reasonable degree of correlation that is evident between energy imparted and health detriment for a wide range of X-ray examinations suggests that **in appropriate circumstances** it can represent a useful practical quantity for estimating the risk to patients.

3. Population Related Dose Quantities

Some dose quantities that are related to different aspects of the possible radiation health detriment to populations exposed to ionising radiation are defined below.

3.1. Genetically Significant Dose (GSD)

Historically a large emphasis has been placed on the assessment of **genetic** hazards from diagnostic radiology. This emphasis has tended to conceal the significant level of somatic risk. However, it is possible to argue that the potential somatic hazards to individual patients can be balanced against the benefits received but that any genetic harm incurred cannot be offset in this direct way. It is therefore important to keep a close watch on the magnitude of the genetic risk from diagnostic radiology particularly since a significant proportion of the fertile population is subjected to X-ray examination every year.

The GSD is defined as that dose which, if given to every member of a population, would produce the same total genetic injury as the actual doses received by the various individuals (UNSCEAR (1972)).

Assuming that the genetic injury resulting from the irradiation of an individual is proportional to the gonadal dose received and to the number of children conceived by the individual after the exposure, and also assuming that no other factors are involved, then:

$$\text{Total genetic injury to population} = \alpha \, \Sigma \, D(i) \; P(i) \qquad (1)$$

where:
P(i) is the child expectancy of the ith individual;
D(i) is the gonadal dose to the ith individual;
α is the constant of proportionality;
and summation is over every individual in the population.

From the definition of GSD, if all individuals in the population received exactly the GSD, then the total genetic injury would be the same as that resulting from the actual doses received as given in equation (1).

Thus:

$$\alpha \, \Sigma \, GSD \; P(i) = \alpha \, \Sigma \, D(i) \; P(i)$$

Hence:

$$GSD = \frac{\Sigma \, D(i) \; P(i)}{\Sigma \, P(i)} \qquad (2)$$

Since it is assumed that genetic injury is caused only by doses to gonadal tissue prior to conception, irradiation of the maternal gonads will not cause genetic damage to an existing fetus. On the other hand irradiation of fetal gonads may be the cause of genetic injury to subsequent generations, and therefore fetuses should be considered as individuals in the enumeration of populations for the calculation of GSD and they should be included in both numerator and denominator of equation (2).

The concept of GSD is a general one, and the GSD from any source of radiation during any period can be considered. However, the remainder of this section concentrates on the annual GSD from diagnostic radiology. In so doing it is convenient to divide the population into groups according to age and sex since these are the major determinants of child expectancy. The individuals within any age-sex group can then be further subdivided according to the types of examination that they underwent during the year in question.

Hence:

$$GSD = \frac{\sum_{jk} N(j,k)\ \bar{D}(j,k)\ P(j,k)}{\sum_{k} N(k)\ P(k)} \tag{3}$$

where:

N(j,k) is the number of individuals in the kth age-sex group who underwent an examination of type j during the year in question.

P(j,k) is the child expectancy of an individual in the kth age-sex group who underwent an examination of type j.

$\bar{D}$(j,k) is the mean gonadal dose received by an individual in the kth age-sex group from an examination of type j.

N(k) is the number of individuals in the kth age-sex group in the population.

P(k) is the child expectancy of an individual in the kth age-sex group.

Usually the child expectancy of an individual will be independent of the type of examination carried out in which case P(j,k) = P(k). However there are a few types of examination like hysterosalpingography where a higher than average rate of infertility is expected, and for which the child expectancy should be correspondingly reduced.

3.2. Somatically Significant Dose (SSD)

A somatically significant (SSD) dose can be defined, in a manner analogous to the GSD, as that dose which, if given to every member of the population, would result in the same total number of fatal cancers as the doses actually received by the individuals exposed.

Note that somatic effects have been restricted to **fatal** cancers only. To be analogous to the GSD, the effect of the finite **life expectancy** of the exposed individuals in reducing the opportunity for the expression of long term radiation induced cancers should be taken into account in the same way that GSD allows for the dependence of genetic effects on the **child expectancy** of the patient. Expression factors E(i,k) which depend on the type of cancer or organ (i) and the age and sex (k) of the patient are required and their derivation is discussed in a later chapter.

A weighting factor $W_s(i,k)$ is also required to express the relative risk of fatal cancer induction per unit dose in each organ (i), and it may also depend on the age and sex (k) of the patient e.g. for breast cancer.

Hence:

$$SSD = \frac{\sum_{ijk} N(j,k)\ \bar{D}(i,j,k)\ W_s(i,k)\ E(i,k)}{\sum_{k} N(k)}$$

where:

i is the organ;

j is the type of investigation;

N is the number of people;

$\bar{D}$ is the mean organ dose;

k is the age/sex group;

W_s is the weighting factor for somatic effects;

E is the expression factor (accounts for probability of expression in relation to life expectancy).

Leukaemia Significant Dose and **Malignancy Significant Dose** are special cases of the above quantity that are restricted to leukaemia and solid cancers respectively.

3.3. Collective Effective Dose Equivalent (S_E)

This quantity combines genetic effects and fatal somatic effects of radiation as in H_E and sums the individual doses for the exposed population.

$$S_E = \sum_{ijk} N(j,k)\ \bar{H}(i,j,k) \cdot w(i)$$

where:

$\bar{H}$ is the mean dose equivalent;

w(i) is the ICRP 26 weighting factor;

Unit is the man sievert (man Sv).

3.4. Per Caput Effective Dose Equivalent ($\bar{H}_E$)

As with GSD and SSD, per caput effective dose equivalent averages the collective dose over the entire population irrespective of the proportion of the population that was irradiated.

$$\bar{H}_E = \frac{S_E}{\sum_{k} N(k)}$$

Unit: sievert (Sv).

These last two quantities do not include a factor for taking account of the reduced opportunity for expression of long term radiation effects for exposed populations which include a significant proportion of elderly people.

The ICRP 26 weighting factors w(i) assume expression factors typical of a population of workers.

It may also be important to recognise differences in sensitivity between the sexes for radiation induced health effects and to include non-fatal cancers and hereditary effects beyond the first two generations in assessing risks to patients. These considerations are excluded from the effective dose equivalent concept of ICRP 26.

4. Recommendations

The dose quantities of ultimate interest for radiological protection of patients in diagnostic radiology are the mean absorbed doses received by radiosensitive organs or tissues in the patient.

These cannot be directly measured.

They are usually derived from measurements in free air or on the surface of the patient or a phantom using empirical or calculated conversion factors.

The following specific recommendations should help to standardize the dose quantities used, but it should be appreciated that there is no single correct way to carry out patient dose measurements and the most suitable quantities to use are often dictated by the dosimetry techniques and level of access to radiology facilities that are available:

1. The quantities "Exposure" and "Air Kerma" should not be used for measurements relating to the doses received by patients from diagnostic X-rays.
2. Instruments calibrated against national primary standards of "air kerma" should be regarded as indicating "absorbed dose to air" when measuring diagnostic X-rays.
3. Organ doses can be derived from either the "Entrance absorbed dose to air, in free air" (2.1.) or the "Entrance surface absorbed dose to muscle with backscatter" (2.2.). The choice of the directly measured dose quantity depends on both the dosimetry technique and organ dose conversion factors, that are available, e.g.
   require 2.1. if:
   (i) measuring absorbed dose to air with an ionisation chamber or TLD in the absence of a patient or phantom; and
   (ii) using TARs to estimate organ doses; or
   (iii) using entrance dose to organ dose conversion factors.
   require 2.2. if:

(i) measuring absorbed dose to muscle with TLDs on the surface of a patient or phantom; and

(ii) using PDD data to estimate organ doses; or

(iii) using entrance surface dose to organ dose conversion factors.

4. The acceptable abbreviation for "Entrance absorbed dose to air, in free air" is "Entrance dose".
5. The acceptable abbreviation for "Entrance surface absorbed dose to muscle, with backscatter" is "Entrance surface dose".
6. A complete description of these quantities should be included in any report that quotes them.
7. Whenever organ doses are quoted the directly measured doses from which they were derived should also be given.
8. A complete description of, or reference to the conversion factors used to derive the organ doses should be provided.
9. If organ doses are combined into a single risk-related quantity (e.g. GSD or effective dose equivalent) the exact method for combining the organ doses should be described and the organ doses themselves should still be quoted.
10. Absorbed doses should be expressed in grays.
11. Dose equivalent should be expressed in sieverts.
12. "Dose-area product" and the "total energy imparted to the patient" are useful quantities when the complexity of the X-ray examinations (e.g. the use of fluoroscopy) precludes the measurement of entrance doses and hence the derivation of organ doses.

5. References

BCRU (1982) "Memoranda from the British Committee on Radiation Units and Measurements. Brit. J. Radiol. 55 375-377.

Bengtsson, L.G. and Jensen, M. (1979) "Significance of photon attenuation for the choice of quantities in X-ray protection". In Application of the dose limitation system for radiation protection. Practical Implications IAEA-SR-36 (IAEA, Vienna).

Cristy, M. (1980) "Mathematical phantoms representing children of various ages for use in estimates of internal dose". NUREG/CR-1159, ORNL/NUREG/TM-367 (Oak Ridge National Laboratory, Tennessee).

Drexler, G. Panzer, W., Widenmann, L., Williams, G. and Zankl, M. (1984) "The calculation of dose from external photon exposures using reference human phantoms and Monte-Carlo methods. Part III: Organ doses in X-ray diagnosis". GSF - Bericht S-1026 (Gesellschaft für Strahlen- und Umweltforschung, München).

Greening, J.R. (1985) "Fundamentals of Radiation Dosimetry". 2nd Edition. Medical Physics Handbook 15 (Adam Hilger, Bristol).

Gustafsson, M. (1979) "Energy imparted in Roentgen diagnostic procedures Acta Radiological Diag. 20 123-144.

Hasman, A., and Groothedde, R.T. (1985) "$CaSO_4$ (Dy) thermoluminescent dosimeters for the determination of gonadal doses. Acta Radiologica Th. Phys. and Biol. 14 369-375.

ICRP (1977) "Recommendations of the International Commission on Radiological Protection". ICRP Publication 26 (Pergamon Press, Oxford).

ICRP (1982) "Protection of the patient in diagnostic radiology". ICRP Publication 34 (Pergamon Press, Oxford).

ICRU (1970) "Radiation dosimetry: X-rays generated at potentials of 5 to 150 kV". ICRU Report 17 (International Commission on Radiation Units and Measurements, Washington).

ICRU (1979) "Average energy required to produce an ion pair". ICRU Report 31 (International Commission on Radiation Units and Measurements, Washington).

ICRU (1980) "Radiation quantities and units. ICRU Report 33 (International Commission on Radiation Units and Measurements, Washington).

Jones, D.G. and Wall, B.F. (1985) "Organ doses from medical X-ray examinations calculated using Monte-Carlo techniques". NRPB-R186 (HMSO, London).

Keriakes, J.G. and Rosenstein, M. (1980) "Handbook of radiation doses in nuclear medicine and diagnostic X-ray, (Chemical Rubber Co., Florida).

Kramer, R. Zankl, M., Williams, G. and Drexler, G. (1982) "The calculation of dose from external photon exposures using reference human phantoms and Monte-Carlo methods. Part I: The male (ADAM) and female (EVA) adult mathematical phantoms". GSF Bericht S-885 (Gesellschaft für Strahlen- und Umweltforschung, München).

Pauly, H. (1978) "Stochastic late effects after partial body irradiation in diagnostic radiology: evaluation of approximate data". Rad. and Environ. Biophys. 15 21-33.

Rosenstein, M. (1976) "Handbook of selected organ doses for projections common in diagnostic radiology". US Dept of Health, Education and Welfare, Bureau of Radiological Health, HEW Publ. (FDA) 76-8031.

Rosenstein, M. (1976) "Organ doses in diagnostic radiology". US Dept of Health, Education and Welfare, Bureau of Radiological Health, HEW Publ. (FDA) 76-8030.

Rosenstein, M. Anderson, L.W. and Warner, G.G. (1985) "Handbook of glandular tissue doses in mammography". HHS Publication FDA 85-8239.

Rosenstein, M., Beck, T.J. and Warner, G.G. (1979) "Handbook for selected organ doses for projections common in pediatric radiology". US Dept of Health, Education and Welfacre, Bureau of Radiologica Health, HEW Publ. (FDA) 79-8079.

Shrimpton, P.C. (1983) "Energy imparted as a measure of radiological hazard to patients from X-ray examinations". In proceedings of the VIIth International Congress on Medical Physics, Helsinki, Medical and Biological Engineering and Computing, Vol. 23.

Shrimpton, P.C. and Wall, B.F. (1982) "An evaluation of the Diamentor transmission ionisation chamber in indicating exposure-area product (R cm^2) during diagnostic radiological examinations. Phys. Med. Biol. 27 871-878.

Shrimpton, P.C., Wall, B.F., Jones, D.G., Fisher, E.S., Hillier, M.C., Kendall, G.M. and Harrison, R.M. (1986) "Doses to patients from routine diagnostic X-ray examination in England. Brit. Jour. Radiol. 59, 749-758.

Shrimpton, P.C., Wall, B.F., Jones, D.G., Fisher, E.S., Hillier, M.C., Kendall, G.M. and Harrison, R.M. (1986) "A national survey of doses to patients undergoing a selection of routine X-ray examination in English hospital". NRPB-R200 (HMSO, London).

Takenaga, M., Yamamoto, O. and Yamashita, T. (1983) "A new phosphor $Li_2B_4O_7$: Cu for TLD. Health Physics 44 387-393.

United Nations Scientific Committee on the Effects of Atomic Radiation (1972) "Ionising Radiation: Levels and Effects". Vol. 1: Levels, New York, UN.

Wall, B.F. and Shrimpton, P.C. (1981) "Deliberation on a suitable quantity and technique for assessing somatic risk for diagnostic radiology". In proceedings of seminar on Patient Exposure to Radiation in medical X-ray diagnosis EUR 7438 (Commission of the European Communities, Luxembourg).

Wall, B.F., Driscoll, C.M.H., Strong, J.C. and Fisher, E.S. (1982) "The suitability of different preparations of thermoluminescent lithium borate for medical dosimetry". Phys. Med. Biol. 27 1023-1034.

OPTIMIZATION OF IMAGE QUALITY CRITERIA AND PATIENT EXPOSURE

C. MACCIA
INSERM U 240
B.P. 48
92263 Fontenay aux Roses Cedex
France

1. Introduction

In many countries, medical exposure of patients represents the greatest man-made contribution to the collective dose imparted to the population.

Many publications have already investigated the various aspects and level of such radiation sources, and both United Nations Scientific Committee on the Effects of Atomic Radiation (UNSCEAR) and Advisory Committee on the Biological Effects of Ionising Radiations (BEIR) have issued very comprehensive reports on this subject. Furthermore, the increasing number of X-ray examinations yearly performed in the most industrialized countries [e.g. from 6.2×10^6 to 55×10^6 examinations between 1957 and 1988 respectively in France (Charpak et al. (1988))], the extension of medical indications together with the introduction of new diagnostic imaging techniques, have continuously stimulated reassessment of such radiation sources., Several dosimetric studies were therefore recently carried out in Europe (Shrimpton et al. (1986), Padovani et al. (1987), Maccia et al. (1988)) in order to assess and rationalize the use of ionising radiations in medical field. All these studies, which basically dealt with the patient dosimetry, have clearly shown strong differences in strategies for conducting some type of X-ray examinations, and provided a baseline of data which could be used to optimize practices, image quality and doses delivered in diagnostic radiology. In taking such a complex problem, which is the optimization of the radiographic image, several methods may be envisaged at the practical level (measuring physical characteristics of image detectors, applying R.O.C. analysis...). Unfortunately the implementation of these methods require, almost always, rather complex infrastructures to be involved (personnel availability, experimental laboratories, interaction with the health care system organization). Now radiologists in

M. Oberhofer (ed.), Advances in Radiation Protection, 265–276.

their daily practice, are faced to the crucial problem of establishing the most relevant diagnosis through the most suitable imaging technique while keeping the patient received dose as low as reasonably achievable. In other terms, they have to look, in a very reliable manner, for a best compromise between the "image quality and dose" using very simple criteria and methods. As a first tentative attempt, a study group of the radiation protection programme of the Commission of the European Communities (CEC), initiated, in 1987, a project on the establishment of quality criteria for diagnostic radiographic images.

The main goal of this project was to provide practitioners with a provisional acceptable list of both *radiological* and *technical* requirements (including dose values) which could be useful to judge the quality of the radiographs routinely undertaken in diagnostic radiology. Through this project, the existing exposure ranges, which have been described by several authors, would be restricted without limiting the choice of the radiological technique. Since conventional X-ray techniques contribute significantly to the total effective dose equivalent that an average individual may annually receive from other man-made radiation sources, the project was orientated towards only those examinations which are still commonly and frequently carried out in every day radiological practice: chest, skull, lumbar spine, pelvis and sacrum, urinary tract, breast.

In order to evaluate the suitability of the requirements listed in a draft-document of the CEC, the group set up a trial which collected information on more than 900 X-ray examinations performed in 10 European countries. A questionnaire was therefore circulated to 24 different European X-ray departments which actively participated in the project by checking, for each examination type, the suggested image quality criteria and commenting on their relevance.

This is believed to be the first time that a trial of this type, involving at the same time radiologists and physicists, and using the same approach, has been carried out in so many different countries practising radiology.

2. Material and Method

From the methodological point of view, two major problems had to be solved in structuring the questionnaire.

The first one dealt with the definition of an acceptable list of medical radiological requirements which referred, for each projection of each examination type, to important characteristic features of the "normal" radiographic image. This task implicitly required that all CEC study group radiologist members agreed, besides any abnormality or pathology, upon the anatomical patterns that should be visible on a given radiograph.

This very positive and fruitful exercise revealed, on one side, the necessity of harmonizing the European radiological terminology and, on the other side, pointed out differences existing in the radiology schools in describing, in a very simple manner, what kind of information should be contained in a "normal" X-ray film.

Concerning requirements of important image details, the questionnaire was limited to provide dimensions of specific normal and abnormal structures projected on the film or image receptor.

The second important problem was to find out a relevant dose quantity which could be easily measured on a patient who undergoes an X-ray examination and, at the same time, could be useful to check the compliance with the quality criteria for diagnostic radiographic images listed in the questionnaire.

The suggested quantity was the "Entrance Surface Dose" (ESD) which has already been demonstrated to be a rather appropriate indicator of the relative risk for the examined patients. It was therefore decided to collect this information using thermoluminescent dosimeter chips stuck on the X-rayed patient's skin.

Three European dosimetry laboratories namely the NRPB (National Radiological Protection Board) in U.K., the GSF (Gesellschaft für Strahlen-und Umweltforschung) in F.R.G. and the USL (Unità Sanitaria Locale) No. 7 in Udine (Italy) were involved in this part of the trial which needed to handle and read out thousands of dosemeters sent throughout Europe. In order to achieve this step, a preliminary dosimetry intercomparison between the mentioned laboratories was made to ensure reliability of measurements.

Finally, patient dose data and completed questionnaires were collected and centralised at the INSERM (France) for the evaluation of the trial.

3. Results

3.1. General Statistics

Despite the rather complex management problems, the trial was successfully completed and a satisfactory number of questionnaires were collected for each X-ray examination category considered. Table 1 summarizes the actual number of patients for whom both quality image criteria and doses were recorded in the questionnaires and gives the number of X-ray departments that participated.

More detailed data are also presented in Table 2 for each projection type of each examination.

TABLE 1. General characteristics of trial.

Examination Type	Number of Countries	Number of X-ray Dept.	Number of Patients	(*) Number of Dose Measurements
Breast	8	15	160	160
Chest	8	16	211	300
Urin. Tract	7	15	155	191
Skull	5	12	117	223
Lumbar Spine	7	14	149	204
Pelvis	6	13	139	134
Total	10	24	931	1212

(*) All projections together.

TABLE 2. General results of the trial.

Examination Type	Reject Rate (%)	Radiographs Meeting All Diagnostic Criteria (%)	Hospitals Meeting Dose Requirements (%)
Breast	5	79	33
Chest (p/a)	2	79	69
Urin. Tract			
(before inject.)	6	62	87
Skull (a/p)	10	74	89
Skull (p/a)	10	65	56
Skull (lateral)	5	67	75
Lumbar Spine			
(a/p)	10	69	50
(lateral)	4	77	69
Pelvis (a/p)	6	77	58
Pelvis (lateral)	5	83	67

First of all, one may notice the reject rate values which compare well with those generally found in the literature (Belletti et al.). It can be, therefore, deduced that there was no bias associated with the trial, either in terms of excessive attention paid by radiographers in performing the examinations, or in terms of particular severity required by radiologist of the quality of radiographic images. This encouraging finding, nevertheless, raises the more general problem of the reject rate level which should always be kept as low as possible in diagnostic radiology to limit unnecessary patient exposures.

Concerning the percentage of radiographs having met all diagnostic criteria, two interesting remarks may be made: first, the resulting average figure, all examinations together, is relatively high, 72%; second, if one adds to that value, examination by examination, the corresponding reject rate figure, the percentage of 100% is never reached. This implicitly means that, some radiographs were judged by the radiologists as acceptable for the diagnosis, while being not complying with all diagnostic criteria. In other terms, radiologists, in their routine practice seem to be spontaneously less restrictive when judging the quality of the radiographic image they produce. This result raises the problem of adapting the list of the image criteria to the every day radiological practice.

On the other hand, if one excludes the breast examination, for which the provisional CEC dose requirement was probably too severe, compliance rates with this latter markedly varied (close to a factor of 2) depending on the projection considered. There exists therefore no significant correlation between the dose score and image quality score as determined by compliance with the "diagnostic requirement".

In the light of these preliminary findings, "hospital by hospital" data analysis was carried out to evaluate the relevance of the image criteria. A "multi-score system" was then developed for each criterion and implemented for a selected number of projections.

3.2. The Image Scoring System

Basically, the idea was to define, for each image criterion, a numerical index capable of reproducing the "step by step" process implicitly followed by the radiologist when checking radiographic image quality. More precisely, a tentative attempt was made to determine schematically the basic elements which underlie the acceptability of the film for diagnosis. In splitting such a process up into different components, one may note that it requires that all potentially visible anatomical structures should be shown on the film (*M* for "medical" component), all abnormal or pathological details should be adequately contrasted (*T* for "technical" component) and, finally, the part of the body projected on the film should correspond to the field size (*P* for "positioning" component). From such a basis, it was therefore decided to translate this process into a multi-numerical scoring system related to the previously mentioned components. A group of practising radiologists actively participated to the establishment of this scoring system. An example of this scoring system for the chest image criteria is given in the Table 3.

TABLE 3. Example of the image criteria scoring system (chest (p/a)).

Image Criteria (p/a)	M(*)	T(**)	P(***)
- Symmetrical reproduction of the thorax	2	2	2
- Reproduction of vascular pattern in the lung periphery	2	2	0
- Reproduction of the costopleural boundary from the apex of the lung to the diaphragm	2	3	1
- Visually sharp reproduction of the peripheral vessels, the border of heart and the diaphragm	1	2	1
- Visualisation of the retrocardial lung and the mediastinum	3	3	1
- Performed at inspiration (min 6 anterior costal arches)	2	2	2
Global score	12	14	7

Scoring system: 0 = irrelevant; 1 = minor; 2 = important; 3 = fundamental.

M(*): Medical Score; T(**): Technical Score; P(***): Positioning Score.

As shown in the table, each criterion has three scores corresponding to the basic components mentioned before and each score may range from 0 (irrelevant component) to 3 (fundamental component).

Supposing the M-T-P scoring system reliable, one may see that the same image criterion, for instance "visually sharp reproduction of the peripheral vessels, the border of heart and the diaphragm" has not the same relevance for the radiologist when medical or technical view point is taken into account. This theoretically implies that, when the M-T-P scores are generally low, the answer to a criterion may be negative without having any impact on the final result of the film, that is to say, its acceptability for the diagnosis. Conversely, when a particular criterion scores higher, for instance "visualisation of the retrocardial lung and the mediastinum", a negative answer strongly affects the acceptability of the X-ray film.

3.3. *The Image Criteria Evaluation*

A comparison between the dosimetric results and the image quality, evaluated using the previously mentioned scoring system, was carried out for all the considered projections. The chest (p/a) and IVU (before injection film) examination results are presented in Figures 1 and 2 respectively.

3.3.1. Chest (p/a). Figure 1 shows the average entrance skin doses histogram, all techniques together, measured for the chest (p/a) projection in 16 hospitals which participated to the trial.

It also gives, for each hospital, the percentage of X-ray films having obtained the maximum global score, i.e. the rate of examinations for which all diagnostic image criteria listed in the questionnaire were met.

Despite the idea that this examination type would technically be the most "standardized" one, rather wide range of doses was measured in the considered X-ray departments with a minimum average dose of 0.16 mGy, and a maximum of 0.95 mGy. In only two out of sixteen hospitals, compliance with all the image criteria, was achieved for all

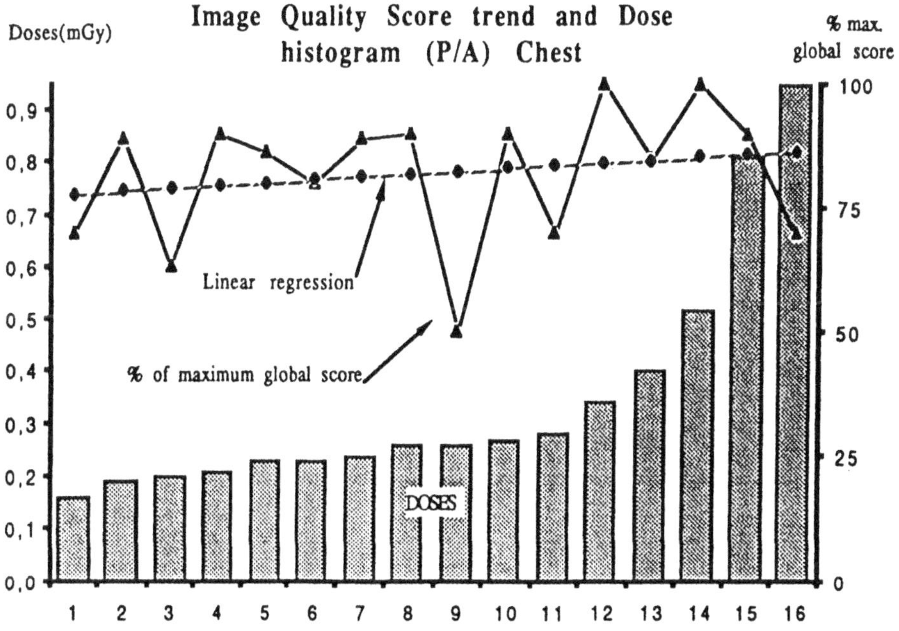

Figure 1. Histogram of doses and Image Quality Score trend. Chest (p/a).

examinations (continuous line), but unfortunately, all the corresponding entrance skin dose values were above 0.3 mGy which corresponds to the dosimetry requirement indicated in the CEC document. Three other hospitals did not meet the same dose requirement but their image quality index was generally lower.

Concerning the linear regression curve trend (dotted line), it clearly shows that there are no significant differences among all the considered X-ray departments: the curve being parallel to the horizontal axe.

This implicitly means that the image quality was supposed to be as reliable as possible, independently of the hospital taken into account.

Unfortunately, such a "rather" good image quality was obtained using different techniques that involved a wide range of doses.

3.3.2. IVU (Before Injection Film). In the Figure 2 results are presented for the IVU examination.

As far as the dose range is concerned, inter-hospital average dose estimates markedly varied from 2.5 mGy to 30 mGy (a factor of 10).

Conversely to the chest (p/a) projection, 2 hospitals out of 15 reached the maximum image score, and 2 other different hospitals failed to meet dose requirements.

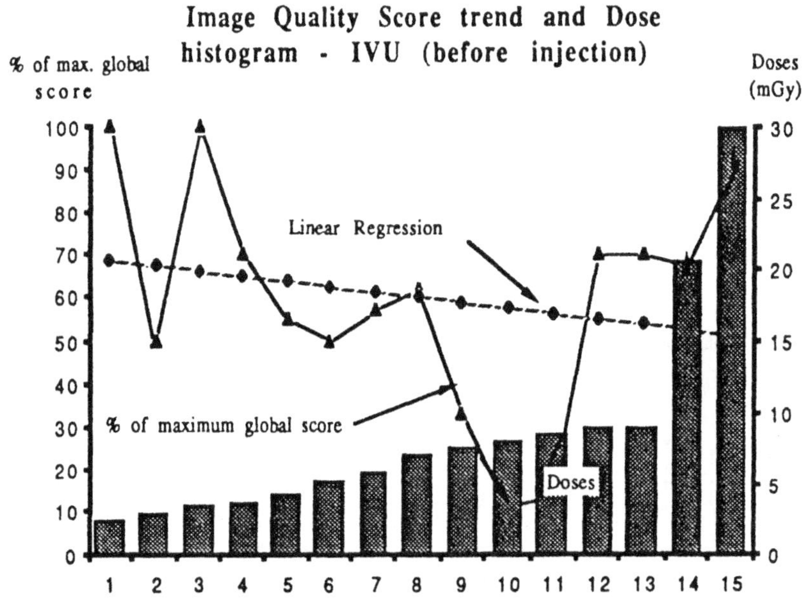

Figure 2. Histrogram of doses and Image Quality Score trend. IVU (before injection).

Despite the fact of trying to keep the image quality comparable among the considered hospitals, one may notice that the best quality of image was generally found associated to the low dose techniques, as clearly demonstrated by the linear regression curve trend.

3.4. Selecting the "Optimal" Technique

In this final step of the evaluation of the image criteria relevance, the scoring system was used to select the "optimal" technique which would correspond to that complying with both radiographic and dosimetric requirements recommended for the trial. In doing so, priority was given to the medical component of the score to reflect, in a more realistic manner, the quality of the radiological information which is essential to guarantee the acceptability of the film. In other words, for all X-ray projections considered, a sub-set was created keeping only those films for which the medical score was equal to the maximum or within 1 of the maximum value. To this sub-set obviously belong all films complying with all image criteria, and some other films for which a minor image criterion, from the medical view point, was not met.

Considering the radiological equipment and the film/screen combinations, four categories were defined in order to compare results obtained with so many different radiological units.

Basically, discrimination was made between the X-ray tables equipped with an Automatic Exposure Control system (AEC) and those manually operated (NO AEC).

Concerning the film/screen combinations, attention was paid to the CEC document requirements.

In average, because of their bad medical score, 20% of all acceptable for diagnosis films were eliminated through this selecting procedure. Furthermore, taking into account the compliance with the dose reference values, the percentage of the discarded films reached 50%. In other terms, looking for the best compromise between the image quality and the dose, led, all projections together, to keep only one film out of two.

Results of this selection are summarized in the Figure 3 and 4 for the chest and for the IVU respectively.

As the Figure 3 indicates, implementing the selecting procedure led to keep 167 high image quality score chest X-rays out of 208 acceptable films corresponding to three groups of technique shown in the "chest pie" chart.

First of all, no people were found to work exclusively manually and using a low speed film/screen class (NO AEC < 200), and very few examinations were found to be carried out using low speed films with the AEC system actually installed (1 X-ray department).

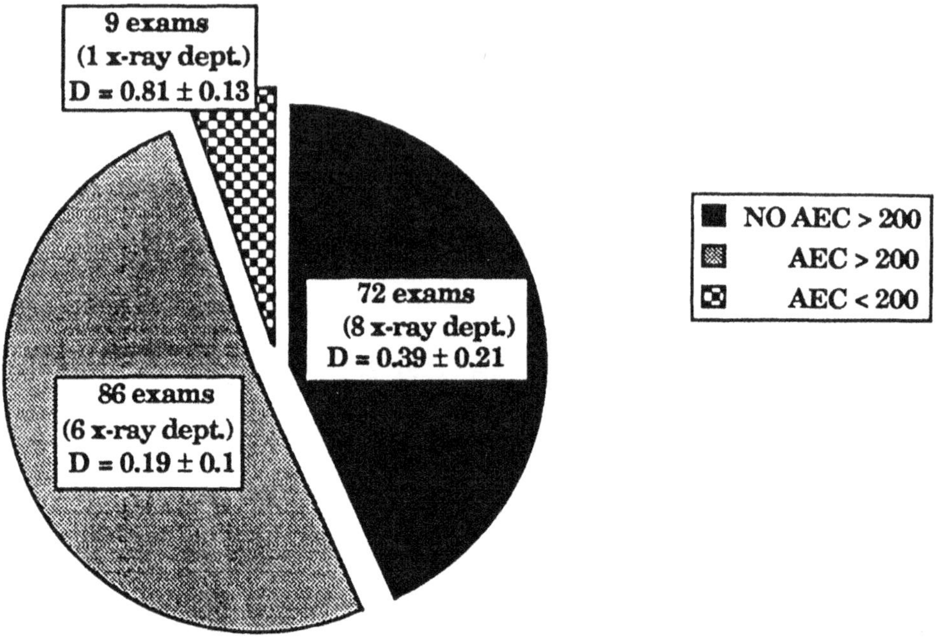

Figure 3. Selecting procedure results for the chest (p/a).

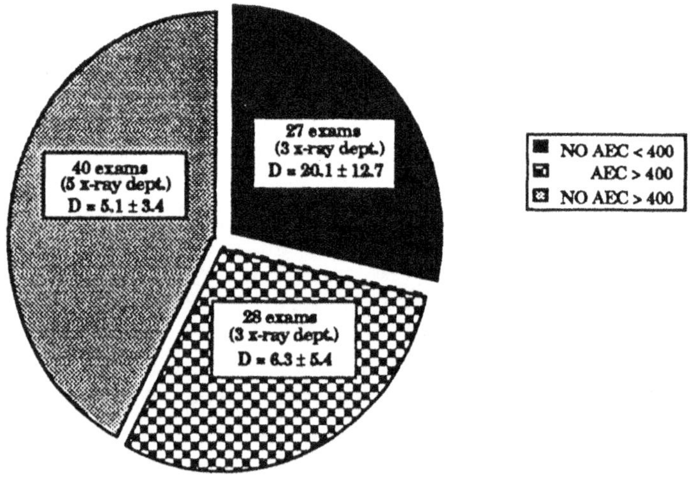

Figure 4. Selecting procedure results for IVU (before injection).

The great majority of the selected high image quality score films were taken either with an operating AEC system or with a manual operating equipment but always with film speed class above 200.

In particular, for these two categories of technique, dose hierarchy was respected with an average figure of 0.19 mGy (AEC system) and 0.39 mGy (NO AEC system) respectively. This clearly shows the strong impact of the AEC techniques in improving dose reductions when the appropriate film/screen combinations are used.

By implementing the same procedure to the IVU, 95 high image quality score X-rays out of 139 acceptable films were selected and three different radiological techniques were identified. Among them, two radiological techniques (AEC $\geq$ 400 and NO AEC $\geq$ 400) were found to be comparable and valuable from both dosimetric and radiological point of view. To the third one belong 27 out of 95 examinations which were found, despite their "good" image quality score, to be performed by using a very high dose-rate technique i.e. low speed film/screen class and manual operating equipment (NO AEC $<$ 400).

Almost all entrance skin dose values (85%) associated to these X-ray films were actually higher than the suggested CEC dose requirements i.e. 10 mGy. From such a basis, and in the context of the selecting procedure, this particular technique, requiring unacceptable and unnecessary dose level, should be avoided if one wants to keep patient doses associated to this complex examination as low as reasonably achievable.

4. Conclusions

The intention of this paper has been to present the main results of a trial set up by a Study Group of the Radiation Protection Programme of the CEC.

In view of harmonising the radiological techniques commonly used in different European countries, the relevance of quality criteria for radiographic images together with dose requirements were checked on about 900 examined patients and a first tentative attempt was made to correlate the image quality with the corresponding patient dose.

Although no evident correlations were found, such an approach was found to be relevant to determine technical parameters which provide a baseline of data concerning practices and doses delivered in diagnostic radiology. The necessity of implementing both quality control and quality assurance in this particular field was finally pointed out as being the most important tool to guarantee the best image quality while keeping patient exposures as low as reasonably achievable.

5. References

Belletti, R., Gallini, R. and Giugni, U. "Operative programme of quality control in the radiodiagnostic field in a general hospital: results and proposals". British Journal of Radiology, Supplement No 18 131-133.

Charpak, Y., Maccia, C., Fagnani, F. (1988) "Etude prospective sur l'imagerie médicale en France". Rapport INSERM No. 51.

Maccia, C., Benedittini, M., Le Faure, C. and Fagnani, F. (1988) "Doses to patients from diagnostic radiology in France". Health Physics, 4, 397-408.

Padovani, R., Contento, G., Fabretto, M., Malisan, M.R., Barbina, V. and Gozzi, G. (1987) "Patient doses and risk from diagnostic radiology in North-East Italy". British Journal of Radiology, 60, 155-165.

Shrimpton, P., Wall, B.F., Jones, D.G., Fisher, E.S., Hillier, M.C. and Kendall, G.M. (1986) "A national survey of doses to patients undergoing a selection of routine X-ray examinations in English hospitals". NRPB-200 (HMSO, London).

EXPERT SYSTEMS FOR QUALITY ASSURANCE IN DIAGNOSTIC RADIOLOGY

B.M. MOORES
Integrated Radiological Services Ltd. (IRS)
Radiation Protection Service
42 Rodney Str. - Liverpool
United Kingdom

1. Introduction

Information is the single most important non biological entity which has contributed towards the development of the human race. The ability to develop, store, transmit and apply information, or knowledge, sets the human race apart from all other species on this planet. With the advent of information technology, with its basis founded on the microchip an information revolution has already commenced. One aspect of this revolution centres around the development of even more powerful computers of smaller and smaller size. This evolution in computing power has created an environment where a belief that computers will be made to mimic the human brain and think for themselves has become almost accepted as inevitable. Thus if human reasoning methods could be analyzed and understood then it may be possible to reproduce them in computer technology and produce some form of artificial intelligence. Hence there exists the feeling that knowledge is the key to computing advance.

As the human race came to have access to more and more information, its analysis and manipulation became increasingly the subject of "scientific method". Consequently information was fitted into a jigsaw of knowledge, piece by piece, with each piece analyzed for its relevance and interconnection to other areas of knowledge. The modern educational process is more concerned with illiminating given parts of the jigsaw of knowledge to successive generations, where its relevance to our present state of knowledge is constantly reviewed. A piece of knowledge may subsequently be discarded if new information proves to be more relevant. This rigid knowledge based educational process has completely blurred the development and assessment of empirical expertise. Consequently we tend to treat rules-of-thumb as a temporary stop-gaps to be replaced when a more formal solution is developed. Unfortunately

M. Oberhofer (ed.), Advances in Radiation Protection, 277–285.

the majority of decisions taken by human beings, even scientists, tend to develop from rule of thumb reasoning. Most significant scientific discoveries appear to be based upon this empirical reasoning with the theory and full understanding of any discovery following afterwards.

This empirical knowledge base is termed know-how and it is rarely put into a written form. Although this know-how may play a major role in human decision making its role cannot definitely be proved since most evidence on the subject also tends to be empirical.

Thus we can consider a three stage process in the overall knowledge field; first comes information which may loosely be termed any non-random input into the human intellect. Information which is deemed relevant or interesting is then stored logically for future use and consideration and constitutes knowledge. Alternatively information may be assimilated and stored for use on an empirical basis. This type of knowledge or know-how is then refined through use, so that a way of doing things evolves, often apparently subconsciously. If someone, "an expert", is asked to describe the execution of some task, then they usually attempt to do so in terms of a logical sequence of events without any redundancy. However the true answer would be obscure since much reasoning takes place below the level of consciousness. Know-how includes short cut strategies, sudden jumps, zig-zag reasoning, analogies with a variety of tests and checks. Its aim is to provide a capacity for quick decisions over a wide range of contexts. This ability was an obvious pre-requisite for the survival of the human race against much larger and stronger predators. Knowledge taught as know-how must have formed the basis of the educational process for a million years in order to prepare successive generations. Unfortunately the virtual suppression in schools of this form of education, beyond the primary school level, must severely handicap child development. This area of development and its lack of presence in conventional education programmes is increasingly being recognized in the business field. Management is beginning to recognise that crucial know-how is what distinguishes the high flyer from the average performer. In fact it is perhaps no accident that many of the most successful business tycoons in the developed world had minimal formal education. There is now a partial recognition that the majority of quality control and safety manuals serve no useful purpose. They are written so as to be defendable and not readable. There has also been one medical experiment which highlights gaps in the formal education process. Twelve general practitioners programmed a computer based on simple rules of thumb which they employed in diagnosing four common conditions. At the finish most agreed that the programmes approximated their own diagnostic practices, but the programmes were entirely different to anything they had ever seen in medical school or elsewhere.

2. Expert Systems

Expert systems consist of computer programmes based upon human knowledge which has been provided by one or more experts in a particular area. The knowledge is then usually presented in a format which employs IF-THEN rules. These rules are accompanied by lists of facts or assertations which may include items of data included in the rules (e.g. "IF blurred and noisy" some relevant measure of blur or noise). All the rules, facts and algorithms are stored in a knowledge base analogous to a data base. The knowledge base may be tested, by various techniques, for logical consistency and apparent completeness. Its accuracy cannot be verified but its value or worth can be compared to the performance of a human expert by being executed against sample problems with a non expert following the rules.

Although the idea of Artificial Intelligence dates back to the early work of Turing in the 1930's it wasn't until the 1960's that it became fully established. Coincidentally the evolution of expert systems approaches to intelligent systems developed against the grain of the accepted work in this field. Whereas most researches attempted to develop a problem solving capacity within a computer, Feigenbaum working at the Carnegie Melon Institute assumed that decision-making knowledge was specific and largely the result of experience. A well known early approach was MYCIN in which, using approximately 400 rules, skilled advice to a physician is provided on the diagnosis and antibiotic treatment of blood diseases. At any point in the consultation the physician may enquire why MYCIN has asked a particular question or how it reached a conclusion. The system replies in terms of rules which are relevant to any enquiry. A sample rule in MYCIN is as follows:

IF 1) the site of the culture is blood, and
   2) the gram stain of the organism is gramneg, and
   3) the morphology of the organism is rod, and
   4) the patient is a compromised host.
THEN there is suggestive evidence (0.6) that the identity of the organism is pseudo-aeruginosa.

Obviously MYCIN is only intended for users who should be familiar with the term pseudo-aeruginosa. Also the certainty factors are measured from 0 to 1 so that (0.6) is indicative but not highly probable.

The MYCIN program is known to possess three weaknesses:

i) the explanations are not quite as good as they appear;
ii) it occasionally reveals an embarrassing lack of common sense e.g. it has little appreciation of the difference between men and women;
iii) the treatment of uncertainty in the knowledge base is known to be logically flawed.

All of the above are intimately related to the nature of the problem being addressed and the accuracy/quality of the knowledge data. As such they are the type of weakness which will be common to each and every expert system.

The fundamentals of an expert system approach to improving human performance may be considered at three levels. First there must be a well defined activity which can be isolated from external influences. In this way the area of expertise and the corresponding knowledge base can hopefully be defined. Secondly the activity must be amenable to know-how programming. There must be a defined framework for the activity and it should be expressible in simple rules. Such as IF-THEN. Finally the system should ideally be usable by those who can benefit most including, possibly, non experts in the field. In other words there will be minimal advantage from applying an expert system approach if only computer experts can employ the final result.

In a know-how program the knowledge must be represented explicitly so that any reasoning from that knowledge can be exercised by both the builders and the users of the programme. This ability to exercise or implement frequently is crucial since it provides a mechanism for recall and refining of the rules of thumb. Empirical knowledge is held below the level of conscious awareness and running a program can aid iterative improvements in the expression of the knowledge. Also, end users can gain a great deal from running a well presented Know-How program with their own choice of data. Repetition contributes greatly to the transfer of knowledge. Alternatively humans rarely read the best written manuals more than once or twice.

The main purpose, therefore, of know-how programming or Expert Systems is to act as an aid or a prompt to someone performing a given task. Under no circumstances are they intended as a replacement for the human and are not necessarily even aimed at the human level of performance, nor even as giving advice. A model has been quoted for these systems of a parrot sitting on ones desk providing pieces of know-how on demand. This is desirable because nobody would attempt to ascribe common sense to a parrot nor transfer to it their own responsibilities for a decision.

In an expert system it is essential to only promise what can be delivered. The fundamentals of an expert system approach outlined previously can be expanded to include the following:

a) the purpose of the application is primarily to share expertise and this purpose governs the design;
b) the application is selected so that it fits the capabilities of the implementation team and their choice of software;
c) the application is limited, sharply focussed and unequivocal;
d) the application relies on well defined pieces of expertise confi-

dently held by an articulate and authoritative expert or experts;

e) the implementation team does not modify or dilute the expert knowledge to fit personal or external constraints;

f) the structure and presentation of the know-how is professionally arranged to fit the subjective criteria of the target end-users, preferably with the help of a sample of those users.

The very nature of the undertaking itself places, at present, many constraints upon the effective application of expert systems approach. Therefore it is worthwhile considering the radiological process within the framework of expert system requirements.

3. The Radiological Process

Diagnostic radiology is a complicated and multi-faceted undertaking which has many links into other aspects of the health care-infrastructure. In fact diagnostic radiology is one of the most widely used medical techniques by all medical disciplines. Operating as it does at the forefront of medical technology it also has major links with industry and commerce. Similarly education and training play an important role in its structure. In spite of its multi-faceted aspect it is possible to consider radiology in terms of three well defined areas of activity:

i) the referral process including a patient case history;

ii) the radiological examination including image production;

iii) the clinical diagnostic process from the imaging data.

Each of these areas of endeavour are of course inter-related and should be considered in any overall quality assurance framework. In fact within the breast screening programme in the United Kingdom quality assurance covers all these different aspects. Review mechanisms and ongoing education and training for radiologists are already built into the operational framework. Of course, this aspect alone is amenable to expert system applications. However within the framework of radiation protection the most obvious area for an expert system approach concerns the radiological examination including image production.

4. Expert Systems in Quality Assurance

An expert systems approach to quality assurance related to the radiographic process is a distinct possibility. Particularly following the Commission of European Communities recent initiative within the field of radiation protection. This initiative is concerned with the development of image quality and patient dose criteria for diagnostic radiology (BIR Report 20 (1989)). For the first time the three key elements relating to the production of a radiographic image have been highlighted and preliminary attempts made to define them. These three elements consist of:

a) quality criteria for the normal anatomy of a particular radiographic view;
b) the maximum entrance surface dose for a particular examination for an average size patient;
c) the radiographic factors which should lead to the production of the necessary image quality criteria at an entrance dose less than the suggested upper value for an average size patient.

These three components adequately define the complete framework for the knowledge base necessary to express how to satisfactorily perform a radiographic examination. Thus one of the more important requirements for the application of an experts systems approach can be met. Mainly a well defined area of activity with minimal links to other areas. The key to meeting this requirement lies in the fact that the image quality criteria are defined solely in terms of the normal anatomy together with some indication of generalised detail. Consequently the patient is in some ways treated as a test phantom in which image quality can be defined in virtually every examination of a particular type for which image quality criteria have been drawn up. Only for those patient examinations in which the disease process is so extensive that it actually interferes with, or masks, the normal anatomy may this approach not apply. Consequently when attempting to apply expert system programme to the third components of the radiological process outlined previously, namely the image reading or clinical diagnostic process it would be necessary to try and relate key diagnostic determinants contained in the patients medical records to any abnormalities present in the radiographic image. However by employing the normal anatomy, as a basis for image quality criteria this link is broken.

The framework for an expert system approach to quality assurance in the radiographic or image production process is concerned with establishing and quantifying any relationships between the presence or not of relevant image quality criteria, the dose employed in an examination and a choice of radiographic factors. Although this framework is somewhat complicated it has at least now been defined.

An initial attempt at exploring the basis of an expert system approach to quality assurance in the radiographic process has already been performed. A preliminary evaluation of the C.E.C. image quality criteria for diagnostic radiographic images was performed in 24 X-ray departments located in 10 European countries. The results of this trial have already been presented (Maccia et al. (1989)) and been discussed at this present course, therefore I do not intend to discuss them in detail. Nonetheless in evaluating the results of this trial it was necessary to try and study and explore the presence or not of individual quality criteria and any relationship to radiographic factors including patient dose.

We consider the P.A. chest examination as an example. The image quality criteria deemed to be of most importance are:

i) symmetrical reproduction the thorax;

ii) reproduction of vascular pattern in the lung periphery;

iii) reproduction of the costopleural boundary from the apex of the lung to the diaphragm.

iv) visually sharp reproduction of the peripheral vessels in the border of the heart and the diaphragm;

v) visualisation of the retrocardial lung and mediastimum;

vi) performed at inspiration (minimum of six anterior costal arches).

In order to attempt to "quantify" the degree of compliance of a particular radiograph in meeting these requirements some form of quantification or numerical index is required. In analysing the C.E.C. trial data Maccia et al. assigned to each individual criteria a medical requirement M in which all potentially visible anatomical structures should be shown on the radiograph, a technical requirement T in which all abnormal or pathological details should be adequately contrasted and a positioning requirement P in which the part of the body projected on the film should correspond to the field size. A multi-numerical scoring system was then applied to each individual criterion with a score allocated to each of the M-T-P requirements. The image scoring system for the P.A. chest is shown in Table 1.

Based upon the presence or not of particular image quality criterion a radiographic image could be crudely quantified in term of quality. In the analysis of the C.E.C. trial data attempts were then made to study the relationship between the image quality score and other aspects of the examination like entrance dose and radiographic technique. This work represents a preliminary step in exploring the framework of an expert system programme. Once this framework has been explored more fully it should be possible to create a logical sequence of intelligence "prompts" for those undertaking particular radiographic examinations which can act as a positive aid in achieving satisfactory image quality at acceptable levels of radiation.

The exact form of this framework and the way it is structured for human interrogation is not clear since a full analysis of the knowledge base needed has not yet been performed. Nonetheless we may highlight some of the possibilities which could be considered. For instance it may be beneficial to interrogate the knowledge base in terms of the image quality criteria, the radiographic factors or alternatively the patient dose. For instance a sequence within the knowledge base for a

TABLE 1. Image criteria scoring system.

Image Criteria (PA)	M	T	P
i) Symmetrical reproduction of the thorax	2	2	2
ii) Reproduction of vascular pattern in the lung periphery	2	2	0
iii) Reproduction of the costopleural boundary from apex of lund to the diaphragm	2	3	1
iv) Visually sharp reproduction of the peripheral vessels, in the border heart and diaphragm	1	2	1
v) Visualisation of the retrocardial lung and the mediastinum	3	3	1
vi) Performed at inspiration (minimum of 6 anterior costal arches)	2	2	2
Total score	12	14	7

Scoring system: 0 = irrelevant, 1 = minor, 2 = important, 3 = fundamental
M = Medical score,
T = Technical score,
P = Positioning score.

particular radiological examination might read:

IF 1) if the patient's height is less than 1.8 metres;
2) if the patient's weight is greater than 80 kg;
3) if the KV employed is less than 85 KV;
4) if a screen film combination of speed class less than 400 is employed;
5) if a grid ratio greater than 10 and grid lines of more than 40 cm is employed.

THEN the patient entrance surface dose will definitely exceed the maximum recommended for an average sized patient (0.95).
It will probably be more than twice this value (0.7).

Alternatively the expert system may be expressed in terms of the relationship between the visualisation of a particular image quality criterion and the radiographic factors. We can see in Table 1 that the preliminary analysis of the C.E.C. trial attempted to relate the importance of the technical aspects of an examination T to the presence of a particular criterion in the final image. Obviously the technical aspects consist of many interrelated factors and much work will need to be undertaken in order to try and clarify the relative importance of given technical factors to the visualisation of particular criterion. This work is presently underway.

5. Conclusion

Know-how or knowledge based or expert systems can play an increasing role in raising the standards of performance of human tasks. Particularly if those tasks are well defined and a knowledge base of suitable refinement can be generated for the task. The field of diagnostic radiology is an extremely interesting field for the application of such techniques offering as it does the possibility for improving the diagnostic process as well as reducing the population radiation dose arising from this activity, which is the largest single man-made source of population dose.

A framework for work now exists in this field which is encompassed by the image quality and patient dose criteria developed by a Working Group of the Radiation Protection Programme of the Commission of European Communities. This framework now contains all the essential elements of the radiological image production mechanism which affect the levels of radiation received by patients undergoing an examination. The inter-relationships between the technical aspects of an examination, the reproduction of relevant clinical structures reproduced on a radiograph and the patient dose need to be examined carefully. However, this analysis should provide the basis for the implementation of an expert system which can aid both radiologists and radiographers to achieve consistently high standards of performance throughout the whole of Europe.

6. References

BIR Report 20 (1989) "Optimisation if Image Quality and Patient Exposure in Diagnostic Radiology" ed by B.M. Moores, B.F. Wall, H. Eriskat and H. Schibilla (British Institute of Radiology, London).

Maccia, C., Wall, B.F., Padovani, R., Shrimpton, P., Husson, B. (1989) "Results of a trial set up by a Study Group of the Radiation Protection Programme of the C.E.C. in Optimization of Image Quality and Patient Exposure in Diagnostic Radiology" ed by B.M. Moores, B.F. Wall, H. Eriskat and H. Schibilla. BIR report 20 (British Institute of Radiology London).

RISK ASSESSMENT FOR THE MEDICAL USES OF IONISING RADIATION

B.F. WALL
National Radiological Protection Board
UK - Chilton, Didcot, Oxon OX11 ORQ
United Kingdom

1. A Measure of Harm

A measure of the harm incurred from an exposure to radiation should take account not only of the probability of all possible types of injurious effect occurring but should also include, if possible, some measure of the severity of the effects. ICRP (ICRP (1977)) has introduced the concept of radiation health detriment G, to express this measure of harm from a source of radiation, given by:

$$G = N \sum_{T} r_T \, s_T \, \bar{d}_T$$

where r_T is the incidence of an adverse health effect in tissue T per unit dose, s_T is a weighting factor to express the severity of that effect and d_T is the mean dose to tissue T received by the N people exposed to the source. The product of these factors is to be summed over all tissues for which deleterious effects of ionizing radiation are known.

At the dose levels generally encountered in diagnostic radiology, the threshold doses for non-stochastic radiation effects such as opacification of the lens of the eye will not normally be approached and stochastic effects only need be considered in most instances. These stochastic effects, for which the probability of occurrence and not the severity is proportional to dose, can be divided into two types; somatic and genetic. The somatic effects are mainly radiation induced cancers appearing in the exposed individuals many years after they have been irradiated. The genetic effects are the result of damage to reproductive cells in the ovaries or testes and appear most frequently as inherited disease in the descendents of the exposed individuals and may not occur for many generations. If the exposed individual is pregnant,

M. Oberhofer (ed.), Advances in Radiation Protection, 287–303.

the fetus may be regarded as an additional individual who may also suffer somatic and genetic radiation effects. For the fetus somatic effects take the form of either abnormal development which is mostly evident as mental retardation or increased probability of childhood cancer.

Factors expressing the induction of these stochastic effects per unit dose (r_T) and a measure of their severity (s_T) are needed to estimate the radiation health detriment G.

2. Choice of Appropriate Induction and Severity Factors

Predictions of the late effects of low level radiation, are based mostly on epidemiological surveys of a limited number of groups of people exposed to considerably higher radiation levels than those experienced in diagnostic radiology. The number of radiation induced effects observed in these highly exposed groups is still very small compared with their natural incidence, leading to large statistical uncertainties in determining the excess level of effects and hence r_T even at high doses. Moreover, there is considerable controversy concerning the appropriate model to adopt for extrapolating from effects observed at high doses to the likely response at low doses. While ICRP assumes that a linear dose-response model prevails over the range of doses experienced by those requiring protection, it uses values of r_T to predict detriment in this dose range that are lower than those observed at substantially higher doses. This is in line with radiobiological experiments that indicate a curvilinear dose-response model (concave upwards), for low LET radiation, over the entire dose range from the level of observable effects down to protection level doses. A similar approach is adopted in this paper.

Some tissues demonstrate a dependence of r_T on the sex or age of the exposed individual. Sex differences in radiosensitivity sometimes reflect differences in the corresponding natural cancer rates. Leukaemia with a slight preponderance in men, thyroid cancer being about three times more prevalent in women and breast cancer almost entirely restricted to women, are cases in point. For other types of cancer a sex dependence in sensitivity is not so evident. In the absence of more reliable data, a sex difference in radiosensitivity will be ascribed only to breast cancer in this paper. The effect of age at exposure on the opportunity for expression of somatic and genetic effects is taken into account by a separate "expression factor" discussed in section 3.

The factor s_T denoting the relative severity of the health effect induced in tissue T will have a value of one if detriment is equated to the risk of induction. If the risk of death from the effect is considered to be a more appropriate measure of the detriment, then s_T would be the fraction of individuals dying from the effect after a prescribed time according to established morbidity and mortality statistics. These are rather crude estimates of the severity of radiation induced disease, taking no account of the degree of incapacitation or suffering prior to death, nor indeed of the significance of any delay between exposure and first appearance of the disease. Considerations such as these are difficult to quantify although concepts like the number of days of life lost have been invoked to refine the measure of severity.

In the context of protecting populations of workers ICRP assumed values of s_T appropriate for fatal cancers when dealing with somatic effects, and values appropriate for the incidence of serious hereditary defects in the first two generations when dealing with genetic effects. This no doubt represents a reasonable compromise in the situations for which effective dose equivalent was defined but may require modification for use in estimating hazards in diagnostic radiology. It might, for example, be preferable to express the detriment from the somatic and genetic effects of radiation separately, with s_T set at unity in each case, rather than attempting to combine them using debatable assumptions regarding their relative severity. Possible effects on the fetus might also be considered separately. This is the approach adopted in this paper.

The human epidemiological data that were available prior to 1988 for establishing values of r_T were reviewed by the United Nations scientific Committee on the Effects of Atomic Radiation (UNSCEAR (1977)) and the Committee on the Biological effects on Ionizing Radiation (BEIR (1980)), an expert scientific advisory committee of the US National Academy of Sciences - National Research Council.

The National Radiological Protection Board has needed estimates of the induction of stochastic health effects for assessments of the consequences of accidental and planned releases of radionuclides from nuclear power stations. The values used in their calculations prior to 1988, based on a balanced judgement of the reports published by the above committees and other more specific data, are discussed by Clarke and Smith in NRPB Report R102 (Clarke (1980)), and by Pochin in his Proof of Evidence in the Sizewell B Inquiry (Pochin (1983)). NRPB report R171 (National Radiological Protection Board (1984)) also contains an analysis of risk coefficients for leukaemia and other cancers that were used to assess the risks to members of the public from emissions and discharges from the Sellafield reprocessing plant for the Sir Douglas Black Inquiry in 1984. Table 1 gives illustrative examples of risk factors for the induction of somatic and genetic effects based largely on these publications.

TABLE 1. Illustrative Risk Factors for Induction of Stochastic Health Effects

Stochastic Effect	Risk of Induction (r_T, 10^{-4} Gy^{-1}) Male	Female	Mortality Rate (%)
Breast cancer	0	100 (<50y)	50
Leukaemia	20	20	100
Lung cancer	20	20	100
Thyroid cancer	100	100	5
Osteosarcoma	5	5	100
Skin cancer (non melanoma)	100	100	1
Other cancers (combined)	50	50	100
Total cancers	295	395	
Genetic effects(*)	200	200	
Effects on fetus			
Malformations(**)	4000		
Fatal childhood malignancies	230		

(*) Incidence of serious hereditary defects in *all* future generations.

(**) When irradiated between the 8th and 15th week of gestation.

Appropriate mortality rates are also shown. Similar mortality rates have been applied in deriving the weighting factors that represent the relative probability of inducing *fatal* cancers in the definition of effective dose equivalent in ICRP Publication 26. The very low mortality associated with thyroid and non melanoma skin cancers provides the justification for treating them as being less severe than other radiation induced cancers.

Risk factors for the somatic effects of radiation are currently being revised due to a reassessment of the doses received by Japanese atomic bomb survivors, a larger period of follow-up for this important exposed population and increasing evidence for a "relative" rather than "absolute" model for most solid cancers. These revisions will probably lead to more cancers being predicted per unit dose, but with a larger delay between exposure and expression than is assumed in section 3 of these notes. The overall effect of these changes on the risks predicted in section 5 is difficult to predict at the present time, but is the subject of further study.

The effects on the fetus have been divided into two types at the bottom of Table 1; malformations and fatal childhood malignancies. Mental retardation following in utero exposure to atomic bomb radiation at Hiroshima and Nagasaki is the only effect of malformations observed in humans at low doses. Recent analysis of the times of exposure of the affected children show an apparent sensitivity for the induction of mental retardation that is largely limited to the period from the 8th to the 15th week of gestation (Otake and Schull (1984)). During this 8 week period the risk factor (0.4 Gy^{-1}) is relatively high compared with the other somatic and genetic risk factors shown in Table 1. There is some evidence for a threshold for this effect at about 300 mGy, but it is not conclusive. The risk factor for *fatal* childhood malignancies on the other hand (about half of which will be leukaemias) is only about 1.5 times the mean induction rate for *fatal* cancers in adult females.

3. Opportunity for Expression of Long-Term Radiation Effects

The induction factors for somatic effects shown in Table 1 assume that the exposed individuals will live long enough for the risk to be expressed, and those for genetic effects assume full potential for parenthood. Many types of X-ray examination are conducted on predominantly elderly and sick patients whose poor prognoses will reduce their probability of suffering the delayed somatic effects of radiation before death from other causes intervenes. The opportunity for expression of the genetic effects of radiation also falls with increasing age in line with the reduced expectation of parenthood.

To estimate the degree of reduction in expression of radiogenic cancers knowledge is required of their pattern of incidence with time following exposure. However, current epidemiological surveys are incomplete in the sense that no exposed human population studied so far has been followed to the point where all its members have died.

Predictions regarding the timing of occurrence of radiogenic cancers depend on assumptions about the way radiation affects natural cancer rates; i.e. are the effects additive or multiplicative? Since natural cancer rates generally increase with age, a multiplicative (relative risk) model will predict more excess cancers in the future than a purely additive (absolute risk) model. The epidemiological studies so far are inadequate to resolve this question completely.

However, in the case of leukaemia (and bone sarcoma induced by ingested ^{224}Ra) the rate of incidence appears to peak between 8 and 15 years after exposure and to return almost to the natural incidence rate after about 30 years. This pattern of incidence can be represented approximately by an additive risk model with a log-normal distribution in time of median value 12.5 y and standard deviation 0.8 as shown in Figure 1.

Solid radiation induced cancers, however, do not generally appear for a few years following exposure, reach a peak after about 20 years and continue to occur for at least a further 20 years. This total period of 40 years since exposure is the longest time for which exposed populations have been studied so far. For the purposes of this paper an additive risk model and a log-normal distribution in time have also been assumed for solid cancers, but the median value has been increased to 25 years with a standard deviation of 0.4. This model is also shown in Figure 1 and provides a latent period of about six years following exposure when no excess cancers are observed, followed by a gradual increase in incidence rate which peaks shortly after 20 years and then gradually declines but has not reached zero even 50 years after exposure.

Demographic data for the population of England and Wales published by the Office of Population Censuses and Surveys (OPCS) include information on the probability of surviving for a further year as a function of age and sex. These probabilities ("life tables") have been combined with the above incidence patterns to derive the curves shown in Figure 2, representing the probability of expression (E_i) for leukaemia and solid cancers as a function of age at exposure and sex. For men the probability is seen to fall to about one half by age 65 for leukaemia and by age 50 for solid cancers, while for women the corresponding ages are about six years older, due to their longer life expectancy.

The opportunity for expression of genetic effects of radiation also falls with increasing age, in line with the reduced expectation of parenthood. Child expectancy data as a function of age and sex are also available from OPCS and can be used to generate similar curves to those above, representing the probability of expression of *genetic* effects as a function of age at exposure and sex. These are shown in Figure 3, where it can be seen that the probability has fallen to one half by age 30 for both sexes.

The risk coefficients used by ICRP in deriving the weighting factors for the quantity effective dose equivalent, assume full opportunity for expression of all fatal cancers, but take account of the reduction in opportunity for genetic effects in a general population. The risk coefficient of 10^{-2} Sv^{-1} to either parent for serious hereditary defects occurring in the first two generations, is reduced to 40%

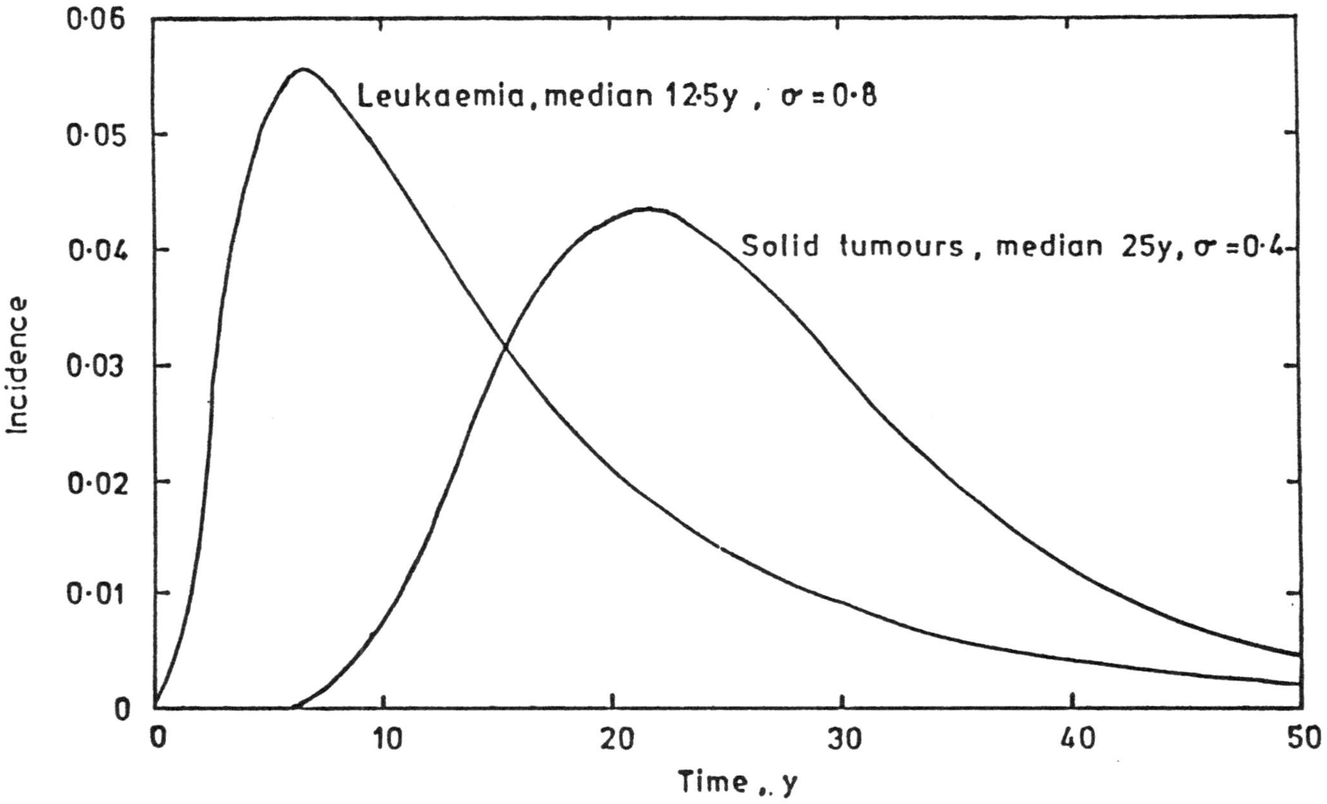

Figure 1. Models assumed for distribution in time of radiogenic cancers, following exposure.

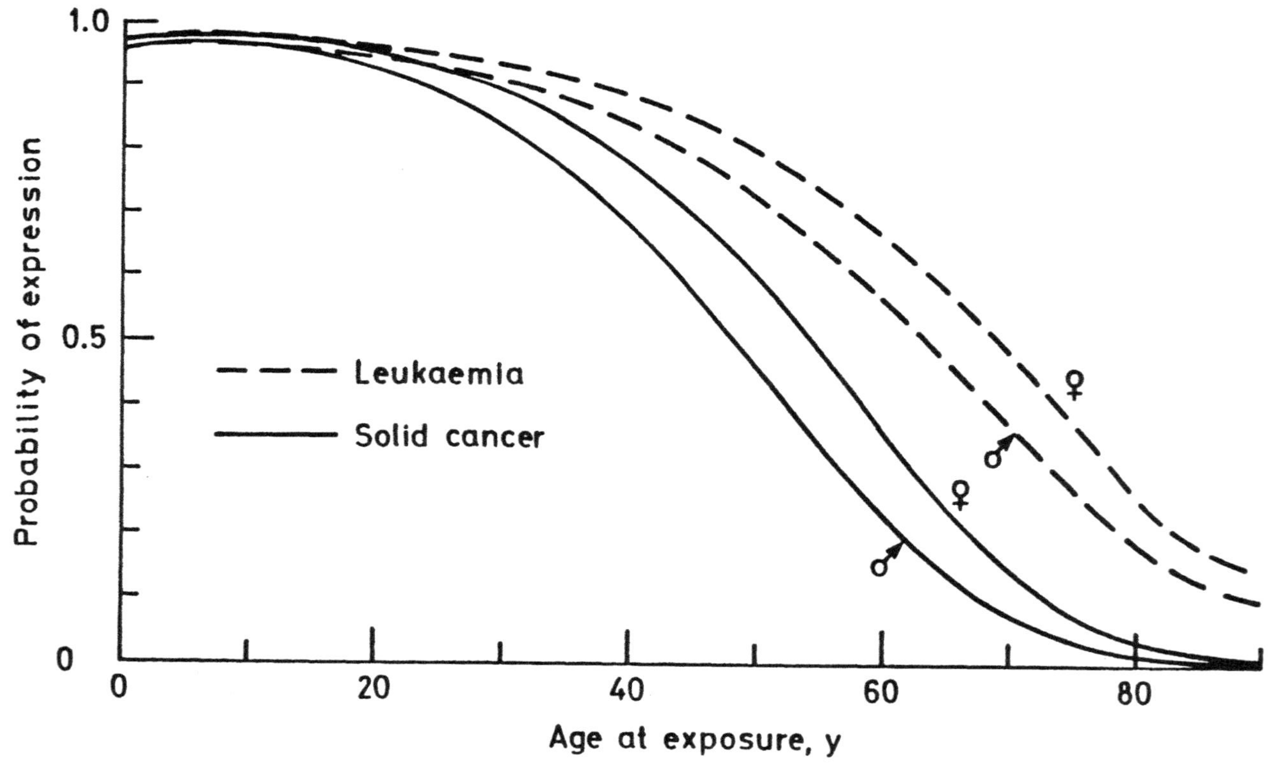

Figure 2. Expression factor curves for leukemia and solid cancers.

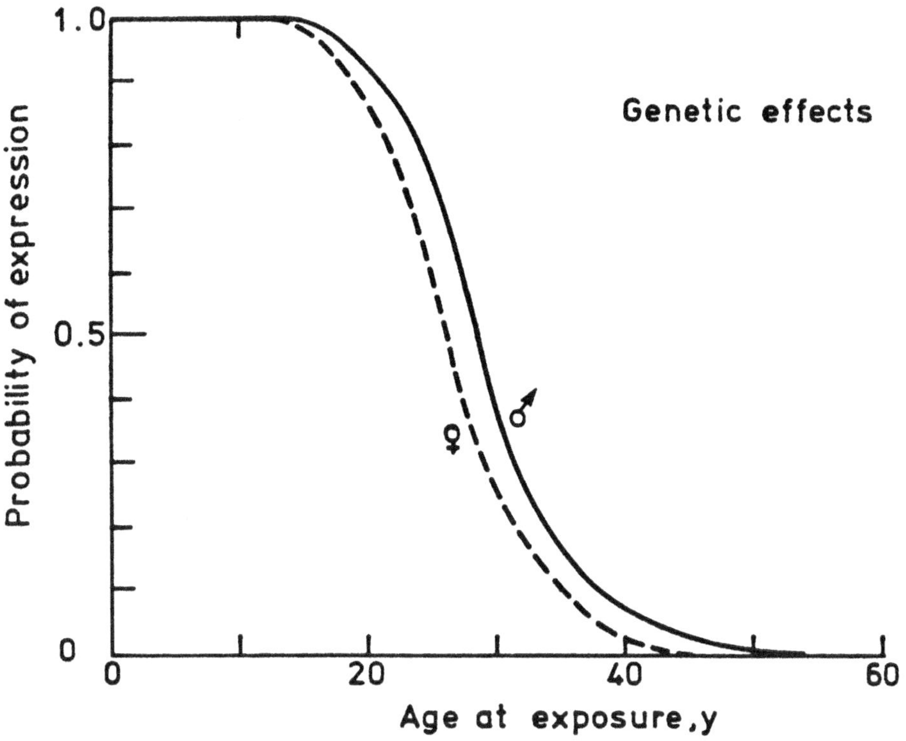

Figure 3. Expression factor curves for genetic effects.

(4.10^{-3} Sv^{-1}) to allow for only 40% of a *working* population being potential parents. The opportunity for expression of both somatic cancer risks and genetic risks will be reduced still further in exposed groups containing higher proportions of elderly people than the average working population.

4. Reduction Factors for Patients

Age distributions for the general population of Great Britain in 1983 and for patients undergoing all types of X-ray examinations combined (except those carried out by dentists) have been taken from (Otake and Schull (1984)) and are shown in Figure 4. The distribution for patients is remarkably uniform between the ages of about 10 and 75 whereas the general population starts to drop rapidly in numbers above 65 for men and 70 for women. Individual types of examination however, demonstrate markedly different age and sex distributions (Otake and Schull (1984)). Examinations of fractures in the limbs, for example, occur most fre-

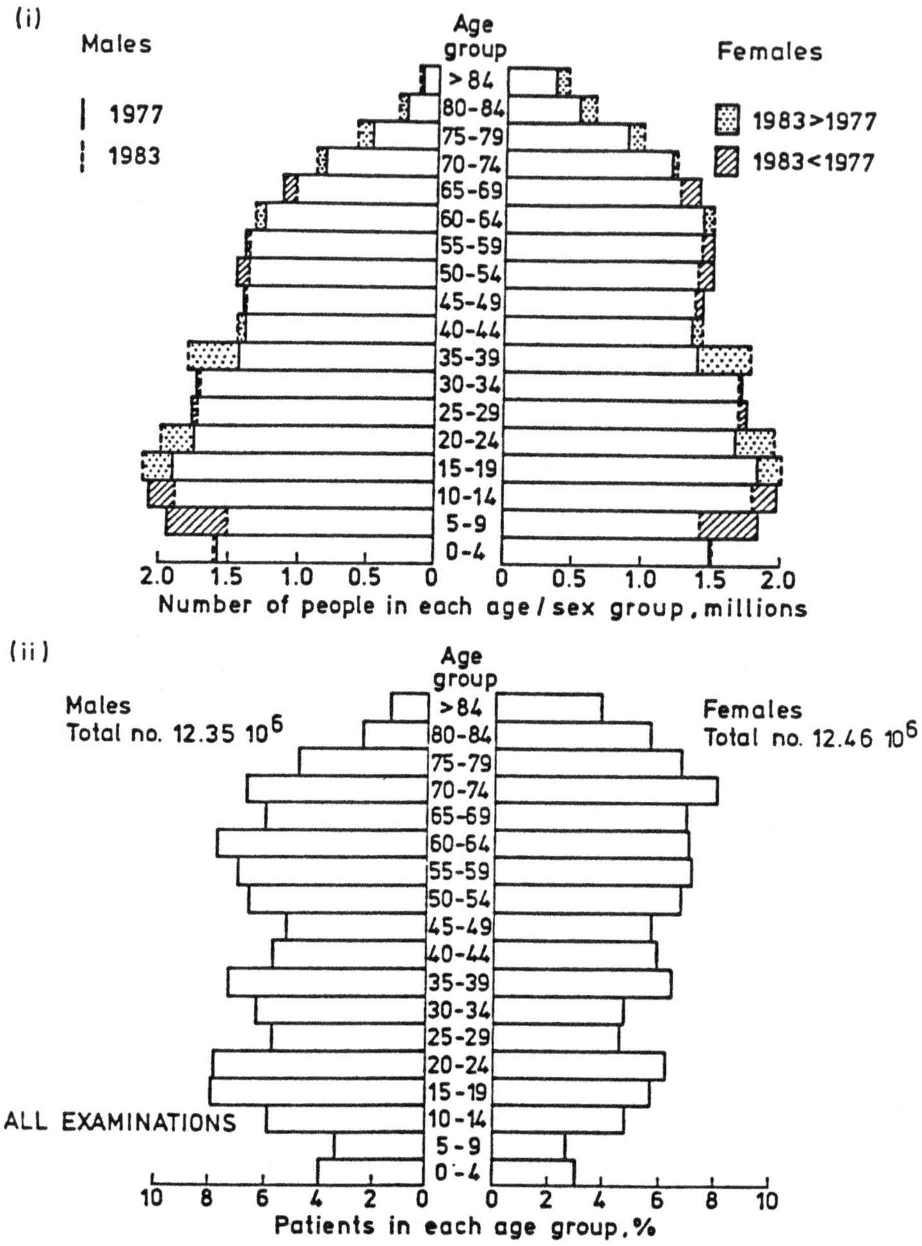

Figure 4. (i) Population pyramid for England and Wales in 1977 and 1983. (ii) Population pyramid for patients undergoing all types of X-ray examination in Britain in 1983, except those conducted by dentists.

quently during accident-prone childhood and adolescence, while examinations associated with the diagnosis of degenerative diseases (e.g. barium enemas) are concentrated in the elderly, as shown in Figure 5.

The reduction in the opportunity for expression of radiation effects in populations of patients can be calculated by averaging the expression factors in Figure 2 and 3 over the appropriate age/sex distributions. I.e.:

$$\text{Reduction factor} = \frac{\sum_i N_i E_i}{\sum_i N_i}$$

where N_i is the number of patients in age/sex groups i and E_i is the corresponding expression factor. Reduction factors calculated in this way are shown in Table 2 for nine types of routine X-ray examination, which between them contribute over 80% of the collective dose to the population of Britain from diagnostic radiology. The factors vary between examinations, depending on the age distribution of the patients, but on average indicate a reduction to about 0.7 for leukaemias, 0.5 for solid cancers and 0.2 for genetic effects. These are substantial factors and a measure of detriment from diagnostic X-ray should take them into account.

TABLE 2. The reduction in opportunity for expression of radiation effects for patient undergoing X-ray examinations.

Examination	Reduction Factor		
	Leukaemia	Solid Cancers	Genetic Effects
Skull	0.78	0.64	0.37
Chest	0.66	0.46	0.18
Thoracic spine	0.70	0.56	0.33
Lumbar spine	0.73	0.55	0.19
Abdomen	0.68	0.52	0.25
Pelvis	0.66	0.47	0.26
IVU	0.70	0.54	0.27
Barium meal	0.70	0.49	0.13
Barium enema	0.61	0.38	0.08
Mean	0.7	0.5	0.2

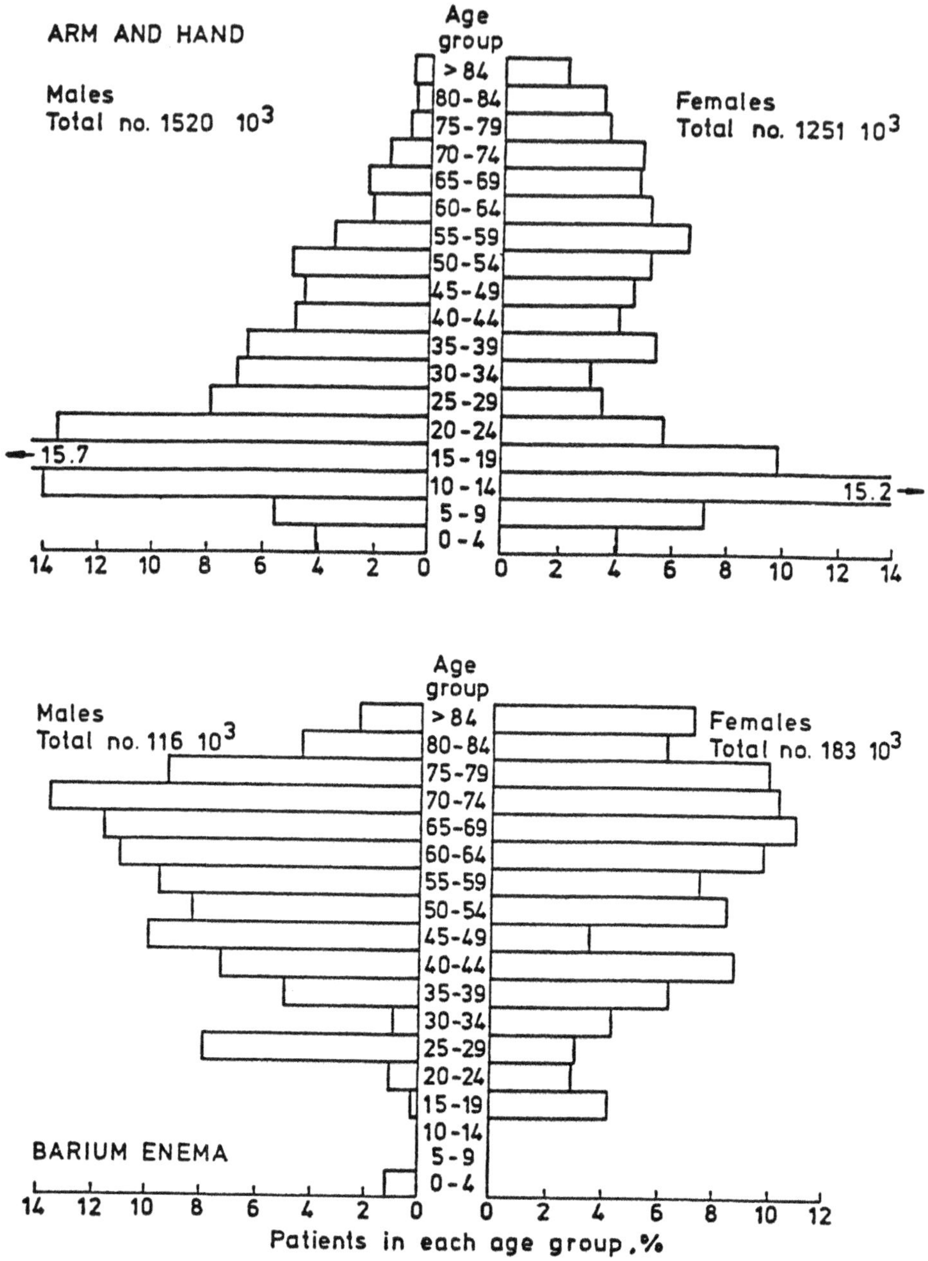

Figure 5. Population pyramids for specified examinations in Britain in 1983.

5. The Average Individual Risk from Routine X-ray Examinations

A more complete expression for the detriment associated with diagnostic X-rays, that takes into account the age and sex of the patients, consequently becomes:

$$G(i) = \sum_{i} \sum_{T} N_i\ E_i\ r_T\ s_T\ \bar{d}_T$$

Since X-ray examinations should be justified individually for each patient, it is more useful to express the detriment as a risk value for an average individual rather than collectively as above.

$$\text{Average risk} = \frac{\sum_{i} \sum_{T} N_i\ E_i\ r_T\ s_T\ \bar{d}_T}{\sum_{i} N_i}$$

and can be regarded as the weighted sum of the probabilities of all adverse health effects occurring in the average patient. This measure of risk will not apply precisely to every individual within the exposed group of patients, but only to an individual with average characteristics (in terms of susceptibility to radiation and life expectancy) and receiving the average exposure.

It is also important to realise that while the effect of age and sex on life expectancy has been taken into account, the data for doing this have been taken from studies on the whole population of England and Wales. Some types of X-ray examination are carried out on severely sick patients whose life expectancies will be less than these national averages. The risk to these patients will be less than values estimated using the reduction factors of Table 2.

Table 3 shows the average individual risks for nine types of routine X-ray examination. The severity factor s_T has been set to unity and the risks have been expressed separately for fatal and non-fatal somatic effects, genetic effects and effects on the fetus, to avoid the problem of quantifying their relative severity. Values of r_T have been taken from Table 1 suitably adjusted by the mortality rate, and the reduction factors have been taken from Table 2. The mean organ doses (d_T) used are those reported by Shrimpton in a British national survey of patient doses (Wall et al. (1986)). The somatic and genetic risks represent a weighted average for the spectrum of patients exposed to each type of examination. The fetal risk is only of relevance to patients pregnant at the time of exposure, and the probability of producing fetal malformation resulting in serious mental retardation is

TABLE 3. Average individual risks for nine types of X-ray examination.

Examination	Probability of Radiation Effect Occurring x 10^{-6}							
	Somatic				Genetic		Fetal	
	Fatal		Non-Fatal		Male	Female	Fatal Cancers	Malformation
	Male	Female	Male	Female				
Skull	1.7	1.7	4.7	4.7	-	-	-	-
Chest	0.27	0.47	0.11	0.32	-	-	-	-
Thoracic spine	7.0	11	13	16	-	-	-	-
Lumbar spine	25	26	11	11	0.2	16	80	1384
Abdomen	9.4	9.5	3.9	4.0	2.0	11	66	1160
Pelvis	3.9	3.9	2.4	2.4	24	6.3	38	660
IVU	26	37	20	30	23	19	82	1430
Barium meal	26	31	15	21	0.8	9.4	83	1440
Barium enema	37	38	20	21	5.4	26	370	6400

further restricted to exposures occurring between the 8th and 15th week of gestation (assuming a linear dose-response relationship with no threshold).

It can be seen that the estimated somatic risks range from less than one in a million for a chest examination to about one in 30,000 for the more complex examinations at the bottom of the Table, for both fatal and non-fatal cancers. Genetic risks rarely exceed the one in 50,000 level for these common types of X-ray examinations. Risks to the fetus, however, can be appreciable for those examinations involving direct fetal irradiation. Barium enemas performed on pregnant patients, for example, are estimated to involve a risk of about one in 3,000 for inducing fatal childhood cancers and a risk of about one in 200 for producing severe mental retardation but only if the examination is made between the 8th and 15th week of pregnancy. The approximate ranges of these risks for the nine selected types of X-ray examination are indicated by the lengths of the boxes in Figure 6.

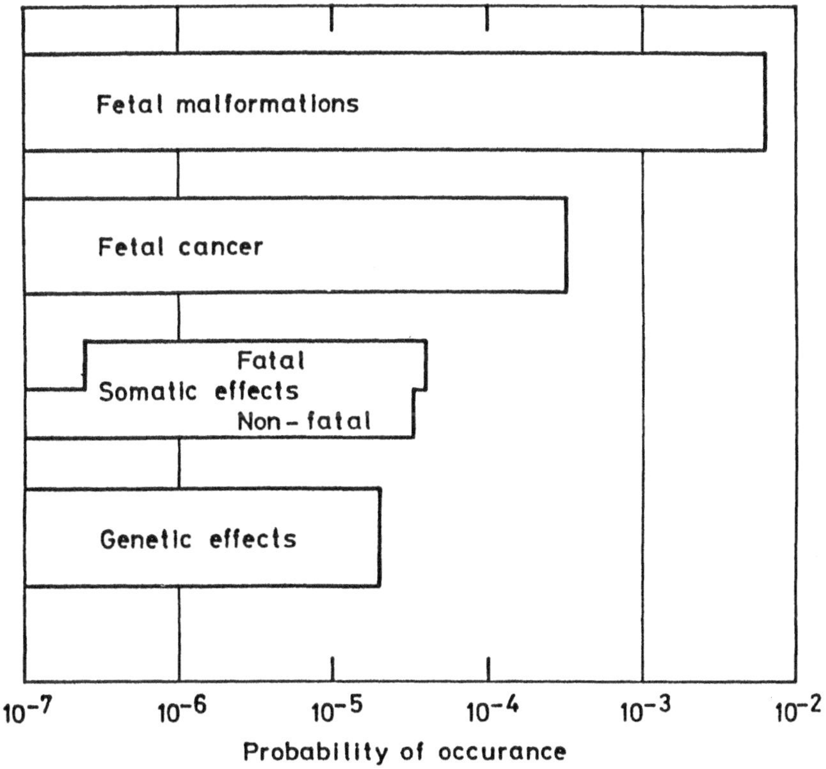

Figure 6. Ranges of risks for nine common X-ray examinations.

6. Discussion

Risk assessments in any field are invariably accompanied by large areas of uncertainty; this analysis is no exception. The probabilities presented in Table 3 provide only an approximate guide to the levels of risk involved in medical X-ray examinations. Different assumptions about the dose-response curve, the timing of appearance of long term effects after exposure, age and sex related differences in susceptibility to radiogenic cancers and the influence of patients' health on their life-expectancy, could all lead to revised risk estimates ranging from zero to at least an order of magnitude greater than the values quoted here.

The significance of uncertainties of this magnitude, however will depend on the level of risk involved. A simple summation of the somatic and genetic risks for each type of examination leads to total probabilities of serious adverse health effects occurring, ranging from $0.5 \cdot 10^{-6}$ for chest examinations to nearly 10^{-4} for intravenous urograms (IVU) and barium anemas. A possible uncertainty of a factor of 10 might affect decisions on the justification and acceptability of an IVU or barium enema but is of far less significance for a chest X-ray.

Broad guidance on the acceptability of risks has been provided by a report on risk assessment by a study group of the Royal Society (Wall et al. (1986)). To quote from the conclusions of the report:

"While it is clearly not possible to set single quantitative guidelines on risk acceptability some broad indicators of the current position can be noted. If the average expectation of life is 70-75 years, then the imposition of a continuing annual risk of death to the individual of 10^{-2} seems unacceptable. At 10^{-3} it may not be totally unacceptable if the individual knows of the situation, enjoys some commensurate benefit, and everything reasonable has been done to reduce the risk. At the other extreme, there are levels of assumed annual risk so low that the manager or regulator can regard them as trivial. The Study Group judges this figure to be commonly about 10^{-6}. It might, in some circumstances, be ten times less (10^{-7}) and an annual risk ten times as great (10^{-5}) in travelling by train does not cause the ordinary traveller any concern.

It is principally in between these upper and lower levels that risk management should consist of comparing risks, detriments, costs and benefits.

Above the upper level, the risks must be reduced or the operations suspended, whatever the cost. Below the lower one even the effort needed to make the comparisons cannot be rationally justified".

It is recognised that X-ray examinations do not strictly constitute a "continuing annual risk" and not all the adverse effects considered in this paper are fatal. However the average person in Britain receives an X-ray examination once every 18 months and in most other European countries they are more frequent. Even if only the fatal somatic effects are considered, all types of examination, with the possible exception of chest X-rays, fall within the upper and lower levels defined by the study group, where efforts to justify and optimise the exposures are required. These efforts must however take account of the high level of benefit perceived by the majority of patients from all types of health care, including diagnostic radiology. Higher levels of risk than normally accepted elsewhere are willingly undertaken by patients seeking relief from the anxiety and pain of real or suspected disease or injury.

7. References

BEIR (1980) "The effects on populations of exposure to low levels of ionizing radiation", (National Academy of Sciences, Washington DC).

Clarke, R.H. and Smith, H. (1980) "Calculations of the incidence of stochastic health effects in irradiated populations", NRPB R102 (HMSO, London).

ICRP (1977) "Recommendations of the International Commission on Radiological Protection", ICRP Publication 26 (Pergamon Press, Oxford).

National Radiological Protection Board (1984) "The risks of leukaemia and other cancers in Seascale from radiation exposure", NRPB-R171 (HMSO, London).

Otake, M. and Schull, W.J. (1984) "In utero exposure to A-bomb radiation and mental retardation; a reassessment", Brit. J. Radiol. 57, 409-414.

Pochin, E.E (1983) "The biological bases of the assumptions made by NRPB in the calculation of health effects", NRPB/P/2 (Rev).

Royal Society Study Group (1983) "Risk Assessment" (The Royal Society, 6 Carlton House Terrace, London).

Shrimpton, P.C., Wall, B.F., Jones, D.G., Fisher, E.S., Hillier, M.C., Kendall, G.M. and Harrison, R.J. (1986) "A national survey of doses to patients undergoing a selection of routine X-ray examinations in English hospitals, NRPB-R200 (HMSO, London).

UNSCEAR (1977) "Sources and Effects of Ionizing Radiation", (United Nations, New York).

Wall, B.F., Hillier, M.C. and Kendall, G.M. (1986) "An update on the frequency of medical and dental X-ray examinations in Great Britain - 1983", NRPB-R201 (HMSO, London).

RADIATION PROTECTION IN INDUSTRIAL RADIOGRAPHY

J. VAZ CARREIRO
Laboratorio Nacional de Engenharia e Tecnologia Industrial
(LNETI - Departamento de Protecçao e Segurança Radiológica)
Estrada Nacional, 10
2685 Sacavem - Portugal

1. Introduction

The increase of uses of ionizing radiation in the different activities even outside of the atomic energy industry calls the attention for the need of the application of good and safe working practices in day to day work with radiation. In fact is reported in the literature (Lushbangh et al. (1980)) that the number of radiological accidents world-wide since 1940 to March 1988 were 296 involving about 25000 persons with significant exposure to radiation and with 69 fatalities due to acute effects. It should be also emphasized that the frequency of accidents due to mishandling of radiation devices has risen significantly in comparison with other type of accidents such as the use of radioisotope or criticalities. Approximately 48% of the 296 accidents occurred are due to the misuse of devices with sealed sources. Many of the victims were not educated people but it should be pointed out that professional radiographers are included in the number of injured persons at significant rate.

It is quite reasonable to say that the use of sealed sources constitutes an issue that requires a special concern from the Governments in order to reduce accidents in the working places and to avoid situations similar to those occurred in Juarez (Mexico), Goiana (Brazil) and Morocco.

The use of ionizing radiation in industry implies a great number of activities such as radiography, fluoroscopy, production control through the application of radioisotopes and X-ray sets in thickness, level and density gauging and static elimination, sterilization and so on. It is our purpose in this lecture to deal only with radiation protection in industrial radiography with special attention for the problems related, as was enhanced above, with the prevention and good practice with sealed sources.

M. Oberhofer (ed.), Advances in Radiation Protection, 305–328.

2. Requirements for Industrial Radiography

2.1. General

The industrial radiography is carried out using gamma radiography devices and X-ray apparatus. The proper design of the requirement, a good planning of the work and a sound radiological protection in operation are very important aspects to the overall safety, avoiding unnecessary over-exposures and the misuse of the radiation sources. A adequate maintenance of the equipment, frequent checks following the instruction of the manufacturers and the rules prescribed by the licensing authorities will contribute to assure a continuous good standards of safety operation.

The requirements that equipment should comply are those referred in the following chapter.

2.2. Gamma Radiography Apparatus

2.2.1. General Characteristics. The gamma radiography apparatus normally used comprise one or more sealed sources, a source holder, a projector (exposure container) and accessories (mechanical system and guide tubes).

The radioactive sources used in industrial radiography are usually sealed sources of cobalt 60, cesium 137 and iridium 192 with a great predominancy of the last mentioned radionuclide. The type and characteristics of these sources are shown respectively in Figure 1 and Table 1.

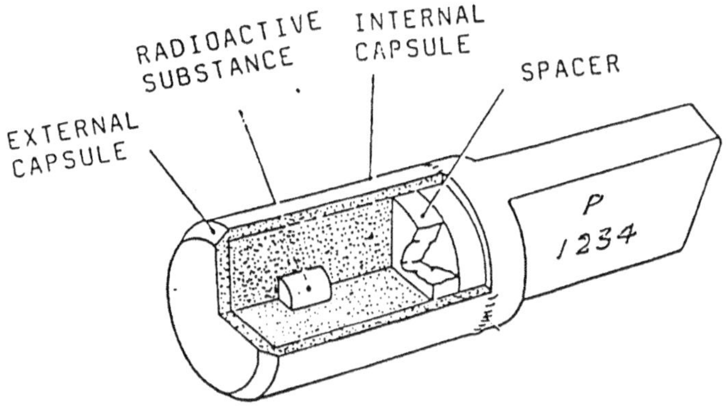

Figure 1. Radioactive source.

TABLE 1. Characteristics of gamma sources used in industrial radiography.

Source	Gamma energy (MeV)	Radio toxicity Group	Half Value Layer (cm)				Exposure
			Steel	Concrete	Lead	Uranium	
Cobalt 60	1.17 - 1.3	2	2.0	6.1	1.2	0.7	1.35
Cesium 137	0.66	3	1.5	4.9	0.7	0.3	0.33
Iridium 192	0.13 - 0.61	2	1.3	4.1	0.6	0.4	0.50

The activity levels for iridium could vary from 0,74 TBq to 7,4 TBq (20 to 200 Ci) and much lower activities for cobalt and cesium sources.

The source holder comprises a receptacle for the source, a cap covering the top of the source, a shielding in rear part and a cable and a fastner for moving and for connection with the ejection cable. In Figure 2 it can be seen, a holder with the radioactive source.

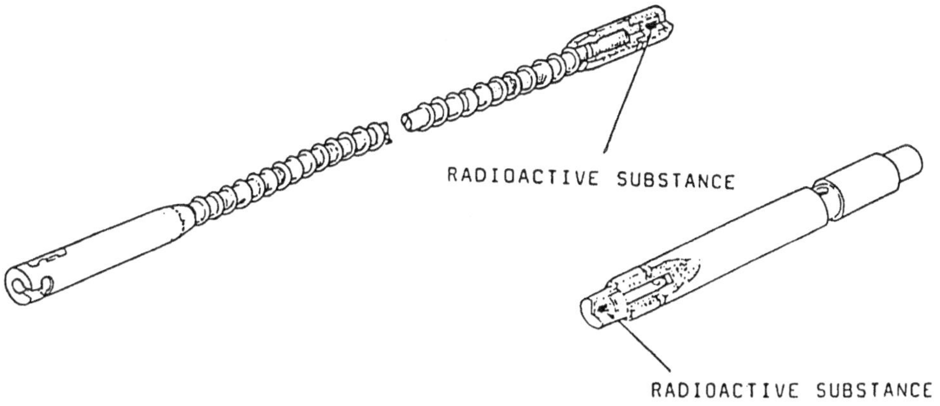

Figure 2. Source holder.

The projector, or exposure container, is a very important piece of equipment from the point of view of radiation protection because it has multiple functions such as an irradiator, a storage container and transport container.

It is common to classify the projectors considering their mobility in the following classes:

Class P: a portable projector designed to be handled by a single person.

Class M: a mobile projector, not portable designed to be moved by a car (or wheels).

Class F: a fixed projector normally used in an enclosure installation.

Considering the characteristics of the ejection facilities the projectors are also classified in three categories such as:

Class 1: without ejection.

The projector enables the emission of the radiation without any movement of source or source holder.

An example can be seen in Figure 3 and is a shutter type container consisting of a block of shielding material at the centre of which the source is located. A portion of the shielding is removed or rotated to expose the source.

Class 2: with ejection, the source holder is linked mechanically to the ejection system.

This type consists of a block shielding with internal tube so that the source when not in use is shielded within the container. The principle is a simplest S-type as it is shown in Figure 3a. The source is exposed when projected out of the container along an external guide tube, as can be seen schematically in Figure 4 and Figure 5.

Class 3: with ejection, the source and holder are not linked mechanically to the ejection system.

It is a type of projector enabling the ejection through a pneumatic, hydraulic or electromagnetic system.

2.2.2. Requirements for Gamma Apparatus.

a) General Specifications

The apparatus for gamma radiography should in general comply with the requirements in the international standard ISO 3999 (ISO 3999 (1977)). Thus the designers and manufacturers should consider primarily the operation conditions of this type of equipment and identify those effects that could have unfavourable incidence in the components of the devices and consequently a diminution of the safety standards. The principal feature of the equipment should be good durability, resistance to corrosion, no damaging effects to extreme temperatures, no radiation damaging effects on non metallic pieces (rubber, plastic, etc.). The prevention of ingress water,

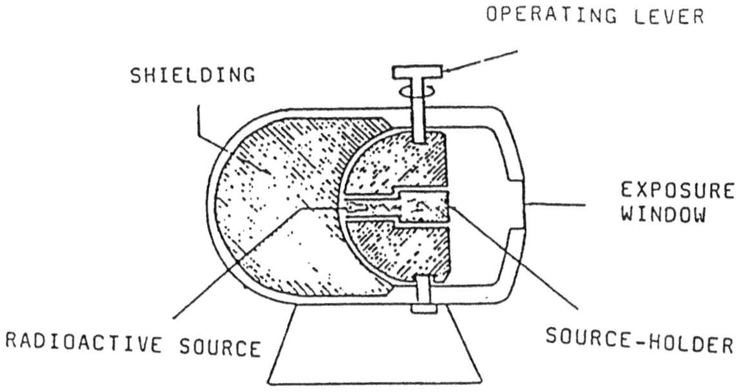

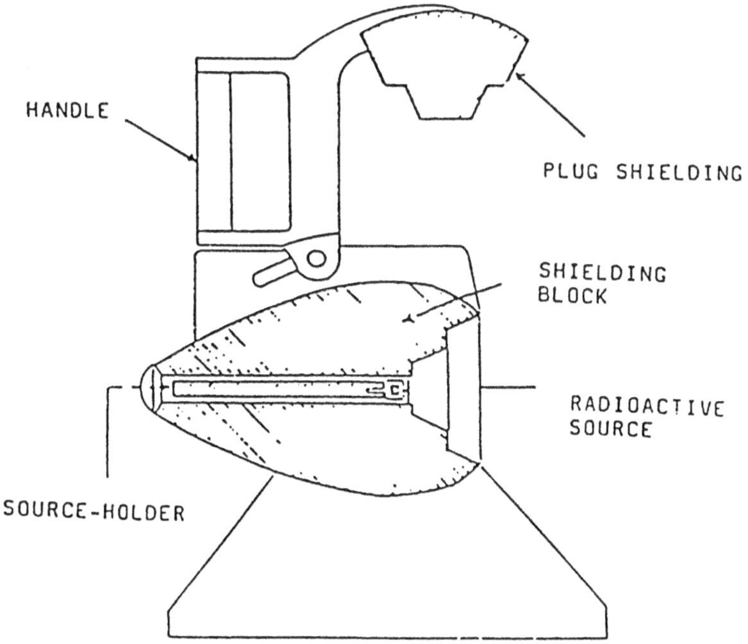

Figure 3. Projection shutter-type exposure containers.

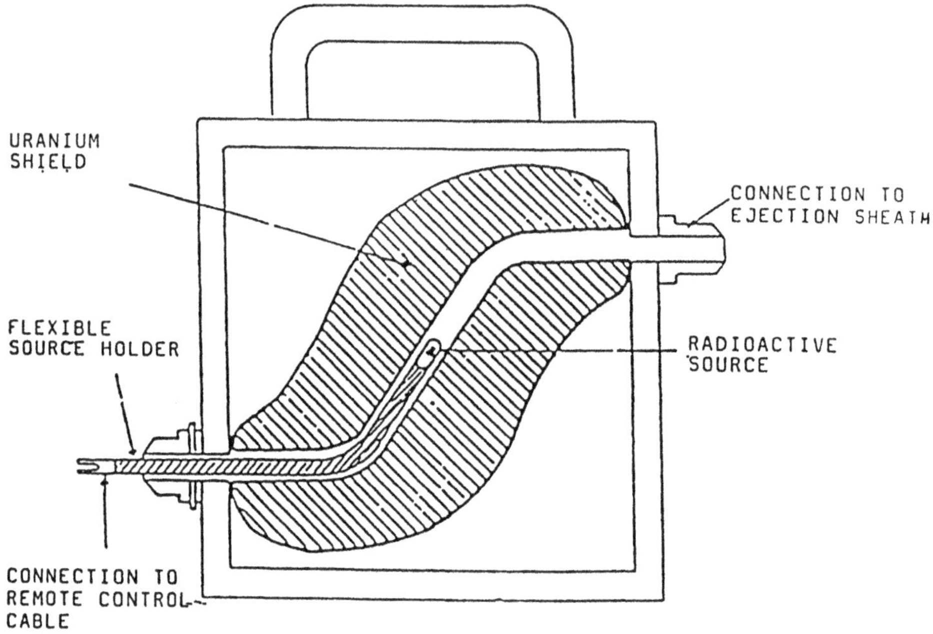

Figure 3a. S bend projection type container.

dust and other matters is also an important parameter to improve safety.

b) Sealed Sources and Holders
The sealed sources should comply with the international standards ISO 1677 "General Requirements For Sealed Radioactive Sources" (ISO 1677) and with ISO 2919 "Sealed Radioactive Source Classification" (ISO 2919).
The source holder should be designed to prevent accidental release of the radioactive substance.

c) Projectors (Exposure Containers)
In what concerns the projectors they should be constructed in order to provide adequate shielding so that the source in storage position and maximum of activity gives outside dose equivalent rates not superior than those referred in Table 2.

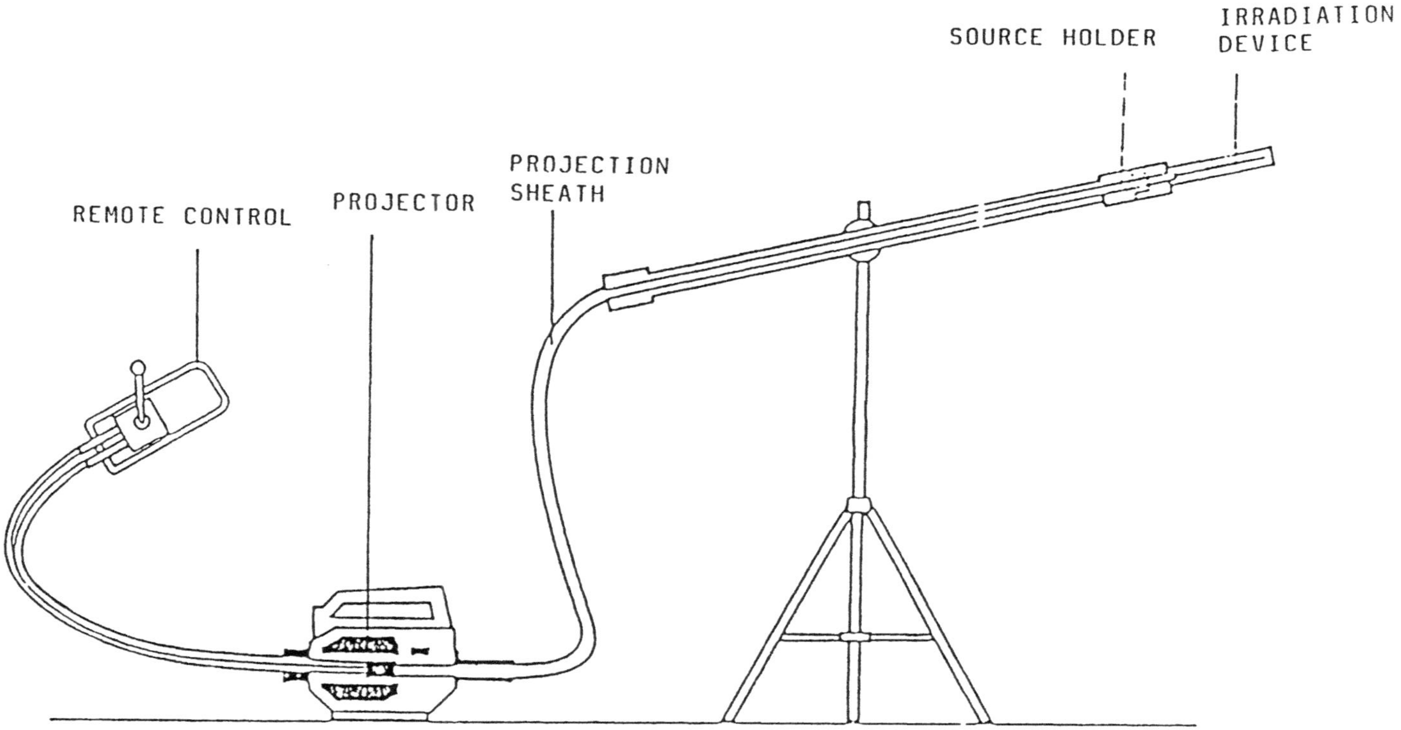

Figure 4. Gamma radiography apparatus.

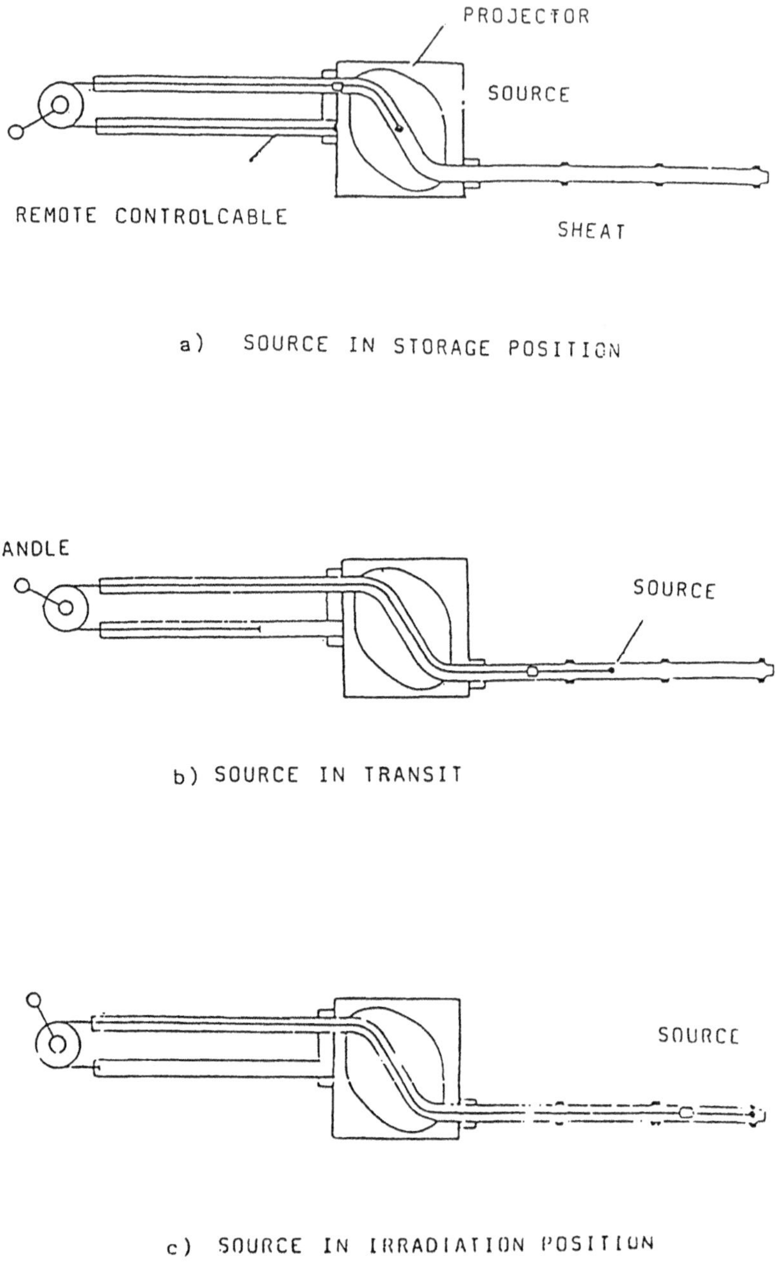

Figure 5. Remote control system.

TABLE 2. Limit of absorbed dose rate in air in the vicinity of the projectors.

Projector	Absorbed Dose Rate in Air (mGy/h)		
	On surface	To 1 m distance	
	Maximum	Mean	Maximum
Class P	1	0.02	0.1
Class M	2	0.02	0.1
Class F	2	0.02	0.1

This table quotes values from "Norme francaise NF-M60=551", (NF M60) which are more restrictive than those cited in ISO 3999 standard.
Furthermore the exposure containers when used as transport containers should comply with the requirements for the type B(U) packages of "Regulations For the Safe Transport of Radioactive Materials" I.A.E.A. (I.A.E.A.)
As an important feature all projectors must be fitted with a safety lock controlled manually by a key which prevents beam emission or ejection.
The locking device should be designed in such a way that even if a fault occurs it cannot prevent the shutting of the beam, or re-insertion of the source.

d) Warning Devices
The projector itself or on the remote control system, should have a warning device in order to indicate clearly the position of the source, let us say, if it is in operation or in storage. Normally the signs are red for operation mode and green when the projector is locked or shut-off.

e) Remote Control System
The projector with a remote control system should be designed in order to cope with a failure in the activation system. It is indispensable that in case of fault, the source automatically returns to the storage position or the shutter is closed. In addition a manual safety device should be provided which allows to bring the projector to the secure mode without exposure to the personnel.
Remote control cables and sheaths should allow a distance of at least 15 meters between the remote control system and the exposure

container.

f) Collimation
Collimators giving different field sizes for different kinds of exposure should be available. The collimators should provide an attenuation corresponding to at least 2 tenth values layers.

g) Tests
The industrial radiological apparatus should be submitted to tests under normal operation conditions or under accident situation taking into account its functions namely the storage, transport and irradiation, according the ISO 3999 standard.

h) Administrative requirements
The apparatus for industrial gamma radiography should be marked with the information concerning the type of source and activity in accordance with ISO 3999 standard.
An instruction manual for each apparatus should be available with the equipment in the site.

2.3. X-Ray Equipment

2.3.1. General Characteristics. The more common X-ray set components are schematically shown in Figure 6 and are the X-ray tube head, the high voltage generator and the control panel. In Figure 7 is shown a X-ray tube composed by a metal target (anode), the filament (cathode) within a glass envelope where the vacuum is produced. There are of course different types of tubes taking into account the type and shape of target and the type of work to be performed. It is important to refer from the point of view of radiation protection there is a protecting housing of the tube with an opening for passage of the radiation beam which could be limited by a diaphragm or localizing cone or have its energy modified by filtration. It is usual to group the X-ray sets considering its mobility in portable, when easily handled by a man, mobile when movable by means of a car and fixed when its mobility is limited to an enclosed installation. If one considers the geometry of the beam, the apparatus could be classified in directional, if the beam has one direction, or panoramic, if the beam makes a 180 angle.

2.3.2. Requirements for X-ray Equipment

a) Tube Leakage
As a requirement the dose equivalent rate from leakage radiation measured at a distance of one meter from the focus should not exceed

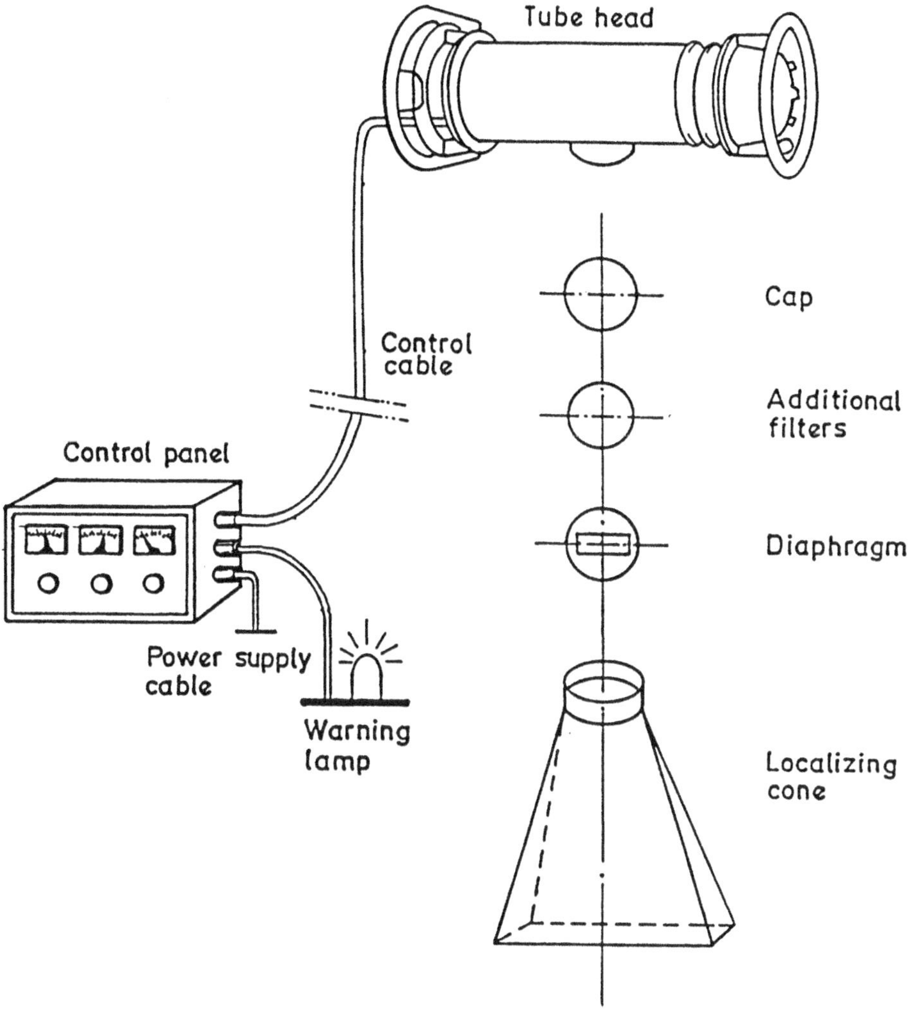

Figure 6. Components of the X-ray equipment.

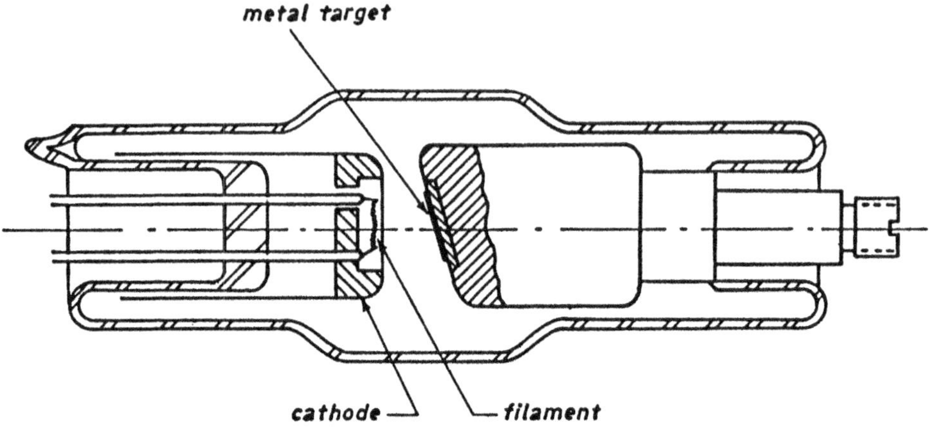

Figure 7. X-ray tube.

the following values considering the applied voltage and maximum tube currents (Nordic recommendations (1987)):

- up to 150 kV : 1000 μSv/h;
- above 150 kV to 200 kV : 2500 μSv/h;
- above 200 kV : 5000 μSv/h.

b) Filtration

Dose rates in vicinity of X-ray sets can be reduced by addition of suitable filtration. So the tube head should provide a total filtration at least equal to the following values corresponding to the maximum voltage (Nordic recommendations (1987)):

- up to 50 kV : no requirement;
- above 50 to 100 kV : 2 mm Al or the equivalent, (1st HVL: 2.5 mm Al/100 kV);
- above 100 up to 200 kV : 3 mm Al or the equivalent, (1st HVL: 0.2 mm Cu/200 kV);
- above 200 up to 300 kV : 4 mm Al or the equivalent, (1st HVL: 0.6 mm Cu/300 kV);
- above 300 kV : 0.5 mm Cu or the equivalent, (1st HVL: 2.2 mm Cu/350 kV).

Equipment with lower total filtration may be used for special purposes and additional changeable filters should be available in order to minimize low energy radiation.

c) Collimation

Collimators should be available to reduce the useful beam and to define the minimum size necessary to the work.

Diaphragms enabling different field sizes of the primary beam should be available and easy to mount. Panoramic X-ray equipment should be supplied with diaphragms enabling directional beam exposure.

d) Control Panel

This unit should comply with strict requirements in order to minimize radiation doses to the radiographers and misuse of the X-ray tube. So the control unit should be provided with:

- a control key activation of the system;
- two independent means (indicators) indicating that X-rays are being generated; one of these indicators should be a reliable red lamp;
- door switches connectors in case of equipment working in enclosed installation;
- easy connections for external working lamps.

Furthermore the control panel should be the only mean for controlling the radiation exposures.

e) Cable

The cable length is an important feature to be considered. In fact the risk of exposure will be minimized if the control panel is placed far from tube head. Length of cable should normally be not less than 20 meters for X-ray apparatus up to 300 kV or even bigger for more powerful equipments.

f) Administrative Requirements

For quick information on the site it is recommended that the tube head be marked distinctly with some specifications such as: maximum tube voltage (kV), maximum tube current (mA), location of focus, anode side, primary beam angle, filtration, etc.

A brief guide, in the language of the country, concerning the operation unit mode, the hazards involved, should always, accompany the equipment.

3. Radiological Protection in Operation

3.1. General Principles

A wide variety of working places in the industrial radiography operation, a great risk of exposure of the personnel and the public, obliges a careful organization of the radiation protection. Two basic principles should be observed, one is the optimization of the irradiation which involves the safety of the equipment and the characteristics of the premises and the irradiation procedures and the other is concerned with the limitation of the individual doses to the workers and the general public.

The optimization should be applied in case of mobile equipment in the equipment itself through the checking of the safety requirements and through operational measures regarding the conditions of irradiation in particular related with the collimation of the gamma and X-ray beams.

In what concerns enclosed installations the optimization measures should be applied to the layout of the premises namely to the shielding aspects of the irradiation rooms.

The limitation of individual doses to the workers are based on the exposure limits currently in force in EEC and obviously they should be observed in order to prevent over-exposure of the personnel.

3.2. Limitation of Individual Doses

The dose limit of 50 mSv/year is a primary limit for the limitation of individual doses. Additional reference levels may be used, as 30 mSv/quarter or 10 mSv/quarter, as guide values in day to day work, enabling radiological protection services by intervention to prevent that annual dose limits are exceeded.

The annual dose limit of 5 mSv should be observed to the members of public who might be exposed to radiation if they are in the vicinity of the site. However for partial body exposures, for instance the hands, dose limits are also established by ECC Directive (Council Directive 84/467).

Table 3 sums up the limits for whole body exposures and partial exposures.

3.3. Classification of Areas

It is a good practice in radiological protection to classify areas mainly when the work is performed with mobile equipment at work site. They are non-regulated areas and regulated areas.

Non-regulated areas: the access of personnel is unrestricted.

Regulated ares: the access is limited in order to protect the public against radiation hazards. Supervised, controlled and high irradiation areas are normally included in this category.

Supervised areas: areas where the workers might receive exposure in excess of one tenth of the occupational limit 5 mSv/year (500 mrem/year). The maximum exposure level allowed should not exceed 0.02 mSv/h (2 mrem/h). In this area workers not directly involved in the radiograph work may stay for some time.

Environmental radiation monitoring must be carried out.

TABLE 3. Annual dose limit for occupational exposure and members of the public.

Body	Dose limits (mSv)	
	Workers	Public
Whole body (uniform irradiation)	50	5
Individual organs or tissues (except lenses of the eye)	500	50
Lenses of the eye	150	15
Hands, legs, forearm, ankle	500	50
Skin	500	50

Controlled areas: areas where the workers might receive exposures in excess of three-tenths of the dose equivalent limit 15 mSv/year (1.5 rem/year). The maximum exposure level allowed is 0.05 mSv/h (5 mrem/h).

In this area the workers assigned to irradiation work may stay. Individual radiation monitoring should be performed and also the environmental monitoring.

High radiation areas: areas where the exposure levels exceed 1 mSv/h (100 mrem/h) and where no worker is allowed to enter during the irradiation period.

The signalization of those areas is very important because the safety rules for industrial gamma radiography require a severe accomplishment of the warning signs. The regulated areas should be marked out with suitable signs giving the necessary information. The supervised areas must be signalized in order to inform about the access restrictions.

The controlled areas must bear the trefoil radiation hazards warning sign while high radiation areas should be marked out with the same sign but expressing "no entry" during the irradiation time. When necessary and specially in enclosed installation visual and audio alarm systems should be installed as additional safety.

3.4. Classification of Workers

In certain way related with the classification of areas also workers should be classified. Thus the workers who are directly assigned to irradiation work are considered occupational workers and the annual dose limit is 50 mSv. It is the case of the operators who are eventually exposed to radiation doses that could exceed 15 mSv/year.

Individual dosemeters must be used by this category of workers and they must be submitted to medical surveillance. Workers occasionally involved in radiation work should be regarded as individuals not receiving more than 15 mSv/year. This must be the annual dose limit. No individual dosimeters and medical surveillance are required.

The other workers should be considered as common workers not engaged in radiation work and so they are regarded as public. The dose limit should be 5 mSv/year.

3.5. Planning

A careful planning of radiography work is essential in order to avoid unnecessary risks.

The hazards analysis and the selection of the equipment are very important issues in order to minimize undue exposures and to prevent serious accidents.

This can be achieved by adequate choice of the following aspects:

a) Selection of the source and its strength in accordance with the type of work.
b) Type of equipment.
c) Duration of exposure.
d) Provision of shielding.
e) Location of the controlled area.
f) Use of collimators.
g) Time of day for operation (out of normal working hours).
h) Exclusion of all non-radiographer personnel from controlled areas.
i) Position of the operator and other personnel.
j) Safety features and warning devices.
k) Provision of radiation monitor and personal dosemeters.
l) Choice of the film and processing control.
m) Backscattering from workpieces.
n) Segregation of radiographic and non-radiographic work.
o) Supervision of each source.

Industrial radiography should be preferably performed in enclosed installations, however there are many situations that justify the use of mobile projectors outside namely the work on site.

A radiography team should always consist in two persons, one experimented radiographer and one assistant except in the case of the work in enclosed installation. As an important rule the team should not operate simultaneously more than one unit.

The operating team should have always available personal dosemeters, one rate meter (at least), and an audible-alarm dosemeter for each member of the team.

Before starting the radiography work the team should ensure that radiography equipment and radiation protection equipment are available and in proper working condition.

3.6. Operation

3.6.1. Enclosed Installations. As was said before, the work in enclosed installations offers less radiation risk than the work on site if the installations are complying with the minimum requirements. Some specifications should be followed during the construction and operation of such type of premises.

An enclosed installation is a permanent enclosure in which the radiation source, X-ray sets and the object to be exposed are placed and when a remote control system to activate the devices exists on the outside.

The following requirements should be taken into consideration:

a) The installation where the equipment is activated electrically must have an interlock system which prevents undue exposures if one of the entrance doors is open before or during the irradiation.
   The re-initiation of the operation should be only possible by manual restart after the door has been closed.
b) An installation with manual radiograph apparatus should have a safety system which either prevents access in the enclosure during the exposure or gives an audible alarm if a door would be open.
c) An enclosed installation should have adequate warning signs defining it as a controlled area or high radiation area and visual-audible devices showing that the irradiation is going on.
d) The dose equivalent rate at 1 metre distance from the outer surface must not exceed 7.5 uSv/h for the maximum working conditions of the equipment.
e) The installation should have an emergency device which allows a person accidentally left in the room to open the door and leave the enclosure. Also the doors not controlled by the radiographer should be lockable from the inside.
f) Enclosed installation should be marked with signs showing the maximum ratings and limitations on the primary beam directions.

In what concerns the operation the radiographer must perform some checks such as:

a) Compliance of the installation with the conditions established for the enclosed installations namely those related with limitation of the use of the radiation sources and limitation of beam directions.

b) To assure that nobody is inside of the premise before starting the irradiation.
c) The exposure is ending in accordance with the termination procedures.
d) Verification, before entering in the installation, that the irradiation is terminated. A dose equivalent rate meter must be always used.

3.6.2. Open Installations. The work in open installations determines that the radiographer follows some general recommendations mainly those concerned with the definition of zones of restricted access and the establishment of barriers.

Some of the more important recommendations are the following:

a) Permanent barriers on the site should be used to minimize the radiation doses and even additional shielding should be used if it turns necessary.
b) A high radiation area where the irradiation takes place should be established by means of physical barriers to prevent the access and to reduce radiation doses. The dose equivalent rate outside of this area should not exceed 50 µSv/h.
c) During the exposure no person is allowed to remain or enter in the high radiation area.
d) During the irradiation a controlled area should be established under the supervision of the radiographer. A dose rate outside of this area must not exceed 7.5 µSv/h.
   Workers associated with radiography work are allowed to pass through the area, however no work place may be occupied during exposure.
   Temporary barriers should be mounted if it is difficult to control the area.
e) The control unit should be well located in order to permit to the radiography team the best supervision to the high radiation and controlled areas.
   At least one of the radiographers must be permanently at the place to take immediate corrective measures if something goes wrong.
f) The dose rate at the position of radiographer must not exceed 20 µSv/h during the irradiation time.
g) The radiographer should not enter in the high radiation area unless he is sure that irradiation is terminated. A portable radiation monitor should always be used.

Earlier the requirements were mentioned needed for improving the safety parameters of the equipment. However the use of the equipment plays an important role in the safety of industrial radiography operation. Thus it is essential that some recommendations, or precautions should be considered during the use of the apparatus in open installations.

Regarding the X-ray equipment:

a) During the preparation or testing of X-ray tube the radiation window should be closed with a lead cap.
b) Diaphragm and collimators should be used to reduce the beam size to the minimum necessary and to minimize the scattered radiation.
c) The use of additional filtration in order to reduce low energies is recommended.
d) When the primary beam is not intercepted by the object to be radiographed, additional shielding must be applied to avoid the increase of the controlled and supervised areas.
e) The warning devices, visual-audible, should be activated during the irradiation period.

Regarding the use of gamma radiograph apparatus:

a) The radiographer should select the activity of the radiation sources to have reasonable exposure times compatible with acceptable doses. In general activities requiring exposure times less than one minute should not be used.
b) The use of high activity sources exceeding:
   - 1500 GBq Ir - 192;
   - 1500 GBq Cs - 137;
   - 400 GBq Co - 60;

   is acceptable only when there is an evident need for performing the work.
   The use of high activities demands special license delivered by national authorities.
c) Collimators should be used to reduce the beam size to the minimum necessary and to minimize scattered radiation.
d) During the preparation of the equipment the radiographer team should pay special attention to the security of the different connections between exposure container, collimator and remote control cables.
e) When a projection sheath is used, it should be as short as practicable.
f) The radiographer must use a radiation monitor in order to confirm that source is in the storage position after the termination of the irradiation.
g) If an emergency occurs the radiographer should activate the contingency plan.
h) The work with high activity sources (superior to those referred in b)) should be performed with two qualified radiographers.
i) The remote control cables and sheath must be long enough to allow distances at least 15 metres between the remote control unit and the exposure container. Emergency equipment (mainly shielding) should be available on the site.

3.7. Transport of the Radiation Sources

The transport of the industrial gamma radiography apparatus is an important aspect from the point of view of radiation protection because it is common to operate this equipment at different work sites outside of the establishment. The risk of an accident during the transport is real and it is necessary to fulfil some rules concerning the type of the packages, and the dose rates involved.

The radioactive sources should be transported in accordance with the requirements stated in "Regulations for the Safe Transport of Radioactive Material" from A.I.E.A. (I.A.E.A.) or specific national regulations.

In general the portable gamma radiography devices containing Ir-192, Cs-137, Co-60 are regarded as type B(U) packagings. This packaging ensures that the dose rate at 1 metre from the surface does not exceed 0.1 mSv/h and the maximum dose rate of radiation at any point of the outer surface is 2 mSv/h. The outer surface of the package is marked with the radiation trefoil symbol and the wording "type B (U)".

3.8. Storage of the Equipment

A storage facility should be provided for the source containers and should be located in areas not adjacents to explosive or flammable materials. The storage room should be lockable, marked out with "access limited". The dose rate at accessible places should not exceed 7.5 μSv/h. A person should be assigned for the responsibility of record keeping and only authorized persons could withdraw the equipment.

Regarding the X-ray equipment no special storage facilities are required other than protection against theft, weather and unauthorized persons.

3.9. Accidents

As it was mentioned in the introduction of this document, from the 296 radiological accidents reported since 1940, 48% occurred with sealed sources. It is reasonable to say that a lot of accidents have not been recorded and it is well known that many fatalities and severely injured persons have occurred in the field of industrial gamma radiography.

The causes of accidents are several but is possible to sum up at least five major causes:

- The source remains exposed;
- the projector was not correctly locked;
- radiation monitoring was not carried out;
- the equipment fails;

- the training of personnel was inadequate.

To cope with the accidents it is necessary to have available a contingency plan. Thus every employer should prepare written emergency procedures covering all type of incidents and the adequate actions to be taken.

A contingency plan should cover the following topics:

1) Immediate actions to be taken by the radiographer in order to prevent excessive radiation doses.
2) Internal and external notification procedures.
3) Accident recovering procedures.
4) Evaluation of the accident, finding causes and evaluation of the consequences.
5) Medical examinations of the persons involved (workers or public).
6) Preparation of detailed report for national authorities.

One of the most common incidents happens when the source fails to return properly to the container or a source becomes separated from the exposure mechanism.

As an example, the actions that must be done by the radiographer are the following:

1) To measure the radiation dose rate with a radiation monitor.
2) To prevent the access of persons to the vicinity of the source. If there is an increase in dose rate the barrier should be moved to position corresponding to 7.5 uSv/h. This means that outside of this area the dose should be inferior to that value.
3) To place warning signs accordingly.
4) To avoid the intervention of non classified persons.
5) To prepare a plan before entering the emergency area.
6) To inform the manager about the situation and the actions prepared for coping with the incident.
7) To implement the emergency plan.
8) To replace the source into the container if possible by means of the emergency equipment keeping the source far, at least to a distance of an arm length.
9) To use a shielding if the source could not be replaced into the container.
10) To call for radiation protection expert assistance.
11) To advise the national authorities.

4. Licensing and Organization Aspects

The use of radioactive sources is an activity that envolves a radiological risk of exposure to radiation. It is a responsibility of the governments to legislate and regulate with adequate laws and codes of practice those activities in order to protect the workers and to prevent the risks to the health of the general public which may eventually arise from work with radiation. The users, or the employers, are obliged to follow the regulations, fulfilling the licensing conditions and complying with requirements prescribed to the installation, equipment and personnel. It is also an important obligation to establish a proper organization for radiation protection with minimum means in order to assure safety conditions in the work and to promote adequate training to the personnel. The process of licensing consists in the issue of license by the national authority which allows the employer to operate an establishment for using and keeping radioactive sources/equipment. The application for the authorization should contain besides the administrative data the following technical data:

a) Description of the premises and type of work to be carried out.
b) Description of the type of sources/equipment and its characteristics.
c) Type of the apparatus to be used.
d) Certificates attesting the compliance with the standards (namely equipment, premises).
e) Organization for radiation protection.

This last item is of utmost importance because it may contribute greatly to good standards of safety if a sound organization is settled and maintained.

A top organization for radiation protection always implies a good training of radiography operators in radiography techniques and basic radiation protection knowledges. Thus it is essential that the licensee has available a safety manual defining all aspects of its radiological protection organization.

The main issues that the safety manual should comprise and clearly specify are the following:

a) Personnel

The licensee should appoint a radiation protection officer with responsibility for all aspects of radiation protection. He should be an expert able to give advice on the issues related with protection on radiography site and should be licensed by the national authority on the subject. The radiographers should have a basic knowledge of radiation protection and skill on the use of apparatus or devices and radiographic techniques. They should have followed a basic training course on radiation protection (see g) training). The assistants should have sufficient information about their jobs and the associated risks involved.

All radiographers and assistants should be classified persons which means they are regarded as occupational workers and are likely to receive a dose of ionising radiation which exceeds three tenths of the annual dose limit.

b) Equipment Tests and Checks
A detailed program for periodic inspection should be carried out at specified intervals following the instructions of the manufacturers. A maintenance service should be performed accordingly taking into account the safety requirements recommended. A record should be kept including any remedial measure taken.
The program which concerns the gamma radiation apparatus should primarily consider the projection sheaths and remote control, the exposure containers with special attention for the shielding and the source holders and source itself. Regarding the X-ray equipment the maintenance work should consider the requirements for the head tube, the control panel, the auxiliary equipment (such as collimators, diaphragms) and for the electrical safety.

c) Personnel Dosimetry
A system for recording the individual doses of the workers should be provided in regular basis following the requirements of the national radiation protection body. Notification procedures should be implemented in order to define the clear actions to be taken.

d) Medical Surveillance
Periodic medical reviews of personnel should be carried out and special medical examination should be performed in case of over-exposures.

e) Control of Sources
An accounting system for controlling the sources should be settled and supervised by an authorized person. The records should be kept for a certain period of time with the data concerning the type of sources, date of receipt, the activity at a specified date, condition of the source (on storage, in operation on the site, at disposal).

f) Contingency Plans
Such a plan should define the emergency procedures covering the incidents or accidents with the sources/equipment.

g) Training
The type and extension of the training will be strongly dependent of the importance of the establishment where the radiography work is carried out. The recommendations of the National Council on Radia-

tion Protection and Measurements (NCRP) point out the need for a continuous process of training defining three distinct phases such as initial training, on-job training and periodic training.
The initial phase is given to all radiographers before they assume responsibility and includes the topics: fundamentals of radiation and radiation protection, fundamentals of radiography, licensing and regulation aspects, radiation emergency procedures and some laboratory exercises. Approximately 80 hours are devoted to this phase.
The phase on-job training will consist in instruction topics related with the practical work involving levels of authority and degree of responsibilities, radiography operating procedures, safety procedures, emergency procedures, records relating to radiation protection program.
The time devoted to this phase will be defined taking into account the importance of work performed in the establishment.
The third phase, periodic training, will depend upon several circumstances such as the modification of the procedures, or the exchange of the equipment and should be considered when necessary on a semi-annual basis.
The employer's safety manual could be not so precise and extensive concerning the training but should clearly define which type of training is supplied to the radiographers and specify the duties and responsibilities of the personnel.

5. Bibliography

Lushbangh, C.C. et al. (1980) "Radiological Accidents". Proc. 5th Congress of IRPA, I.A.E.A. - CN 51-92.

Apparatus for gamma radiography "Specification ISO 3999".

General requirements for sealed sources "ISO 1677".

Sealed radioactive sources classification "ISO 2919".

Norme Francaise "Radioprotection, Appareils de Radiographie Gamma". NF M60-551.

Regulations for Safe Transport of Radioactive Materials I.A.E.A. Safety Series N. 6.

Nordic recommendations on radiation protection in industrial radiography (1987).

Council Directive 84/467/Euratom on the basic safety standards.

Radiation Safety Training Criteria for Industrial Radiography NCRP Report N. 61.

EMERGENCY AWARENESS, PREPAREDNESS AND INTERVENTION IN RADIOLOGICAL ACCIDENTS (Lessons learnt from the Goiânia radiological accident)

M. OBERHOFER
Commission of the European Communities
DG XII - Joint Research Centre Ispra
I - 21020 Ispra (Va)
Italy

1. Introduction

The Chernobyl nuclear accident in 1986 and the radiological accident in Goiânia, Brazil, in September 1987 highlighted the importance of emergency planning and preparedness for both, nuclear disasters and radiological accidents. It was in particular the Goiânia accident (IAEA (1988)) which was the reason for inserting the topic of this lecture into the course-programme, because of the discussions which this accident had triggered on the importance of awareness and preparedness for **non**-nuclear radiation accidents, though in 1984 a large scale Co-60 contamination case in Mexico (CNSNS-IT-001 (1985), NUREG 1103 (1985)) had already drawn the attention of emergency planners to the importance of increased awareness and preparedness for non-nuclear accidents.

2. The Goiânia Accident

What had happened in Goiânia?

In the beginning or mid (the exact date is not known) of September 1987 a 50.9 TBq (1) (1375 Ci) Cs-137 teletherapy source was stolen from an abandoned private hospital by some young people and dismantled in a scrap-dealer's yard in the centre of Goiânia city with a population of 1 million in the State of Goiâs, Brazil, about 200 km South-West of Brasilia (IAEA (1988), Oberhofer and Bacelar Leao (1988)). During the dismantling of the therapy unit the source was ruptured with the highly radioactive cesium-chloride of the source being liberated, contaminating the scrap-dealer's yard and people there with consequent dispersion of the radioactivity to outside areas and finally large parts of the

(1) T = Tera, 1 TBq = 10^{12} Bq.

M. Oberhofer (ed.), Advances in Radiation Protection, 329–339.

city. The severity of the case was enhanced by the fact that the radioactive material glowed in the dark and was considered particularly precious with deliberate distribution of it among family and friends of the scrap-dealer. Most of the material was extracted from the source by means of a screwdriver, given away in little pieces as presents and for using it in finger-rings and/or bracelets. The worst happened when some of the material in form of luminous powder served for rubbing it onto the body of a six year old girl performing nocturnal dances in front of her family. On this occasion the girl also incorporated with food part of the material with an activity in the order of 1 GBq (2) (27 mCi), the highest body burden of Cs-137 ever recorded. Soon after the accident the first symptoms of radiation sickness appeared starting with the scrap-dealer's family when members of the family had accumulated between 1 and 7 Gy (3) (100 to 700 rad). Initially these sicknesses (vomiting and diarrhoea) were treated as tropical diseases. Only when the dealer's wife suspected that the illnesses had some connection with the luminous material she decided to put the source into a plastic bag and take it by means of a city omnibus to a Public Health Office, placing it in front of the medical doctor on service with the statement, that some gaseous emission from the piece of metal might be the reason of the fact that all of her family became ill and are going to die.

In the meantime more people became ill in different areas of the city and underwent treatment in various medical practices and/or hospitals.

Suspecting that some radiation might be involved, the medical doctor contacted a hospital were he knew that some radiation measuring instrument might be available. When a physicist of this hospital then approached the doctor's studio with a radiation detection device it soon became evident that apparently a very severe radiation accident had occurred within the city. Soon it became clear that the accident must have had as a consequence the contamination of large parts of the city and the suspect arose that the contamination might not only be limited to the city. When the medical doctor then informed the national authorities, the Brazilian Commission of Nuclear Energy (CNEN), a vast action started to secure the remnents of the Cs-137 source, to isolate the contaminated areas, to find out what had happened and to bring the situation under control with the final outcome that altogether 120,000 were controlled for contamination, out of which 121 had been contaminated internally, 21 had received doses between 1 and 7 Gy (100-700 rad), all of them hospitalised in Rio de Janeiro and Goiânia, 4 persons died within 4 weeks because of severe overexposure (haemorrhage

---

(2) G = Giga, 1 GBq = 10^9 Bq.

(3) Gy = Gray, 1 Gy = 100 rad.

and infection), also the six year old girl, and 24 others had radiation burns, with one having lost his right arm by amputation. The extent and degree of environmental contamination was such that some 200 individuals had to be evacuated from 41 homes. Seven residences and various associated buildings had to be demolished and top soil had to be removed from many areas making use of heavy building equipment. Altogether, 3500 m^3 of radioactive waste were produced within 3 months and piled-up in waste drums and containers in a large temporary waste disposal area about 15 km outside Goiânia city.

3. Accident Awareness

The Goiânia case is a very good example of dealing with the subject of this lecture, firstly because of the actual scenario and secondly because of the lessons learnt from the accident. Also, the Goiânia accident has drawn the attention again to **non**-nuclear accident situations, for which much less efforts have been made to develop plans to deal with them, compared to nuclear (reactor) accidents, like the one at Chernobyl, for example. This later accident had increased these efforts on both national and international level.

Awareness of the likelihood of *radiological* accidents is a prerequisite for being prepared and for promptly counteracting with efficiency in case of such accidents.

Accidents awareness requires:

- a complete registration of radioactive sources throughout a country;
- an inquiry on the uses of these sources:
- a detailed analysis of the hazards associated with the use of the sources and potential accident situations and;
- regular inspections.

In connection with the Goiânia accident already in the first phase of the accident handling doubts on the observation of these basic premises for being able to successfully deal with the accident had come up, when it was difficult:

- to verify that the therapy-source had been registered at the right place for its intended use;
- when it turned out that the authorities had not been informed on its actual location after transfer of the Oncology Department from the abandoned and practically demolished private hospital to another new construction;
- and when later investigations revealed the fact that the "old" therapy-unit with its source had been completely forgotten by the responsible medical doctors and a hospital physicist. For this latter rea-

son the doctors and the hospital physicist were accused and had to go to court.

Accident awareness is amplified by studying possible accident scenarios. Abandoning of a therapy-unit and/or source is definitely such a scenario. In Safety Series no. 91 of the IAEA (IAEA (1989)) giving recommendations of emergency planning and preparedness for accidents involving radioactive materials, used in medicine, industry and teaching, possible accident scenarios are listed up, the one discussed here as an example included. Quotation:

a) A radioactive source may be misplaced, lost or stolen. An obvious example might be a gamma radiography source container and source found to be missing. One problem here is that the container may get into the possession of those who through ignorance may decide to dismantle or otherwise interfere with it thereby exposing themselves and possibly other to an unshielded source.

b) A radioactive source becomes unshielded as a result of a failure during routine operations. Again, gamma radiography provides an example - after making an exposure, an operator finds it impossible to retract the source into its container.

c) A radioactive material may be dispersed. An example could be a vial containing a radioactive solution that leaks out during storage. Another example would be a violent release of radioactive substances from a radiochemical facility.

End of quotation.

In all these cases, there is a possibility of uncontrolled exposure of persons unless appropriate protective measures are taken.

One impact of the Goiânia radiological accident was that it definitely has contributed to an increased accident awareness of many authorities, in particular in Brazil, CNEN for example, the Brazilian nuclear energy authority.

This authority immediately started to check to what extent radioactive sources in the country, medically and industrially used, had been registered on a state and/or on a national level - not an easy task for a country large as Brazil, as it turned out!

Accident awareness requires a special attitude towards potential accidents, because in many cases accidents derive from unpredictable causes (equipment malfunctions, fire, theft, mechanical damage, human error, etc.). Accident awareness should be present at a local level, where radiation sources are made use of, on licensing authorities-level, state and national authority level.

In all cases accident awareness needs to be stimulated by regular surveys, reviews and special meetings on this subject.

4. Accident Preparedness

Accident awareness is not the only prerequisite to combat accidents successfully. Thorough planning for emergency response is needed for handling accidents with success. Successful interventions need a careful preparation of such interventions on all possible levels.

Important aspects in any nuclear or radiological emergency planning are according to (IAEA (1989)) page 17:

a) identification of organizations and people who are going to be responsible for taking action in the event of an emergency;
b) establishment of procedures for the early detection of an emergency situation;
c) plans for notification of the appropriate authorities that an emergency situation has developed;
d) identification of appropriate monitoring equipment;
e) development of a system of intervention levels based on parameters that can be measured;
f) identification of the medical assistance required;
g) recognition of the requirement of appropriate training.

A very useful scheme for effective emergency planning and preparedness is given in IAEA Safety Services no. 91 (IAEA (1989)) and is reproduced in Figure 1 for consultation.

That emergency plans for any accident with ionising radiation, nuclear and radiological, must have an adequate degree of flexibility in order to cope with accidents of different nature and different levels of severity again demonstrates the Goiânia accident, where casual circumstances led to the discovery of the accident and where initially only one organization responded to the accident promptly, CNEN in Rio de Janeiro, with the gradual intervention of other organizations.

Often detailed plans cannot be followed because the source of radiation has not been under control, as was the case in Goiânia, where ad hoc decisions had to be made to bring the situation under control. Usually such a situation requires improvisation by specialists capable of dealing with the first phase of the accident with a minimum of assistance and equipment and the capability to develop as fast as possible an intervention plan for further actions, as well as getting organizations and/or authorities involved with more possibilities to successfully intervene at the place of the accident.

To what extent the degree of accident preparedness of some authority can affect the handling of the accident can be demonstrated using Goiânia as an example.

CNEN was prepared for an accident like the one at Goiânia, because it had expertise in the field and a well equipped specialized laboratory at Rio de Janeiro, the Instituto de Radioprotecao e Dosimetria

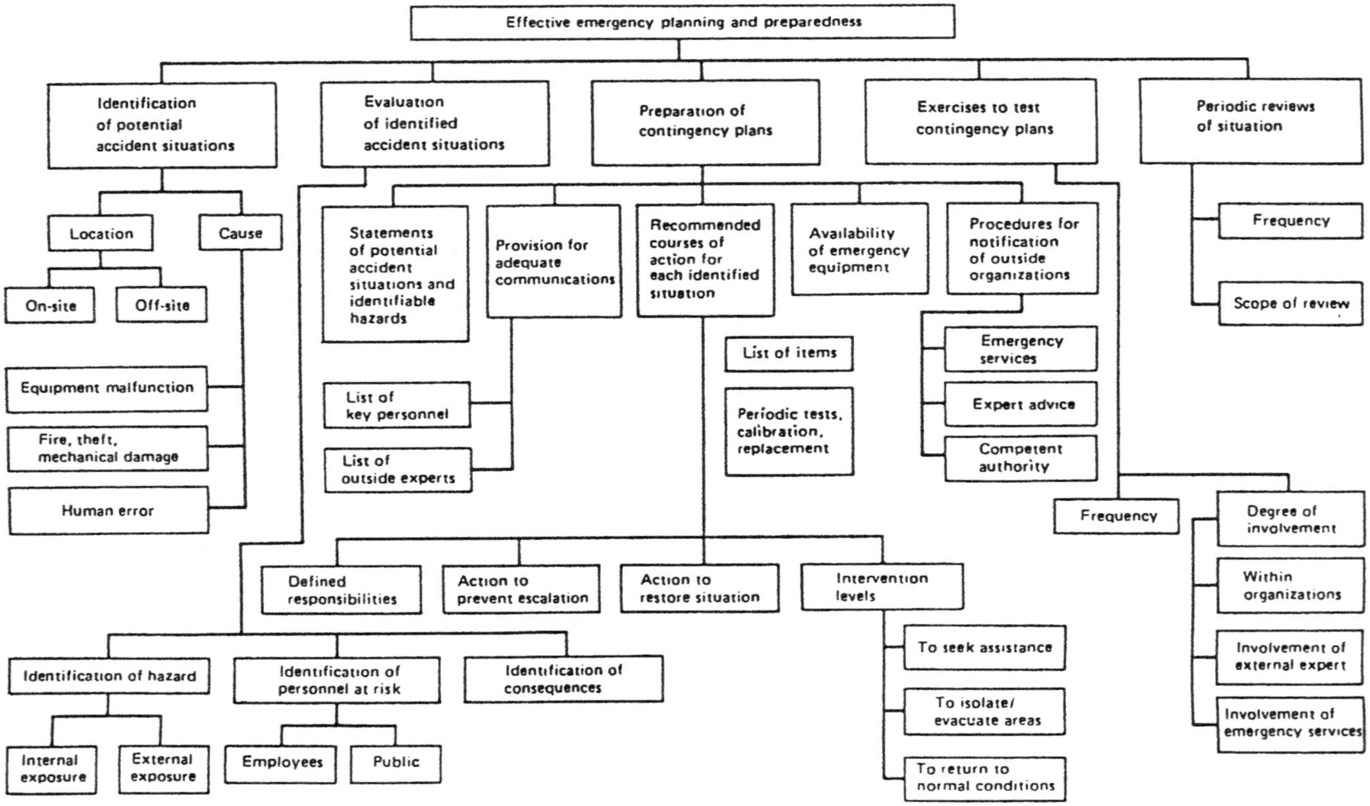

Figure 1. Effective planning and preparedness scheme.

(IRD) and more nuclear laboratories throughout the country, unfortunately all very far away from the place of the accident (Rio, 1348 km away), complicating fast intervention. Nevertheless, CNEN succeeded to respond immediately to the telephone call from Goiânia and air-jetted experts to the scene in order to analyse the situation and initiate the first counter-measures, starting with setting up a local emergency organization. After having activated local police and military, more specialized personnel were flown into Goiânia for identification of persons who needed medical assistance as a consequence to radiation injuries, identify the contaminated areas in the city, isolate these areas and for large scale population monitoring. The response was fast and effective, thanks to the Sherlock Holmes skill of most of the members of the intervention teams.

CNEN was less prepared with regard to the availability of a sufficient amount and suitable protective means for its own personnel, with regards to radiation detection equipment due to the extension of the accident with several monitoring-teams working around the clock and, with regards to equipment for immediate reduction of contamination spread, which was initially practically hampered by heavy rainfalls.

With respect to protective means the situation was so severe that radiation monitors operated in contaminated areas with initially no protective clothing at all and with tennis shoes. No masks were available for protection against airborne radioactivity. Decontamination procedures had to be delayed because of lack of waste containers, remote handling equipment and heavy machines for demolishing houses and for earth movements.

Another problem in the beginning was transport of the intervention personnel within the city, for which city taxis had to be used. Radiation protection control of the intervention personnel was another problem initially, which gradually improved with increasing organizational efforts and efforts to procure all the missing items. Weeks passed before everything was completely under control and several months to have the city cleaned up. By the end of the year (1987) the city was officially declared free of contamination.

CNEN was also prepared to deal immediately with the accident victims at a specialized Naval hospital in Rio de Janeiro, which had been designated much earlier to receive casualties in the event of a nuclear power plant (Angra-I) accident. Another hospital was rapidly adapted for receiving less injured and contaminated persons in Goiânia City. However, at both hospitals improvisation was necessary and preparedness was not sufficient for avoiding contamination of aeroplanes, ambulances and even hospital personnel.

Another important aspect of radiological accident preparedness is the capability for fast assessment of radioactivity incorporation and dose-assessment, which in the Goiânia case was very much the case with its excellent radioprotection teams at the CNEN laboratories at Rio, mentioned above. One difficulty of course was to get the samples, often very radioactive, to this place, 1348 km away from Goiânia.

A further important aspect of the Goiânia accident handling was the lack of co-ordinated dissemination of information through the mass-media to the public, which exacerbated the public fears of radiation (radiophobia) and the consequential demand for information seriously hampered the efforts of those trying to deal with the accident.

The lessons learnt from this is, that in similar cases emergency plans must contain arrangements for the provision of adequate information of the mass-media, which must not interfere with the recovery operations.

In this connection also the need and provisions for psychological advice must be mentioned, not only for members of the public, but eventually also for participants of the personnel directly involved in the intervention actions in highly contaminated areas.

The Goiânia accident and similar accidents have shown that radiological emergency preparedness must very much consider instrumental aspects.

Corresponding passages in the IAEA Safety Series no. 91 (IAEA (1989)) with some omissions read as follows:

- "In order to assess the radiological hazard involved during all stages of the management of the accident, radiation measuring instruments are required. The user, the local emergency personnel and the competent authority should have, or have access to, the appropriate sets of instruments in accordance with established plans.
- The type and nature of the radiation monitoring instruments to be used are dependent on the nature of the radioactive material involved and the type of measurement to be taken.
- The proper maintenance of radiological equipment is very important. It should be regarded, together with periodic calibration, as an integral part of preparedness for an emergency.
- Emergency response personnel and the competent authority's emergency teams, who should be prepared to respond to emergencies of various types and magnitudes, should have access to instruments suitable for all envisaged circumstances. Users, however, need have only those instruments that are necessary for their specific materials. Thus the operator of a ^{60}Co irradiation device has no need for alpha monitors".

In particular proper instrument calibration needs to be mentioned here. Laboratory calibration is very often not sufficient for having the correct instrument reading in a field situation with much scatter-radiation around.

Laboratory instruments casually do not withstand rough handling conditions in emergency situations outside the laboratory and may become obsolete because of malfunction, an experience made in the Goiânia accident operations with a number of commercial radiation protection instruments.

In radiological accident preparedness studies considerations should also be made with regards to the dose-occupancy of persons being part of emergency teams, military and fire-brigade personnel, as well as persons involved in the restoration activities. Control for internal contamination (incorporation of radionuclides) is common practise in accidents with radioactivity release into the atmosphere. Portable equipment may be useful for this purpose.

In the case of contamination with radioactive material environmental radioactivity measurements will be needed. Such measurements and/or controls under circumstances, depending on the type of accident, require an extensive effort and need to be planned and hence should be part of any accident preparedness programme.

In many instances part of an accident preparedness contingency plan is the provision of one or several mobile intervention units, properly equipped for first accident handling.

Such equipment should comprise:

- radiological monitoring instrumentation,
- sampling devices,
- eventually metereological instruments,
- personal protective equipment,
- accessories like ropes, radiation and contamination signs, plastic sheets and rags, remote handling tongs etc., just to mention the most important items.

Radio communication systems should be available.

Often complete mobile laboratories are provided by such plans and actually should be ready for use, if needed.

This latter request requires proper maintenance.

In the Goiânia example this was not the case and no such a unit, though available in Rio de Janeiro, was operative. It was also learnt that in some instances mobile units need to be air-lifted and must satisfy special requirements for such lifts.

And last but not least, special consideration in emergency plans must be given to the disposal of radioactive waste, which, if not carefully planned, can delay clean-up operations. This was the case in Goiânia, where the waste disposal problem, mostly because of political implications in connection with the selection of a proper temporary waste disposal area, was the reason for retarded contamination removal up to nearly two months with consequential continued contamination spread and accumulation of unnecessary large waste volumes.

In this context questions of the handling of deceased contaminated victims also arose, the burial of which may cause severe problems, with the Goiânia accident having given a typical example, where the burials of four contaminated persons were impeded by the aroused population.

5. Maintaining and Updating of Emergency Plans

A last word must be said with regards to maintaining and updating of emergency plans, which have been elaborated for certain conditions. These conditions may change and require modified actions in case of an accident. If such changes are not provided for in time, severe difficulties may arise in following some emergency plan and eventually cause confusion with evident consequences for the emergency interventions.

In agreement with (IAEA (1989)) on the same argument the following advices are given:

1. For each plan a person should be appointed to be responsible for maintenance and modification. Lists of names and telephone numbers should be updated whenever personnel and organizational changes require it, but in any event, at least once every six months. For this purpose, such lists should be kept separate from the plan itself. The person responsible for maintaining a plan should participate as an observer in the exercises in order to be able to feed back the experience to provide improvements to the plan.
2. Provision should be made, at the minimum, for an annual comprehensive review and updating of emergency plans. This review should take into account the lessons learnt from exercises and actual accidents.
3. An indication of the specific distribution of the emergency plans should form part of each plan. All organizations involved in the overall emergency response organization should receive copies of plans and changes to them. A mechanism should be established for those on the distribution list to acknowledge receipt of such changes.

6. References

Accidente por contaminaciòn con cobalto-60, México, 1984 Rep. CNSNS-IT-001 (1985).

Contaminated Mexican steel incident, Rep. NUREG 1103 (1985).

Emergency Planning and Preparedness for Accidents Involving Radioactive Materials Used in Medicine, Industry, Research and Teaching (1989). IAEA Safety Series no. 91. Recommendations. IAEA.

Oberhofer, M. and J.L. Bacelar Leao (1988) The Radiological Accident at Goiânia, in Radiation Protection in Nuclear Energy, Proceedings of an International Conference, Sydney, April 18-22, 1988, IAEA-CN-51/91.

The Radiological Accident in Goiânia, IAEA, STI/PUB/815, Vienna (1988).

SELECTED TOPICS IN NON-IONIZING RADIATION PROTECTION

M. GRANDOLFO
Istituto Superiore di Sanità
Laboratorio delle Radiazioni
Viale Regina Elena, 298
00161 Roma - Italy

1. Introduction

In recent years there has been an increase in the development and use of equipment that produces nonionizing radiant energy, and the question has been raised as to whether adequate measures are being taken to guard the user and the general public from its possible adverse effects. In contrast to ionizing radiation, radiation of longer wavelengths usually interacts with human tissue primarily by generating heat. For want of a better collective term "nonionizing radiation" (NIR) is used to encompass this group of electromagnetic radiation and also ultrasound. Nonionizing radiation pervades the entire environment but, except for the narrow spectrum of visible radiation, it is unperceived by any of the human senses unless its intensity becomes so great that it is felt as heat. Differences in wavelength, even within a single type of radiation, are particularly important when evaluating hazards from exposure to NIR. The ability of the radiation to penetrate into the human body and the sites of absorption will depend on this characteristic and will differ from one type of radiation to another.

In developed countries there has been a remarkable growth in the number of processes and devices that utilize or emit NIR. They find an ever increasing use in industry, engineering, telecommunications, medicine, research, education and the home. This gives rise to a number of questions. How serious are the problems linked with NIR, what are their dimensions, and what acute and chronic effects on the human body are involved? Is there sufficient knowledge of occupational risks and public health hazard? How can exposure be reduced? Are national protection standards adequate and, if not, how can better regulations be drafted and enforced to reduce exposure? Because of the rapidly developing technology and the associated health implications, there is a need to develop international cooperation in the use of NIR and mea-

M. Oberhofer (ed.), Advances in Radiation Protection, 341–360.

sures to prevent over-exposure. Moreover, governments will be expected to intensify the establishment of rules and regulation and means of enforcing them.

When considering the subdivisions of the NIR spectrum, it is important to remember that different energy absorption mechanisms, and thus effects, occur in exposed humans, because one is dealing with radiation having vastly different wavelength and frequency ranges. The ranges covered by electromagnetic non ionizing radiation are: ultraviolet radiation (wavelengths from 100 nm to 400 nm), visible light (wavelengths from 400 nm to 700 nm), infrared radiation (wavelengths from 700 nm to 1 mm), radiofrequency (wavelengths from 1 mm to 3 km or frequencies from 300 GHz to 100 kHz) including microwaves (wavelengths from 1 mm to 1 m or frequencies from 300 GHz to 300 MHz), and extremely low frequencies (frequencies above zero and up to 300 Hz). A gap in the frequencies 300 Hz to 100 kHz exists since little occupational exposure to electromagnetic fields occurs in this frequency range and scant data on its biological effects are available. From a pragmatic biomedical point of view the passage of energy through matter in the form of ultrasonic vibrations is also considered as non ionizing radiation.

The International Radiation Protection Association (IRPA) formed a working group on non-ionizing radiation in 1974; at the 1977 IRPA Congress in Paris, this working group became the International Non-Ionizing Radiation Committee (IRPA/INIRC).

The IRPA/INIRC in cooperation with the Environmental Health Division of the World Health Organization (WHO), has undertaken responsibility for the development of health criteria documents in NIR. These form part of the WHO Environmental Health Criteria Programme, which is sponsered by the United Nations Environment Programme (UNEP). These criteria then become the scientific data base for the development of exposure limits and codes of practice.

2. Extremely Low Frequencies (ELF)

2.1. Epidemiological Studies

In 1979 Wertheimer and Leeper published their study that came to the then remarkable conclusion that children resident near high current configurations (transformers; large gauge, high-current power lines; power substations etc.) had a two-to-three-fold higher incidence of cancer than children resident near low current configurations. This study, although having a number of limitations common to many epidemiological studies, was sufficiently well conducted that interest was raised and a large number of subsequent studies have resulted.

Table 1 gives a summary of the studies on cancer incidence and occupational exposure to electromagnetic fields. Of the 20 studies, about half suggest an association between exposure to these fields and an increased incidence of leukaemia. The other studies suggest an association or an associated increased incidence of other cancers.

Table 2 summarizes the studies on cancer incidence and population exposure to electromagnetic fields. Only 3 of the 11 studies suggest an increased leukaemia risk. It should be noted that the category of study design is indicated in Tables 1 and 2. The proportionate mortality ratio (PMR) or proportional incidence ratio (PIR) studies provide the most limited and potentially flawed data, case-control (CC) studies provide better results, while the retrospective follow-up (RFU) studies are the most methodologically sound. One notes that most of the RFU studies do not suggest an increased incidence of leukaemia with electromagnetic field exposure.

Coleman and Beral concluded from their review of the literature that the most consistent finding is that electrical workers appear to be at increased risk of leukaemia, especially acute myeloid leukaemia, but the effect is small. When they combined the results of eleven separate investigations it suggested an 18% increase in the risk of leukaemia (risk ratio (RR) = 1,18, 95% confidence interval (CI) 1.09-1.29) which is partly or wholly due to a 46% increase in the risk of acute myeloid leukaemia (RR = a 46% increase in the risk of acute myeloid leukaemia (RR = 1,46, 95% CI 1.27-1.65)). They explained that it is not clear whether this increase is specific to certain types of work within the electrical industry. Nor is it possible to determine from the available data if the increase in leukaemia is due to electromagnetic fields or to other factors to which electrical workers are exposed. Coleman and Beral also found no clear association between cancer risk and residence near sources transmitting electricity, although some data suggest that there may be small increases in leukaemia in those living very close to the sources.

Savitz concluded that "The findings of human studies of 50/60 Hz electric and magnetic fields are not sufficiently indicative of health hazards to warrant exposure limitation, nor is there high quality, convincing research to exonerate these exposures as threats to human health" and "Given this array of incomplete information, there is not an empirical basis yielding an answer to the critical question of human health consequences of 50/60 Hz electric and magnetic field exposures".

"Using standard levels of scientific proof, the argument that these fields cause cancer, reproductive damage, or other health effects falls far short of convincing. From the perspective of public health protection, however, one might ask whether the suggestions of health effects raised in some studies have been convincingly negated by

TABLE 1. Cancer incidence and occupational exposure to electromagnetic fields.

Authors	Workers	Study	Cancer Risk
Wiklund et al., 1981	Telecom	PMR	No cancer risk
Milham, 1982, 1985b	Electrical	PMR	Increased leukaemia
Wright et al., 1982	Electrical	PIR	Increased leukaemia
McDowall, 1983	Electrical	PMR/CC	Increased leukaemia
Coleman et al., 1983	Electrical	PIR	Increased leukaemia
Vagero & Olin, 1983	Electrical	RFU	No leukaemia risk Increased pharangeal, respiratory cancer
Swerdlow, 1983	Electrical	PMR	Increased eye melanoma
Pearce et al., 1985	Electrical	CC	Increased leukaemia
Lin et al., 1985	Electrical	CC	Increased brain tumours
Milham, 1985a	Amateur radio	PMR	Increased leukaemia
Gilman et al., 1985	Miners	RFU	Increased leukaemia
Vagero et al., 1985	Electrical	RFU	No leukaemia risk Increased urinary cancer Malignant melanoma
Calle & Savitz, 1985	Electrical	PMR	No leukaemia risk
Olin et al., 1985	Electrical	RFU	Increased malignant melanoma
Stern et al., 1986	Electrical	CC	Increased leukaemia
Tornquist et al., 1986	Electrical power	RFU	No leukaemia risk No brain tumour risk
Flodin et al., 1986	Electrical	CC	Increased leukaemia
Thomas et al., 1987	Electrical	CC	Increased brain tumours
McLaughlin et al., 1987	Electrical	PIR	No brain tumour risk
Lin, 1987	Electrical power	CC	Increased cancer risk

TABLE 2. Cancer incidence and population exposure to electromagnetic fields.

Authors	Subject	Study	Cancer Risk
Wertheimer & Leeper,	Children resident	CC	Increased leukaemia
Fulton et al., 1980	Children resident near HCCs	CC	No increased leukaemia
Wertheimer & Leeper, 1982	Adult resident near HCCs	CC	Increased leukaemia
Coleman et al., 1985	Persons resident near HCCs	CC	No increased leukaemia
Meyers et al., 1985	Children resident near HV	CC	No increased cancer
Rodvall et al., 1985	Person resident near HV	CC	No increased cancer
Tomenius, 1986	Children resident near HV	CC	Increased CNS tumours Increased leukaemia
McDowall, 1986	Persons resident near HV	CC	No increased leukaemia
Savitz, 1986b	Children resident near HCCs	CC	Increased cancer Increased leukaemia
Stevens et al., 1986	Persons resident near HCCs	CC	No increase acute non-lymphocytic leukaemia
Coleman & Bell, 1987	Persons resident near HV lines and sub-stations	CC	No leukaemia risk

superior research. The answer is that they clearly have not. The social and political choices relating to this issue fall outside the bounds of science, but there is a major role for further epidemiologic study in providing the basis for those decisions".

The World Health Organization concluded, following a review of the cancer registry and epidemiological studies that "these associations cannot be satisfactorily explained by the available theoretical basis for carcinogenesis by ELF electromagnetic fields. The preliminary nature of the epidemiological evidence, and the relatively small increment in reported incidence, suggest that, although these epidemiological data cannot be dismissed, there must be considerable further study before they can be accepted".

To date, chronic low-level exposure to 50/60 Hz fields has not been established to increase the risk of contracting cancer. If it is established in the future, it would in all likelihood be an extremely weak carcinogen.

2.2. Exposure Standards

On 3rd May, 1989, interim guidelines on limits of human exposure to 50/60 Hz electric and magnetic fields, developed by the International Non-Ionizing Radiation Committee (INIRC) of the International Radiation Protection Association (IRPA), have been approved by the President of IRPA on behalf of the Executive Council of IRPA, subject to minor editorial amendments.

Analysis of the scientific data does not provide any evidence that exposure at present day levels has a public health impact that would require corrective action. However, it is apparent that there are gaps in our knowledge and that further data need to be collected to determine if any health risk exists. It thus appeared that there was a need to develop an interim guideline on exposure based on an objective analysis of currently available knowledge.

The complete text of the guideline will be available only after its publication in Health Physics, and as part of a book published by Pergamon Press giving a compilation of all IRPA/INIRC guidelines.

The limits of exposure to 50/60 Hz electric and magnetic fields recommended in the guidelines are presented in Table 3.

The basic criterion is to limit current densities induced in the head and trunk by continuous exposure to 50/60 Hz electric and magnetic fields to no more than about 10 mA/m^2.

Continuous occupational exposure during the working day should be limited to root mean square (rms) unperturbed electric field strengths not greater than 10 kV/m. Short-term occupational exposure to rms electric field strengths between 10 and 30 kV/m is permitted, provided the rms electric field strength (kV/m) times the duration of exposure (hours) does not exceed 80 for the whole working day.

Continuous occupational exposure during the working day should be limited to rms magnetic flux densities not greater than 0.5 mT. Short-term occupational whole body exposure for up to two hours per workday should not exceed a magnetic flux density of 5 mT. When restricted to the limbs, exposures up to 25 mT can be permitted.

Members of the general public should not be exposed on a continuous basis to unperturbed rms electric field strengths exceeding 5 kV/m.

TABLE 3. Exposure limits to 50/60 Hz electric and magnetic fields recommended by IRPA.

Exposure Characteristics	Electric Field Strength kV/m (rms)	Magnetic Flux Density mT (rms)
OCCUPATIONAL		
Whole working day	10	0.5
Short term	30 (a)	5 (b)
For limbs	-	25
GENERAL PUBLIC		
Up to 24 hours/day (c)	5	0.1
Few hours/day (d)	10	1

Notes:

(a) The duration of exposure to fields between 10 and 30 kV/m may be calculated from the formula $t \leq 80/E$ where t is the duration in hours per work day and E is the electric field strength in kV/m.

(b) Maximum exposure duration is 2 hours per work-day.

(c) This restriction applies to open spaces in which members of the general public might reasonably be expected to spend a substantial part of the day, such as recreational areas, meeting grounds and the like.

(d) These values can be exceeded for a few minutes per day provided precautions are taken to prevent indirect coupling effects.

This restriction applies to open spaces in which members of the general public might reasonably be expected to spend a substantial part of the day, such as recreational areas, meeting grounds and the like. Exposure to fields between 5 and 10 kV/m should be limited to a few hours per day. When necessary, exposures to fields in excess of 10 kV/m can be allowed for a few minutes per day, provided the induced current density does not exceed 2 mA/m^2 and precautions are taken to prevent hazardous indirect coupling effects. It should be noted that buildings in a 5 kV/m external field have a field strength lower by more than an order of magnitude inside the building.

Members of the general public should not be exposed on a continuous basis to unperturbed rms magnetic flux densities exceeding 0.1 mT. This restriction applies to areas in which members of the general public might reasonably be expected to spend a substantial part of the day. Exposures to magnetic flux densities between 0.1 and 1.0 mT (rms) should be limited to a few hours per day. When necessary, exposures to magnetic flux densities in excess of 1 mT should be limited to a few minutes per day.

3. Radiofrequency and Microwaves

Although concerns about microwave (MW) and radiofrequency (RF) effects and possible hazards first arose in highly developed countries, the problem has now become universal. The number of devices, such as radars, radio relays or repeaters, and television and radio broadcasting stations, which, by design, emit electromagnetic waves into the atmosphere, is growing from day to day all over the world.

Environmental and workplace surveys demonstrate that the main source of significant chronic exposure is occupational, above all in the following activities: broadcasting, telecommunications, radar, satellite communication, metallurgy, food processing, RF-heat sealer operation, scientific research and medicine.

3.1. Advances in RF Dosimetry

Gandhi (1989) has recently shown that the high electric fields sanctioned in the American National Standards Institute (ANSI) guideline for LF to HF frequencies will result in significant RF currents flowing through the human body resulting in high Specific Absorption Rates (SARs) in smaller cross-section areas of the body such as the leg and the ankle region. Based on measurements with standing human subjects for plane-wave fields, induced currents in the order of 628-780 mA and resulting angle-section SARs as high as 182-243 W/kg are projected for 1.75 m tall individuals for the ANSI guideline for the frequency band 3-40 MHz. Using electromagnetic scaling concepts, SARs as high as 371 and 354 W/kg are projected for ten- and five-year-old children, respectively, for f = 50.7 and 62.5 MHz. These values are considerably larger than 8 W/kg for "any 1 g of tissue" assumed in the ANSI guideline and will in fact result in high rates of heating as measured at the surface of the ankle section.

Commonly encountered ungrounded objects such as car, van, bus, etc. will develop open-circuit voltages in the order of several hundred volts exposed to ANSI-recommended electric field of 614 V/m for the frequency band 0.3 to 3 MHz. Upon touching such vehicles, large currents may flow through the human body that are considerably in excess of those needed for perception, pain and even burns in some cases. For example, the current flowing through the hand of a human upon holding the door handle of an ungrounded automotive van is estimated to be 879 mA, resulting in a local SAR in the wrist of about 1045 W/kg.

It has been shown by Gandhi (1989) that the current I flowing through the feet of a standing grounded human being is given by:

$$I/E = 0.108\ h^2 f \tag{1}$$

where E is the plane-wave incident electric field (assumed vertical) in V/m; h is the height of the individual in meters; and f is the frequency in MHz. It is interesting to note that the current in Eq. (1) can be considered as though all the fields falling on an area 1.936 h^2, or approximately 5.93 m^2 for a human of height 1.75 m, were effectively passed through the human body. Similar results have also been reported by a number of authors. Eq. (1) is valid to a frequency of 40 MHz for a 1.75 m human. Because of the f^2 dependence, currents as high as 13.0 mA/(V/m) have been measured at 40 MHz for adult humans. This finding implies a current, I, of nearly 800 mA for the ANSI recommended E-field of 61.4 V/m at this frequency. Anatomical data have been used to estimate the effective areas for current flow for the various cross sections of the leg. The effective area A_e is estimated by the equation:

$$A_e = (A_c\sigma_c + A_1\sigma_1 + A_m\sigma_m)/\sigma_c \tag{2}$$

where A_c and A_1 and A_m are the physical areas of the high-water-content and the low-water-content tissues, and of the region containing red marrow, of conductivities σ_c, σ_1, and σ_m, respectively. Noting that most of the ankle cross section consists of low-conductivity bone or tendon, an effective area of 9.5 cm^2 is calculated for this cross section for a human adult even though the physical cross section is in the order of 40 cm^2. The SARs have been calculated from the equation:

$$J^2/\sigma_c\varrho = (I/2)\ A_e^2\sigma_c\ \varrho \tag{3}$$

where ϱ is the mass density of the tissue, taken to be 10^3 kg/m^3. The current taken for each of the legs is I/2. A fairly large ankle-section SAR of 243 W/kg is projected for a standing adult of height h = 1.75 m at a frequency of 40 MHz (Figure 1). This is, of course, considerably higher than the ANSI guideline of 8 W/kg for any 1 g of tissue.

Gandhi (1989) quantified the currents that would flow the feet with rubber- or leather- soled shoes. The currents for subjects wearing leather-soled shoes and rubber-soled shoes are, respectively, about 90-95 percent and 60-80 percent (the larger values have been measured for higher frequencies) of those for bare feet. The SARs in Figure 1 must

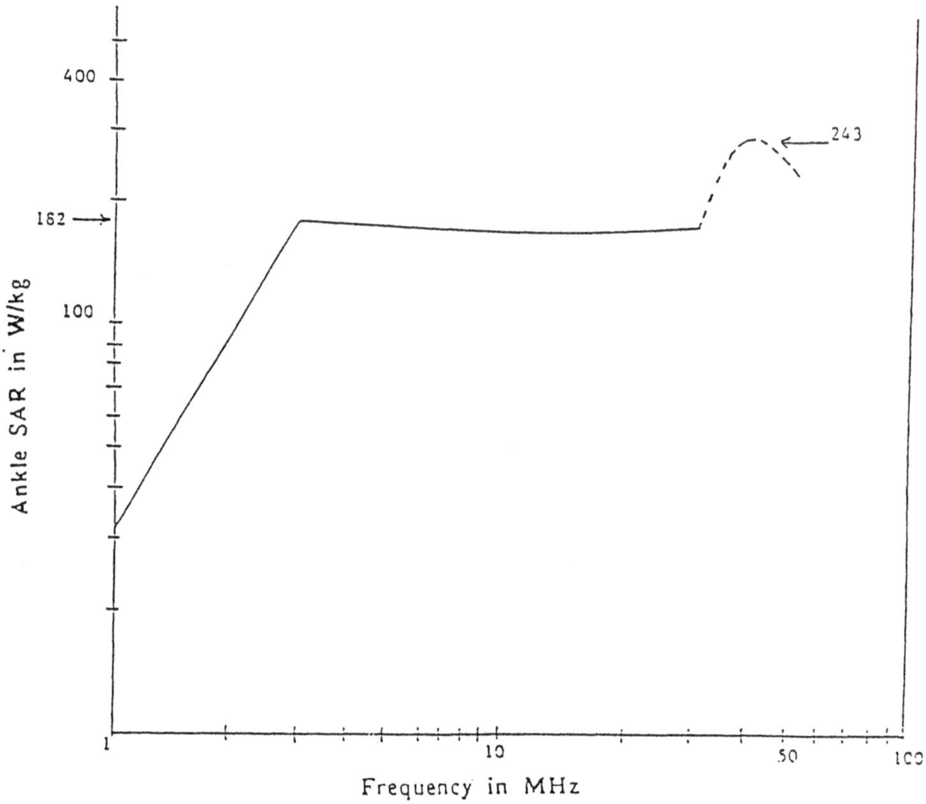

Figure 1. Ankle SAR for an adult male (h = 1.75 m) for electromagnetic fields recommended in the ANSI-1982 RF safety guideline (from Gandhi, 1989).

therefore be multiplied by the square of the appropriate current reduction factors.

The problem of RF shock and burns for human contact with commonly encountered metallic bodies has been discussed in the literature. Ungrounded objects such as car, van bus, fence, metallic roof, etc., will develop open-circuit voltages on the order of several hundred volts when exposed to the ANSI-recommended electric fields of 614 V/m for the frequency band 0.3 to 3 MHz.

From the foregoing it is obvious that substantial current densities and concomitant SARs may be set up in the body extremities such as the ankle section for freestanding individuals (no contact with metallic bodies) or the wrist section for conditions of contact with the metallic bodies, for electric-field limits suggested in the ANSI C95.1-1982 RF safety guideline. Gandhi et al. measured the surface temperature elevation of the wrist and the ankle sections for a healthy human subject at room temperature for a variety of RF currents and estimated SARs in the frequency band 1-50 MHz. A couple of representative graphs on the temperature of several locations of the ankle and wrist sections are given as Figures 2 and 3. As a result of the experiments that have been performed, the observed highest rates of temperature increase $\Delta T/\Delta t$ (for point 1 of the ankle section - see Figure 2, for point (a) of the wrist section - see Figure 3) are given by the following best-fit relationships:

$$\Delta T/\Delta t = 0.0045 \times SAR \ ^\circ C/min \ \text{(ankle section)}$$

$$\Delta T/\Delta t = 0.0048 \times SAR \ ^\circ C/min \ \text{(wrist section)}$$

where the SAR is in W/kg. Since ankle-section SARs in the order of 182-243 W/kg and wrist-section SARs as high as 1045 W/kg have been projected previously for the E-fields recommended in the ANSI C95.1-1982 safety guide, fairly high rates of temperature increase are therefore anticipated.

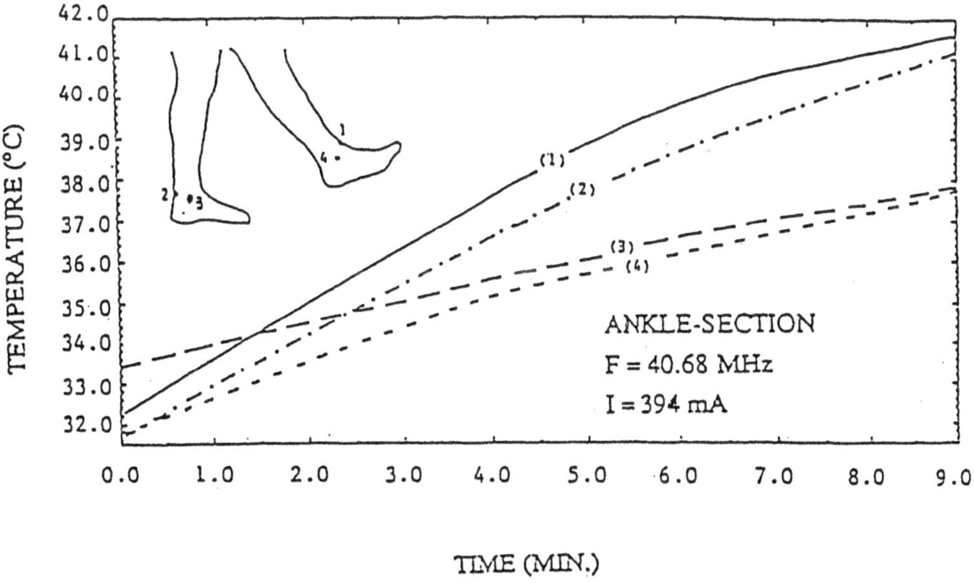

Figure 2. Temperature of several locations at the surface of the ankle section under condition of RF current through the leg. Estimated SAR in high-water-content tissues = 239 W/kg (from Gandhi, 1989).

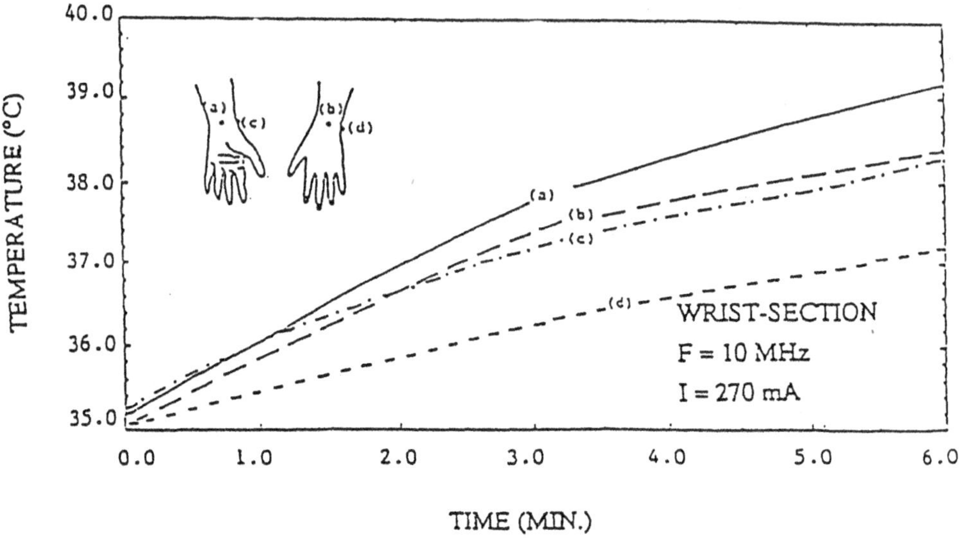

Figure 3. Temperature of several locations at the surface of the wrist section under condition of RF current through the hand. Estimated SAR in high-water-content tissues = 134 W/kg (from Gandhi, 1989).

3.2. Exposure Standards

Among the many factors that go into the development of an exposure standard, the selection of a good scientific biological effects data base plays quite obviously the most important role: different scientific approaches have produced different philosophies of protection guidelines and thus different exposure limits.

In recent years there have been many comprehensive reviews of the scientific basis for setting standards intended for the control of human exposure to non ionizing electromagnetic fields, and here the ground will not be covered again in detail other than to present some general ideas on the debate which has accompanied the recent approval of new national and international safety standards, which in turn proves the need for a new data base as useful input for a definite risk assessment.

These new exposure standards are based on what is known about the frequency-dependent nature of radiofrequency radiation energy deposition in biological systems and the current knowledge of biological effects, and in general provide an added margin of safety over what was previously used.

The inherent risks to health from RF/MW exposures are directly linked to the absorption and distribution of energy in the body, and the absorption and distribution are strongly dependent on the size and orientation of the body and the frequency and polarization of the incident radiation.

The American National Standards Institute (ANSI) has recently completed a review of literature and screened only those reports that produced positive findings, were reproducible, and supplied adequate dosimetric information. Evaluations of these selected reports, in terms of whether or not the biological effects constituted a hazard to health, were made.

Behaviour in experimental animals was found to be the most sensitive indicator of an adverse health effect (e.g. convulsion activity, work stoppage, work decrement, decreased endurance, perception of the exposing field and aversion behaviour). On the basis of their review, ANSI concluded that acute (less than 1 hour) exposure to electromagnetic energy that is deposited in the whole body at an average specific absorption rate (SAR) of less than 4 W/kg does not produce an adverse health effect in experimental animals. However, because prolonged exposure (days and weeks) may cause damage, a tenfold reduction in the permissible SAR (i.e., 0.4 W/kg) was involved.

The ANSI proposals, shown in Table 4, apply to both occupational and general population exposures and cover the frequency range 300 kHz-100 GHz. Exposures are to be averaged over 6 minutes, and levels in excess of those in Table 4 are allowed if it can be demonstrated that the mean specific energy-absorption rate is less than 0.4 W/kg and the peak rate in any part of the body does not exceed 8 W/kg. Emissions from devices radiating less than 7 W and at frequencies below 1 GHz are exempt. Above 1 GHz devices will be treated on a case by case basis.

TABLE 4. ANSI proposals for limiting population and workers exposures to radiofrequency and microwave radiations.

Frequency range (MHz)	Power density (W/m^2)	Electric Field strength (V/m)	Magnetic Field strength (A/m)
0.3-3	1,000	632	1.58
3-30	$9,000/f^2$	1,897/f	4.74/f
30-300	10	63.2	0.158
300-1,500	f/30	$3.65f^{0.5}$	$9.12x10^{-3}f^{0.5}$
1,500-100,000	50	141.4	0.353

f = frequency in MHz.

Exposures slightly in excess of the proposed limits are not expected to be harmful but are regarded as undesirable. In uncontrolled situations it is suggested that exposure should be as low as reasonably achievable (ALARA principle).

A basic criterion of USSR protection standards is that they should be set at levels where there is no change in the state of health or well-being of exposed individuals. In practice, this is taken to be at levels well below those at which there are any perceptible effects of exposure.

There have been reports from the USSR that the exposure of workers to microwaves at levels between about 0.1 W/m^2 and a few W/m^2 has led to *functional disturbances in the nervous system with a predominantly asthenic condition,* including depression, impairment of memory, inability to make decisions, etc. It is claimed that these symptoms are reversible when the exposure ceases and it is noteworthy that they are nonspecific to microwave exposure. There have been similar reports from other eastern European countries but careful studies have failed to confirm these claims.

It appears, however, that to a large degree, the apparent differences in US and USSR standards are based not on actual factual information but on differences in basic philosophy that appear in the areas of industrial hygiene.

Since 1970 the Commission's Directorate-General for Employment, Social Affairs and Education of the European Communities has concerned itself with the influence of non-ionizing radiation on human health.

In its Sixth General Report for 1972 the European Parliament called on the Commission to establish basic standards for health protection against the danger of microwave radiation.

The first meeting of a Working Group in this field took place on March 1973. The Working Group drew up a proposal for a Commission recommendation. At the Commission's request the recommendation was converted into a proposal for a Directive and was submitted to the Council on 26 June 1980. This proposal was published in the Official Journal of the European Communities on 26 September 1980 and is entitled *Proposal for a Council Directive laying down basic standards for the health protection of the workers and the general public against the dangers of microwave radiation.* The European Parliament and the Economic Social Committee gave favourable opinions on 19 June and 25 March 1981 respectively, and the Council of Ministers instructed the Working Group on Social Affairs to prepare an opinion on this Proposal for a Directive. The structure of the Proposal for a Directive on microwave radiation was modelled on that of the Directive on ionizing radiation. The Working Group on Social Affairs had already approved several Council Directives on the protection of workers against the hazards related to chemical, physical and biological substances at the workplace and in July 1983 the Social Affairs Working Group suggested that the proposal for a Directive on microwave radiation be aligned with this model. The Commission modelled the Directive to incorporate the Council's request, with a tendency to accept that occupational exposure should not exceed a SAR of 0.4 W/kg averaged over any period of six minutes and over the whole body, or a SAR of 4 W/kg for one gramme of tissue. The discussion on which degree of conservatism had to be applied in deriving the working limits of electric and magnetic field strengths and power density from these SAR values never reached a conclusion and, currently, no directive is expected for at least 5 years.

IRPA has recently proposed new criteria and standards for limiting the exposure of workers and the general population to radiofrequency and microwave radiations.

Occupational exposure to RF radiation at frequencies below and up to 10 MHz should not exceed the levels of unperturbed rms electric and magnetic field strengths given in Table 5, when the squares of the electric and magnetic field strengths are averaged over any 6 minute period during the working day, provided that the body-to-ground current does not exceed 200 mA, and that any hazard of RF burns is eliminated according to the recommendations stated below. A conservative approach consists in limiting pulsed electric and magnetic field strengths as averaged over the pulse with to 32 times the values given in Table 5.

TABLE 5. IRPA occupational exposure limits to radiofrequency electromagnetic fields.

Frequency range (MHz)	Unperturbed rms electric field strength (V/m)	Unperturbed rms magnetic field strength (A/m)	Equivalent plane wave power density (W/m²)	(mW/cm²)
0.1-1	614	1.6/f	-	-
>1-10	614/f	1.6/f	-	-
>10-400	61	0.16	10	1
>400-2,000	$3f^{0.5}$	$0.008f^{0.5}$	f/40	f/400
>2,000-300,000	137	0.36	50	5

f = frequency in MHz.

Note: Hazards of RF burns should be eliminated by limiting currents from contact with metal objects. In most situations this may be achieved by reducing the E values from 614 to 194 V/m in the range from 0.1 to 1 MHz and from 614/f to $194/f^{0.5}$ in the range from >1 to 10 MHz.

Exposure of the general public to RF radiation at frequencies below and up to 10 MHz should not exceed the levels of unperturbed rms electric and magnetic field strengths given in Table 6, provided that any hazard of RF burns is eliminated according to the recommendations stated below.

TABLE 6. IRPA general public exposure limits to radiofrequency electromagnetic fields.

Frequency range	Unperturbed rms electric field strength	Unperturbed rms magnetic field strength	Equivalent plane wave power density	
(MHz)	(V/m)	(A/m)	(W/m^2)	(mW/cm^2)
0.1-1	87	$0.23/f^{0.5}$	-	-
>1-10	$87/f^{0.5}$	$0.23/f^{0.5}$	-	-
>10-400	27.5	0.073	2	0.2
>400-2,000	$1.375f^{0.5}$	$0.0037f^{0.5}$	f/200	f/2,000
>2,000-300,000	61	0.16	10	1

f = frequency in MHz.

In view of the lack of a complete understanding of the biophysics of living systems, and the controversies concerning the non-thermal electromagnetic field interactions, no predictive theory based on non-thermal considerations is possible. Empirical data must serve for establishing exposure limits. According to Schwan, non-thermal effects may play a significant role at frequencies below 0.1 MHz. No adequate data on biological effects at these frequencies exist, except for the extremely low frequency range. Therefore it seemed to be prudent to limit the lower frequency end of the guidelines to 0.1 MHz. If compelling practical reasons (wide application at a particular frequency) exist, a limit for a narrow range or a particular device (emission standard) operating at a frequency below 100 kHz can be set on the basis of empirical studies.

4. Optical Radiation

Optical radiation covers the ultraviolet (UV), visible and infrared (IR) part of the electromagnetic spectrum. The most important source of occupational exposure to optical radiation for all people working out of doors is natural radiation from the sun, which covers the whole spectrum. Visible light at intensities normally encountered from artificial sources does not appear to pose a risk to health, but hazards can arise from certain high-intensity light sources such as projector bulbs, spotlights and floodlights.

Occupational health hazards from exposure to man-made sources of IR and UV radiation have been known for many years. They occur in a number of situations such as working near high temperature sources (IR and UV), use of electric arcs in metal-welding and cutting (UV), fabrication, control, and occupational use of fluorescent and mercury vapour lamps (UV). Conventional IR and UV radiation is used in medicine for thermography (IR) and phototherapy (UV), and new applications are being introduced.

In response to UV irradiation, both acute and long-term effects may occur. The acute effects are best known from over-exposure to sunlight and consist of sunburn and eye symptoms, while the late effects are skin aging and skin cancer.

IR radiation results in superficial burning and heat stress. The latter problem has to be considered in relation to humidity and ambient temperature of the work environment. Cataract formation was an important long-term sequela of IR irradiation but modern industrial practices have almost eliminated this problem.

Damage from intense sources of visible radiation is confined to the retina and occurs because the light is concentrated on this tissue by the refractive properties of the front of the eye.

There is general agreement on exposure limits for UV and IR but those for visible light are still being developed. Adequate means of protection against all wavelengths of optical radiation exist, and protective devices have been in use for many years.

4.1. Laser Devices

Lasers have already found widespread technological application in metal working, drilling, precision measurements, applied optics (holography), telecommunications, guidance systems, scientific research, dentistry, medicine and military uses.

The health risks from lasers are related to both wavelength and exposure levels, and are identical to those described above for incoherent optical radiation. However, because lasers are more intense sources they constitute the major industrial hazard of optical radiation. The exposure limits for far IR and UV are identical for the eye and skin; however, in the visible range the refractive power of the eye means that for any given exposure the retina is at higher risk of damage.

Most laser damage to the retina results in small lesions with a corresponding loss of visual information. This is only apparent to the victim if the damage is in the macula (centre of the visual field). Exposure limits for laser are essentially the same as those for optical radiations.

Protection is based first upon classification of lasers according to hazard potential, i.e., some lasers are safe, and, secondly, upon system safety and beam controls so that harmful radiation is totally confined within the system. Third, in cases where radiation cannot be confined, personnel protection is undertaken.

5. Ultrasound

Ultrasound is sound (a mechanical vibration phenomenon) having a frequency above the range of human hearing (typically above 16 kHz) which, unlike electromagnetic radiation, requires a medium through which to propagate.

Exposure to ultrasound can be divided into two distinct categories: airborne and liquid-borne. Exposure to airborne ultrasound occurs in many industrial applications such as cleaning, emulsifying, welding, and flaw detection and through the use of consumer devices such as dog whistles, bird and rodent controllers, camera rangefinders, and commercial devices such as intrusion alarms. Liquid-borne exposure occurs predominantly through medial exposure in diagnosis, therapy, and surgery.

As with any other physical agent, ultrasound has the potential to produce adverse effects at sufficiently high doses. In addition, biological effects of unknown significance have been reported under laboratory conditions at low exposure levels. However, the health risks that may be associated with biological effects at the levels of ultrasound currently encountered in occupational, environmental, or medical exposure have not yet been defined.

Though, at present, there is no evidence of adverse health effects in human beings exposed to diagnostic ultrasound, its rapidly increasing use during pregnancy is still of special concern in view of the known susceptibility of the fetus to other physical and chemical agents.

A number of human-exposure limits for airborne acoustic radiation and for diagnostic ultrasound have been recently proposed.

6. General References

ANSI (1982) "American National Standard-Safety Levels with Respect to Human Exposure to Radio Frequency Electromagnetics Fields, 300 kHz to 100 GHz", Report ANSI C95.1-1982, American National Standards Institute, New York.

Czerski, P. (1988) "Extremely low frequency (ELF) electric fields, biological effects and health risk assessment". In: Repacholi, M.H., ed. Non-Ionizing radiations. Physical characteristics, biological effects and health hazards assessment. proceedings of a INIR Workshop, Melbourne; p. 255-271. Available from Australian Radiation Laboratory, Yallambie, Victoria, Australia.

Durney, C.H., Massoudi, H., and Iskander, M.F. (1986) "Radiofrequency Radiation Dosimetry Handbook", 4th ed., Report SAM-TR-85-73, USAF School of Aerospace Medicine, Brooks Air Force Base, Texas.

Elder, J.A., and Cahill, D.F., eds (1984) "Biological Effects of Radiofrequency Radiation", US Environmental Protection Agency, Research Triangle Park.

Franceschetti, G., Gandhi, O.P., Grandolfo, M., eds. (1989) "Electromagnetic biointeraction: mechanisms, safety standards, protection guides". New York and London: Plenum Press.

Gandhi, O.P. (1989) "Advanced in RF dosimetry: their past and projected impact on the safety standards". In: Franceschetti, G.; Gandhi, O.P.; Grandolfo, M., eds. Electromagnetic biointeraction: mechanisms, safety standards, protection guides. New York and London: Plenum Press.

Grandolfo, M. (1986) "Occupational exposure limits for radiofrequency and microwave radiation", Appl. Ind. Hyg., 1:75.

Grandolfo, M., Michaelson, S.M. and Rindi, A., eds (1983) "Biological Effects and Dosimetry of Nonionizing Radiation: Radiofrequency and Microwave Energies", Plenum Press, New York.

Grandolfo, M., Michaelson, S.M., Rindi, A., eds. (1985) "Biological effects and dosimetry of static and ELF electromagnetic fields". New York and London: Plenum Press.

Grandolfo, M., Vecchia, P. (1989) "Existing safety standards for high voltage transmission lines". In: Franceschetti, G.; Gandhi, O.P.; Grandolfo, M., eds. Electromagnetic biointeraction: mechanisms, safety standards, protection guides. New York and London: Plenum Press.

IRPA (1988) "Guidelines on limits of exposure to radiofrequency electromagnetic fields in the frequency range from 100 kHz to 300 GHz", Health Physics, 54:115.

IRPA (1989) "Interim guidelines on limits of exposure to 50/60 Hz electric and magnetic fields", Health Physics, in press.

Michaelson, S.M., and Lin, J.C. (1987) "Biological Effects and Health Implications of Radiofrequency Radiation", Plenum Press, New York.

Repacholi, M.H. (1988) "Carcinogenic potential of extremely low frequency fields". In: Repacholi, M.H., ed. Non-ionizing radiations. Physical characteristics, biological effects and health hazards assessment. Proceedings of a INIR Workshop, Melbourne; p. 303-315. Available from Australian Radiation Laboratory, Yallambie, Victoria, Australia.

Savitz, D.A., Wachtel, H., Barnes F.A., John, E.M., Tvrdik, J.G. (1988) "Case-control study of childhood cancer and exposure to 60 Hz magnetic fields". Am. J. Epidemiol. 128:21-38.

Tenforde, T.S., Kaune, W.T. (1987) "Interaction of extremely low frequency electric and magnetic fields with humans". Health Physics 53:585-606.

Wertheimer, N., Leeper E. (1979) "Electrical wiring configurations and childhood cancer". Am. J. Epidemiol. 109:273-284.

WHO/IRPA (1979) "Ultraviolet Radiation", Environmental Health Criteria 14, WHO, Geneva.

WHO/IRPA (1981) "Radiofrequency and microwaves", Environmental Health Criteria 16, WHO, Geneva.

WHO/IRPA (1982) "Ultrasound", Environmental Health Criteria 22, WHO, Geneva.

WHO/IRPA (1982) "Laser & Optical Radiation", Environmental Health Criteria 23, WHO, Geneva.

WHO/IRPA (1984) "Extremely Low Frequency (ELF) Fields", Environmental Health Criteria 35, WHO, Geneva. WHO/IRPA, 1987, "Magnetic Fields", Environmental Health Criteria 69, WHO, Geneva.

SAFETY ASPECTS IN MAGNETIC RESONANCE DIAGNOSTIC EQUIPMENT

M. GRANDOLFO
Istituto Superiore di Sanità - Laboratorio delle Radiazioni
Viale Regina Elena, 298
I-00161 Roma
Italy

1. Introduction

Magnetic resonance medical devices, which include magnetic resonance imaging (MRI) and in-vivo spectroscopy (MRS), offer unique diagnostic possibilities and applications of these devices are being explored and expanded. During the imaging procedure, lasting up to tens of minutes, the patient lies on a table and all parts of the body are exposed to strong static magnetic fields, time-varying magnetic fields and radiofrequency (RF) radiation. Rapidly switched gradient fields are superimposed on the static field to allow spatial resolution to be obtained.

The present generation of devices for MRI and in vivo MRS use static magnetic fields with flux densities less than 2.5 tesla (1 T = 10^4 gauss). Most of the existing machines for whole imaging contain large-volume superconducting magnets that produce highly stable, uniform fields of 0.6 T or less. Several of the newer whole-body imaging units use fields up to 2.0 T, and fields up to 2.5 T are currently used for in vivo spectroscopy. Higher field levels are being considered for future magnetic resonance units.

The magnetic field gradients that provide spatial information defining the location of magnetic nuclei are less than 10 mT/m, and generally lie in the range of 1-5 mT/m in the existing MRI devices. The rapid switching of gradient fields leads to a time variation of the magnetic flux density (dB/dt) which is less than 3 T/s in the current imaging devices, with typical values in the range of 0.5-1.5 T/s. Higher values of dB/dt are anticipated in future MRI units using faster rates of data projection.

The frequencies of the RF fields used in current MRI devices are a function of the static field strength and the Larmor frequencies of resonant frequencies of H1, C13, Na23, P31 and K39 are, respectively,

M. Oberhofer (ed.), Advances in Radiation Protection, 361–376.

42.576, 10.705, 11.262, 17.235 and 1.987 MHz. Under the near field imaging conditions used for MRI, the absorption of RF energy is primarily associated with the magnetic component of the field.

2. Static and ElF Magnetic Fields

There are three established physical mechanisms through which static and ELF magnetic fields interact with living matter: magnetic induction, magnetomechanical effects, and electronic interactions.

The first mechanism is relevant to both static and time-varying fields, and originates through the following types of interaction:

a) Electrodynamic interactions with moving electrolytes.
   Both static and time-varying fields exert Lorentz forces on moving ionic charge carriers, and thereby give rise to induced electric fields and currents. This interaction is the basis of magnetically-induced blood flow potentials that have been studied with both static and time-varying ELF fields. It is also the physical basis of the weak induced potentials that provide sensory directional cues to elasmobranch fish as they swim through the static geomagnetic field.
b) Faraday currents
   Time-varying magnetic fields induce currents in living tissues in accordance with the Faraday law of induction. Available evidence suggests that this mechanism may underlie the visuosensory stimulation that produces magnetophosphenes and other effects on electrically excitable tissues. In addition, indirect evidence suggests that rapidly time-varying magnetic fields may exert effects on a variety of cellular and tissue systems by inducing local currents that exceed the naturally occurring levels. This effect may be the basis for the wide spectrum of biological perturbations that have been observed with pulsed magnetic fields, such as those used clinically for bone fracture reunion.

The two types of mechanical effects that a static magnetic field exerts on biological objects are:

a) Magneto-orientation
   In a uniform static field, both diamagnetic and paramagnetic molecules experience a torque, which tends to orientate them in a configuration that minimizes their free energy within the field. This effect has been well studied for assemblies of diamagnetic macromolecules with differing magnetic susceptibilities along the principal axes of symmetry. Included in this class of macromolecules are the arrays of photopigments in retinal rod disc membranes.
b) Magnetomechanical translation
   Spatial gradients of static magnetic fields produce a net force on

paramagnetic and ferromagnetic materials that leads to translational motion. Because of the limited amount of magnetic material in most living objects, the influence of this effect on biological functions is negligible.

Certain classes of chemical reactions involve radical electron intermediate states in which interactions with a static magnetic field produce an effect on electronic spin states (electronic interactions). It is possible that the usual lifetime of biologically relevant electron intermediate states is sufficiently short that magnetic field interactions exert only a small, and perhaps negligible, influence on the yield of chemical reaction products.

In addition, to the mechanisms of magnetic field interactions for which there is direct experimental evidence, several other mechanisms have been proposed, on theoretical grounds, in an effort to explain various biological effects that have been reported to occur in static and ELF fields of very low intensity. However, it must be emphasized that many proposed mechanisms have not been subjected to direct experimental tests.

Some organisms possess a sensitivity to static magnetic fields with low intensities comparable to that of the geomagnetic field (about 50 μT). Phenomena for which there is substantial experimental evidence of sensitivity to the earth's field include:

a) direction finding by elasmobranch fish (shark, skate, and ray);
b) orientation and swimming direction of magnetotactic bacteria;
c) kinetic movements of molluscs;
d) migratory patterns of birds; and
e) waggle dance of bees.

In addition, a number of *in vitro* studies have been made of magnetic orientation in assemblies of macromolecules, including retinal rod outer segments, muscle fibres, photosynthetic systems (chloroplast grana, photosynthetic bacteria, and *Chlorella* cells), halobacteria purple membranes and various synthetic liquid crystals and gels. Certain classes of chemical reactions that involve a radical electron intermediate state may also be sensitive to static magnetic fields of moderate intensity (< 10 mT).

For static magnetic fields with flux densities of less than 2 T, there exists a body of experimental data that indicates the absence of irreversible effects on many developmental, behavioural, and physiological parameters in higher organisms. Broadly summarized, available evidence suggests that the following 9 classes of biological functions are not significantly affected by static magnetic fields at levels up to 2 T:

a) cell growth;
b) reproduction;

c) pre- and post-natal development;
d) biolectric activity of isolated neurons;
e) behaviour;
f) cardiovascular functions (acute exposures);
g) the blood-forming system and blood;
h) immune system functions;
i) physiological regulation and circadian rhythms.

Studies on workers involved in the manufacture of permanent magnets in the USSR indicated various subjective symptoms and functional disturbances including irritability, fatiguem headache, loss of appetite, bradycardia, tachycardia, decreased blood pressure, altered EEG, itching, burning, and numbness. However, lack of any statistical analysis or assessment on the impact of physical or chemical hazards in the working environment significantly reduces the values of these reports. Although the studies are inconclusive, they suggest that, if long-term effects occur, they are very subtle, since no cumulative effects are evident.

Recent epidemiological surveys in the USA have failed to reveal any significant health effects associated with long-term exposure to static magnetic fields. A study of the health data on 320 workers in plants using large electrolytic cells of chemical separation processes, where the average static field level in the work environment was 14.6 mT, indicated slight changes in the white blood cell picture (still within the normal range) in the exposed group compared with the 186 controls. None of the observed changes in blood pressure or blood parameters was considered indicative of a significant adverse effect associated with magnetic field exposure.

The prevalence of disease among 792 workers at the US National Accelerator Laboratories, who were exposed occupationally to static magnetic fields, was compared with that in a control group consisting of 792 unexposed workers matched for age, race, and socioeconomic status. The range of magnetic field exposures was from 0.5 mT for long durations to 2 T for periods of several hours. No significant increase or decrease in the prevalence of 19 categories of diseases was observed in the exposed group relative to the controls.

Workers exposed to large static magnetic fields in the aluminium industry were reported to have an elevated leukaemia cancer risk. For persons directly involved in aluminium production, there is no clear evidence, at present, indicating the responsible carcinogenic factors within the work environment.

It can be concluded that available knowledge indicates the absence of any adverse effects on human health due to exposure to static magnetic fields up to 2 T. It is not possible to make any definitive statements about safety or hazard associated with exposure to fields above 2 T. From theoretical considerations and some experimental data, it could be inferred that short-term exposure to static fields above 5 T may produce significant detrimental effects on health.

For time-varying magnetic fields in the ELF frequency range, few systematic studies have been carried out to define the threshold field characteristics for producing significant perturbations of biological functions.

Time-varying magnetic fields generate internal electric currents. For example, 3 T/s can induce current densities of about 30 $\mu A/m^2$ around the perimeter of the human head. Induced electric current densities can be used as the decisive parameter in the assessment of the biological effects at the cellular level.

In terms of a health risk assessment, it is difficult to correlate the internal tissue current densities with the external magnetic field strength. However, assuming worst-case conditions, it is possible to calculate, at least within one order of magnitude, the magnetic flux density that would produce potentially hazardous current densities in tissues. The following statements can be made on induced current density ranges and correlated magnetic flux densities of a sinusoidal homogeneous field, which produce biological effects from whole-body exposure:

a) Between 1 and 10 mA/m^2 (induced by magnetic fields above 0.5-5 mT at 50/60 Hz, or 10-100 mT at 3 Hz), minor biological effects have been reported.
b) Between 10 and 100 mA/m^2 (above 5-50 mT at 50/60 Hz or 100-1000 mT at 3 Hz), there are well established effects. Facilitation of bone fracture reunion has been reported.
c) Between 100 and 1000 mA/m^2 (above 50-500 mT at 50/60 Hz or 1-10 T at 3 Hz), stimulation of excitable tissue is observed and there are possible health hazards.
d) Above 1000 mA/m^2 (greater than 500 mT at 50/60 Hz or 10 T at 3 Hz), extra systoles and ventricular fibrillation, i.e., acute health hazards, have been established.

From the available data on human exposure to time-varying magnetic fields, it can be concluded that induced current densities below 10 mA/m^2 have not been shown to produce any significant biological effects. In the range of 10-100 mA/m^2 (from fields higher than 5-50 mT at 50/60 Hz), biological effects have been established, but these induced current densities from short-term exposure (few hours) may cause minor transient effects on health. The health consequences of exposure to these levels for many hours, days, or weeks are not known at present. Above 100 mA/m^2 (greater than 50 mT at 50/60 Hz), various stimulation thresholds are exceeded and hazards to health may occur.

3. Radiofrequency

Among the safety problems is the possible production of heat in the body due to the radiofrequency power deposited during the procedures (Budinger (1988), Grandolfo (1987)). Recent developments indicate that higher pulse peak powers and closer spacing of RF pulses are needed to optimize the clinical efficiency of the MRI devices (Bottomley (1982), Rzedzian and Pykett (1987), Wehrli (1988)). With increasing static magnetic field strength the RF frequency has to increase, and that also contributes to an increase of specific absorption rate (SAR) in the patient's body. Thus, the tolerated SAR and the associated thermal load may be identified as the prime safety consideration and the limiting factor in clinical magnetic resonance applications.

The primary physical interaction of time-varying magnetic fields with living systems is the induction of electric fields and currents in tissues. In the RF frequency range presently used for medical MRI, approximately 90% or more of the absorbed energy results from tissue currents induced by the magnetic component of the field. The total energy dissipation is a function of frequency, incident RF power density, exposure duration, coupling between the RF coil and the specimen, and several properties such as the mass density, size, and orientation relative to the field polarization. Tissue heating is also influenced by the rate of heat loss via thermal conduction and convection mechanisms in the exposed subject (Adair and Berglund (1987)).

In accordance with Faraday's law of induction, the relationship between the induced electric field intensity, E, and the time rate of change of the magnetic flux density is given by:

$$\vec{E} \cdot \vec{dl} = -d/dt\, \vec{B} \cdot \vec{dS} \qquad (1)$$

where the line integral is around a closed curve and $\vec{dl}$ is a differential element of length along the curve; $\vec{B}$ is the magnetic flux density and $\vec{dS}$ is a differential area element directed normal to the surface enclosed by the curve over which the line integral is taken.

Recently the SAR has been evaluated (Grandolfo et al., in press) in a realistic model of the human torso, composed of thirteen different tissues, using the impedance method. For this purpose, the torso was replaced by a model of cubic cells, each one fully characterized, from the electrical point of view, by its impedances along each direction. The latter were calculated from the known dielectric constant and conductivity of the tissue or mixture of tissues comprising the cell.

In this way, the torso was approximated by a 3-D network of impedances. Applying eq.(1) to each cell face and requiring the current continuity for each loop in the network, we obtained a system of equations which could be solved for the induced currents in each cell's loop. From the currents, power deposition in individual cells was easily calculated. The knowledge of percent distribution of tissues in cells also allowed us to evaluate SARs in different organs together with the whole-trunk average SAR. More details on the calculation technique are given by DeFord and Gandhi (1985).

To apply the impedance method, the torso was enclosed in an ideal parallelepiped divided into 18x12x23=4,968 cubic cells of 2.6 cm edge. The distribution of different tissues in each cell was the same as in earlier papers (Gandhi et al. (1984), DeFord and Gandhi (1985), Orcutt and Gandhi (1988)) and was obtained through a digital scanning of atomic diagrams of cross sections of the torso.

The numerical calculations give the RF power deposition resulting from a continuous exposure. However, it is a common feature of all technical procedures of MRI that RF pulses, rather than continuous waves, are mainly applied; these pulses are the source of the detectable signal, and therefore they are the basis of any useful data acquisition. Presently the imagers can be programmed to produce several pulse sequences and the patterns of RF power deposition strongly depend upon the chosen sequence, which can vary largely in pulse number and type. The numerical data are therefore to be scaled for the duty cycle, D, in order to get SARs for the realistic imaging procedures.

As a first application of the calculation method, we applied the impedance method to the case of an exposure similar to that hypotized by Bottomley and Edelstein (1981).

Through eq.(1), these authors estimated the surface absorption of RF power for an average-sized human head and torso imaged by nuclear magnetic resonance. The torso, which is of interest for comparison with our results, was approximated by a homogeneous cylinder of radius R. The electrical properties and the mass density of the tissue inside the cylinder were assumed equal to those of the muscle. The calculation was performed over a circular loop with radius R intersected by a spatially uniform, time-varying magnetic field perpendicular to the loop and parallel to the torso axis. The field was supposed to be applied as a series of $\pi/2$ RF pulses of sinusoidal waves of frequency f, with pulse duration t and pulse repetition period T.

Under these circumstances, it can be shown (Bottomley and Edelstein (1981)) that the SAR in a tissue of electrical conductivity σ and mass density ϱ is given by:

$$SAR = 6.80 \times 10^{-16}\ \sigma\ f^2 R^2 / \varrho\, t T \qquad (2)$$

In our calculations, a torso of maximum cross section 46.8x31.2 cm and 59.8 cm height was supposed to be exposed to a radiofrequency magnetic field, B_{rf}. In order to get results comparable with those presented by Bottomley and Edelstein, we chose radiofrequency $\pi/2$ pulses of duration 5 ms and repetition period 100 ms, i.e. a duty cycle $5x10^{-2}$ as normally used for instance in fast imaging. In MRI procedures the ϑ pulse duration must satisfy the condition:

$$2 \pi \gamma \; B_1 t_\vartheta = \vartheta \tag{3}$$

where B_1=0.5 B_{rf} is the magnitude of the rotating component of the RF field. Taking into account this restriction, the magnetic field strength B_{rf} used in the calculation was 1.81 A/m (2.35 μT).

The average SARs in individual tissues versus frequency are shown in Figure 1.

In order to allow a better evaluation of the average SAR in the whole trunk as a function of the frequency, we performed a best-fit analysis. Over the whole frequency range of our calculations, we got a $f^{2.07}$ dependence, while the frequency dependence over the reduced frequency range 13.56-64 MHz is better described by a $f^{2.15}$ law.

As already mentioned, the impedance method is a powerful tool to evaluate not only power deposition in the whole trunk, but also its distribution among the individual tissues and organs. The particular situation at 30 MHz is presented in Figure 2. The average SAR in the skin is about 7 mW/kg, to be compared with the value of about 20 mW/kg given by Bottomley and Edelstein for the surface SAR in this case.

The discrepancy can be accounted for by the different dimensions of our realistic torso and the simplified model adopted by Bottomley and Edelstein: the radius of the latter equals the maximum cross dimension of our torso, allowing higher values of SAR in the outer tissues. The distribution of power deposition within every horizontal cross section shows in fact an increase with the distance from the center (Gandhi and DeFord (1984)), as expected due to the larger current loops. That also explains the high value of SAR in the skin compared with all other tissues with the exception of the lung.

To consider different patterns of RF excitation, and to compare our data with more realistic models, we focused our attention on other procedures in which one pulse type dominates within the pulse sequence, typically the spin-echo techniques. In these techniques one $\pi/2$ pulse is followed by one or more π pulses but, taking into account the dependence upon the flip angle, the $\pi/2$ pulse has only a quarter of the energy of a π pulse and thus the former can be neglected also with respect to a single π pulse. Adopting this restriction, we applied the calculation procedure to the spin echo and the inversion recovery methods, following some examples presented by Vollmann and Kuhn (1985) and listed in Table 1.

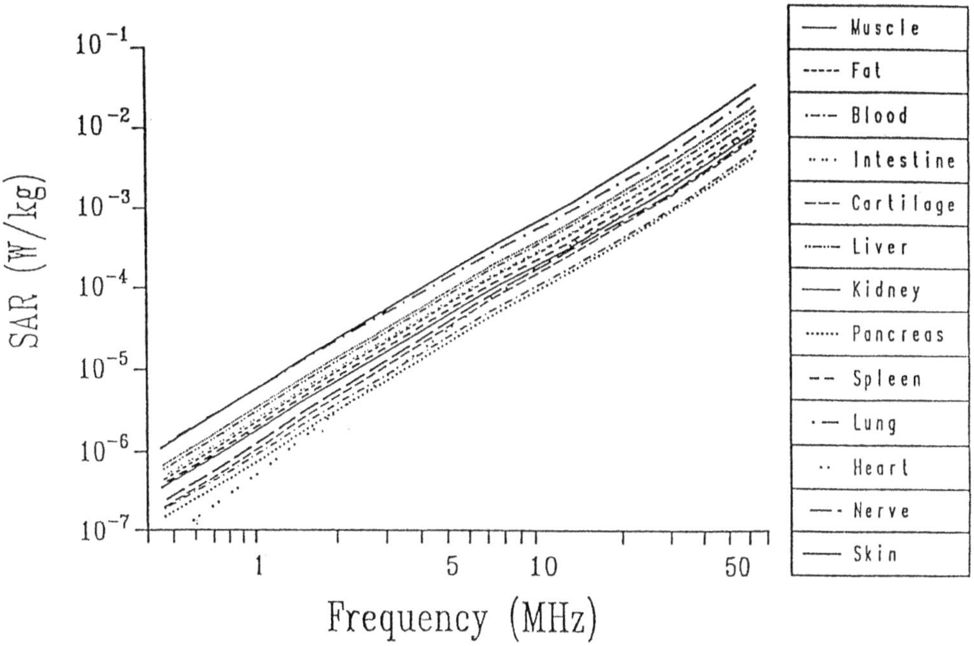

Figure 1. Frequency dependence of SAR for different tissues in a human torso exposed to π/2 pulses of duration 5 ms and repetition period 100 ms.

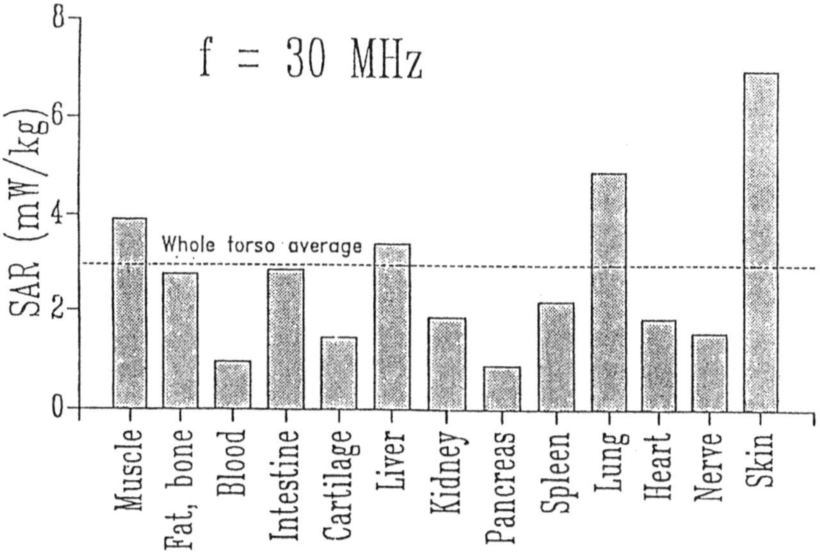

Figure 2. Average SARs in individual tissues and in the whole torso. The figure refers to π/2 pulses of duration 5 ms and repetition period 100 ms(B_o=0.7 T).

TABLE 1 - Whole-torso average SARs (W/kg) for some realistic MRI procedures evaluated by different models and calculation methods.

MRI procedure	Vollmann and Kuhn	Roeschmann	Present work
Spin echo, 1 echo, 1 π pulse t=1 ms, T=1 s B=1.5 T, f=64 MHz	0.026	0.028	0.030
Fast imaging, 1 echo, 1 π/2 pulse t=5 ms, T=0.1 s B=0.7 T, f=30 MHz	0.0029	0.0027	0.0029
Spin echo, 1 echo, 1 π pulse t=2.35 ms, T=0.5 s B=1.5 T, f=64 MHz	0.022	0.024	0.026
Inversion recovery, 4 echoes, 5 π pulses 12 slices t=0.78 ms, T=1 s B=1.5 T, f=64 MHz	2.03	2.14	2.33

The results of our calculations confirm the validity of the impedance method as a useful tool for the evaluation of power deposition in different organs for a number of exposure conditions.

On the other hand, the fairly good agreement of our whole-torso data with those of other authors indicate the reliability of simplified models for an approximate evaluation of RF power deposition.

The impedance method, however, allows evaluation of SAR distribution inside the body, and to locate possible hot spots. For all the exposure conditions we analyzed, the maximum SAR was found in the skin, as expected due to the larger current loops. Relatively high values of SAR were found also in the lung, which therefore appears to be a critical organ.

However, SARs of similar magnitudes were also found for other organs, indicating that significant power deposition may occur also in the internal regions of the body (Bottomley et al. (1985)).

These results support the suggestion (Roeschmann (1987)) that the texture of biological tissue consisting of heterogeneously sized regions of relatively high values of permittivity and conductivity, separated by regions with considerably lower values, allows the RF magnetic field to penetrate much deeper into the body than predicted by homogeneous tissue models. Recent studies (Meyer and Vollmann (1984), Lee and Cho (1985)) actually indicate that RF penetration limits are not likely to degrade the image quality up to 100 MHz, whereas the RF power deposition in patients seems to be more evenly distributed over the exposed body volume than currently assumed (Roeschmann (1987)).

4. Exposure guidelines

Exposure guidelines for MRI and MRS issued in the United States, United Kingdom, and the Federal Republic of Germany are summarized in Table 2. Recommended limits were given for the static magnetic fields, the time-varying magnetic fields, and the RF fields used in MRI devices. These guidelines are in close agreement on a limit for the static magnetic field of 2 to 2.5 T, but significant differences exist in the recommended exposure limits for time-varying magnetic fields and for RF fields.

A limit of 3 T/s for the time-varying field was recommended in the U.S. guideline (National Center for Devices and Radiological Health) based on a review of the known interaction mechanisms and biological effects of time-varying magnetic fields. In contrast, the U.K. guideline (National Radiological Protection Board) was set at 20 T/s (rms) based on the estimate that a time-varying magnetic field of this magnitude would induce a maximum current density of less than 0.3 A/m^2 (rms) in any part of the body. This current density was judged to be safe on the basis that it is approximately one order of magnitude less than the threshold for producing cardiac fibrillation. The limit on dB/dt in the U.K. guideline is permitted to increase as the reciprocal square root of the pulse duration, i.e., $2/t^{0.5}$ in units of T/s (rms), for pulses shorter than 10 ms. This specification was based on various measurements of the stimulus strength versus duration relationship for human responses to electric current, as reviewed in an earlier section of this chapter. The guideline of the Federal Republic of Germany (F.R.G.) for time-varying magnetic fields is not extensively justified on the basis of laboratory data. However, the recommended limits on induced current density and voltage gradient are reasonably consistent with the U.S. limit of 3 T/s, since the maximum current density induced in the human body by a magnetic field with this time rate of change would be approximately 30 mA/m^2. The German guideline also permits the induced

TABLE 2 - Exposure guidelines in clinical MRI.

Source	B(tesla)	Time varying fields	RF fields
National Center for Devices and Radiological Health, USA 1982	2.0	2 T/s	Specific absorption rate <0.4 W/kg (whole body) or <2 W/kg per g of tissue
National Radiation Protection Board, U.K.: 1983	2.5	20 T/s (rms) for pulses >10 ms; $2/t^{0.5}$ for pulses <10 ms (t in s)	Body temperature rise of <1°C (specific absorption rate <0.4 W/kg (whole body) or <4 W/kg per g of tissue
Federal Health Office, F.R.G.: 1984	2.0	Induced current <30 mA/m^2 or induced field <0.3 V/m for pulses >10 ms; induced current <0.3/t mA/m^2 or induced field <3/t mV/m for pulses <10 ms (t in s)	Specific absorption rate <1 W/kg (whole body) or <5 W/kg per kg of tissue (except for the eyes)

current density limit to increase as the reciprocal of the pulse duration, i.e. 0.3/t in units of mA/m^2, for pulses less than 10 ms.

The recommended exposure limits for RF fields given in all three sets of guidelines are designed to avoid significant regional or whole-body heating. The maximum whole-body SAR value recommended in the U.S. guideline is consistent with recommendations by the American National Standards Institute, whereas the F.R.G. limit is 2.5 times higher. In both sets of MRI exposure guidelines, the recommended limit for absorbed RF power in localized tissue regions is increased by a factor of 5 relative to the whole-body SAR limit. The U.S. guideline, however, specifies the local SAR per gram of tissue, whereas the F.R.G. guideline specifies this quantity per kilogram of tissue exclusive of the eyes. The U.K. guideline for the whole body SAR limit is consistent with the U.S. recommendation, but the SAR limit per gram of tissue is twice that given in the U.S. guideline.

Both the U.K. and F.R.G. guidelines discuss the risks of imaging patients with implanted pacemakers, vascular clips, or large prosthetic devices. In addition, the U.K. guideline states that it would be prudent to exclude women in the first trimester of pregnancy from MRI imaging procedures.

The one recommendation that relates to staff members operating MRI equipment is the specification in the U.K. guideline that the Stanford Linear Accelerator limits on static magnetic field exposure should be applied to these personnel. Under these limits, exposure of the whole body is not to exceed 0.02 T for prolonged periods, and exposure of the arms and hands is limited to 0.2 T. For short periods of less than 15 min duration, these exposure limits are increased to 0.2 T for the whole body and 2 T for the arms and hands.

Recently, IRPA (in press) issued a document on the protection of patient undergoing magnetic resonance examinations. According to this document no adverse health effects are expected allowing exposures of patients up to the following limit values.

Static Magnetic Fields

No adverse health effects are expected from exposures of the head and/or trunk to fields up to 2 T, or from exposure of the limbs to fields up to 5 T.

Exposures of the head and trunk to fields above 2 T require an assessment of the potential adverse effects versus the likely benefit to the patient. Short-term whole body exposures to fields of 5 T may pose health hazards, in particular for people with cardiovascular diseases, including high blood pressure. In view of the above, monitoring of the patient's cardiovascular function must be undertaken whenever exposures of the head or trunk above 2 T, and of limbs above 5 T occur. Until more information is available, all exposures should be kept below 10 T.

Time-varying Magnetic Fields

Clinical experience indicates that no adverse health effects are expected when the rate of change of magnetic flux density does not exceed 6 T/s (Czerski and Athey (1987)). It should be noted, however, that patients with changes in the electrocardiogram (ECG) indicative of abnormalities in conduction may be particularly susceptible to exposure to magnetic fields. Therefore an assessment of cardiac function should be made when exposures above 6 T/s are contemplated. An ECG should be made prior to an MRI examination, and the patient's cardiovascular function should be monitored during exposure above 6 T/s.

Radiofrequency Fields

For whole body exposures or exposures to the head and trunk no adverse health effects are expected if the increase in body temperature does not exceed 1°C, except for infants, pregnant women, and persons with cardiocirculatory impairment, in whom it is desirable to limit temperature increases to 0.5 °C.

In practice, no tissue will increase in temperature at a rate more than 1°C/hour for each W/kg of power deposition. In persons (infants, aged persons, patients with local or systemic cardiovascular impairment) in whom the thermoregulatory mechanisms are less efficient than in a healthy human adult, signs of distress may appear at whole body average SAR of 2 W/kg after ten minutes of exposures (Budinger et al. (1985)). Taking the modifying effects of blood flow and environmental heat exchange into account, no adverse effects are expected in persons without cardiovascular impairment exposed to 2 W/kg for one hour. In persons with cardiovascular impairment, 1 W/kg for one hour is an acceptable level, provided proper monitoring is instituted. For exposures shorter than 60 minutes the same level can be attained by limiting the exposure to 120 or 60 W min/kg respectively, provided that the rate of energy deposition does not induce intolerable initial rates of increase in temperature.

Allowing for exposures of the head, the trunk, or extremities alone, no adverse health effects are expected, if the RF fields are such that the absorbed energy (the product of the SAR and exposure time) does not exceed:

- 60 W min/kg averaged over the head, or
- 120 W min/kg averaged over the trunk, or
- 180 W min/kg averaged over the extremities,

provided that the SAR does not exceed:

- 4 W/kg averaged over the head, or
- 8 W/kg averaged over the trunk, or
- 12 W/kg averaged over the extremities.

For exposures of infants or persons with cardiocirculatory impairment a reduction of the above values by a factor of two is recommended. For a RF emitting surface coil the average should be taken over the sensitive volume.

6. General References

Adair, E.R., Berglund, L.G. (1987): "Some thermoregulatory consequences of magnetic resonance imaging". In Schmidt, KH (ed): "Safety assessment of NMR clinical equipment". Stuttgart and New York: Georg Thieme Verlag, pp. 12-21.

Bottomley, P.A. (1982): "NMR imaging techniques and applications: a review". Rev. Sci. Instrum. 53: 1319-1336.

Bottomley, P.A., Edelstein, W.A. (1981): "Power deposition in whole-body NMR imaging". Med. Phys 8:510-512.

Borromley, P.A., Redington, R.W., Edelstein, WA, Schenck, J.F. (1985): "Estimating radiofrequency deposition in body NMR imaging". Magn. Res. in Medicine 2:336-349.

Budinger, T.F. (1988): "Safety of NMR in vivo imaging and spectroscopy". In Budinger, T.F., Margulis, A.R.: "Medical magnetic resonance: a primer-1988". Berkeley: Society of Magnetic Resonance in Medicine, Inc., pp. 327-343.

DeFord, J.F., Gandhi, O.P. (1985): "An impedance method to calculate currents induced in biological bodies exposed to quasi-static electromagnetic fields". IEEE Trans Electromag Comp 27:168-173.

Durney, C.H., Massoudi, H., Iskander, M.F. (1985): "Radiofrequency radiation handbook" (Fourth Edition). Report USAFSAM-TR-85-73, prepared by the University of Utah for USAF School of Aerospace Medicine, Brooks Air Force Base, Texas 78235.

Gandhi, O.P., Deford, J.F. (1988): "Calculation of EM power deposition for operator exposure to RF induction heaters". IEEE Trans Electromag Comp. 30:63-68.

Gandhi, O.P., DeFord, J.F., Kanai, H. (1984): "Impedance method for calculation of power deposition patterns in magnetically induced hyperthermia. IEEE Trans Biomed Eng. 31:644-651.

Grandolfo, M. (1987): "Radiofrequency power deposition in medical NMR equipment". In Schmidt, K.H. (ed.): "Safety assessment of NMR clinical equipment". Stuttgart and New York: Georg Thieme Verlag, pp. 50-60.

Grandolfo, M., Vecchia, P., Gandhi, O.P. (in press): "Calculation of radiofrequency power deposition in magnetic resonance imaging of the human studies. Bioelectromagnetics".

Hutchinson, JMS (1988): "NMR image reconstruction methods". In Maraviglia, B. (ed.): "The physics of NMR spectroscopy in biology and medicine. Proceedings of the International School of Physics E. Fermi, Varenna. Course no. 100, 8-18 July 1986". Amsterdam, North Holland.

IRPA (in press): "Protection of the patient undergoing magnetic resonance examinations". Health Physics.

Lee, S.Y., Cho, ZH (1985): "Attenuation and phase delay of electromagnetic field in NMR imaging with surface coil". In "Fourth annual meeting of the society of magnetic resonance in medicine". (Book of abstracts). London, pp. 1011-1012.

Mansfield, P., Morris, P.G. (1982): "NMR imaging in biomedicine". New York Academic.

Mathur-De Vré, R. (1987): "Safety aspects of magnetic resonance imaging and magnetic resonance spectroscopy applications in medicine and biology: II - recommendations for MR exposures in clinical practice". Arch B Med. Soc., Hyg., Med. Tr. & Med. Leg. 45:425-438.

Meyer, W., Vollmann, W. (1984): "Fundamental limit in MR tomography: RF penetration depth in heterogeneous tissues". In "Third annual meeting of the society of magnetic resonance in medicine". (Book of abstract). New York, pp. 523-526.

Orcutt, N., Gandhi, O.P. (1988): "A 3-D impedance method to calculate power deposition in biological bodies subjected to time varying magnetic fields. IEEE Trans Biomed Eng 35:577-583.

Repacholi, MH (1986): "Limits of human exposure to magnetic fields". In Bernhardt, J.H. (ed.): "Biological effects of static and extremely low frequency magnetic fields". Munich: Medizin Verlag Muenchen, pp. 171-180.

Roeschmann, P. (1987): "Radiofrequency penetration and absorption in the human body: limitations to high-field whole-body nuclear magnetic resonance imaging". Med. Phys. 14:922-931.

Rzedzian, R.R., Pykett, I.L. (1987): "Instant images of the human heart using a new whole-body MR imaging system". AJR 149:245-250.

Vollmann, W., Kuhn, M.H. (1985): "Throughput limitations in clinical MR imaging due to RF heating". In Hopf, M.A., Bydder, G.M. (eds.): "Magnetic Resonance imaging and spectroscopy". Proceedings of the 1st Congress of the European Society of Magnetic Resonance in Medicine and Biology, Geneva, October 5-6, 1984". Geneva: European Society of Magnetic Resonance in Medicine and Biology, pp. 292-298.

Wehrli, F.W. (1988): "Rapid scan magnetic resonance". In Budinger, T.F., Margulis, A.R.: "Medical magnetic resonance: a primer-1988". Berkeley: Society of Magnetic Resonance in Medicine, Inc., pp. 61-81.

9 780792 312321

GPSR Compliance

The European Union's (EU) General Product Safety Regulation (GPSR) is a set of rules that requires consumer products to be safe and our obligations to ensure this.

If you have any concerns about our products, you can contact us on ProductSafety@springernature.com

In case Publisher is established outside the EU, the EU authorized representative is:

Springer Nature Customer Service Center GmbH
Europaplatz 3
69115 Heidelberg, Germany

Batch number: 10809439

Printed by Printforce, the Netherlands